Rural and urban hydrology

M. G. Mansell

⊥T Thomas Telford

Published by Thomas Telford Publishing, Thomas Telford Ltd, 1 Heron Quay, London, E14 4JD.

www.thomastelford.com

Distributors for Thomas Telford books are
USA: ASCE Press, 1801 Alexander Bell Drive, Reston, VA 20191-4400, USA
Japan: Maruzen Co. Ltd, Book Department, 3-10 Nihonbashi 2-chome, Chuo-ku, Tokyo 103
Australia: DA Books and Journals, 648 Whitehorse Road, Mitcham 3132, Victoria

First published 2003

A catalogue record for this book is available from the British Library

ISBN: 0 7277 3230 7

© M. G. Mansell and Thomas Telford Limited, 2003

Throughout the book the personal pronouns 'he', 'his', etc. are used when referring to 'the engineer', 'the designer', etc. for reasons of readability. Clearly, it is quite possible these hypothetical characters may be female in 'real-life' situations, so readers should consider these pronouns to be grammatically neuter in gender, rather than masculine.

Typeset by Alex Lazarou, Surbiton, Surrey
Printed and bound in Great Britain by MPG Books, Bodmin, Cornwall

The voice of my ancestors said to me,
The shining water that moves in the streams and rivers is
not simply water, but the blood of your grandfather's grandfather.
Each ghostly reflection in the clear waters of the lakes tell
of memories in the life of our people.
The water's murmur is the voice of your great-great-grandmother.
The rivers are our brothers. They quench our thirst.
They carry our canoes and feed our children.
You must give to the rivers the kindness you would give
to any brother.

Based on the words of Chief Seattle to the American Government who wanted to buy his people's land (from *Brother Eagle, Sister Sky* by Susan Jeffers).

Preface

There are already several good textbooks on hydrology, so the obvious question is why produce yet another. The original motivation behind this book came from a concern about the lack of textbooks dealing with urban drainage. Flooding in urban areas, for example, is one of the most immediate impacts which hydrology has on the daily life of many people. There is now at least one book (Butler, D. (2000) *Urban Drainage*. Spon, London) on this subject, but on further reflection it seemed that the real problem was that the link between urban and rural hydrology was not being made. Essentially the same hydrological processes of attenuation, evaporation, infiltration and other losses occur in both urban and rural areas. The only real difference is in the relative importance of these processes and the time-scale involved; in urban areas the time-scale is usually minutes while in rural areas it is more often hours or days. Moreover, most 'urban' areas include extensive areas of parks and gardens etc., which behave in much the same way as rural catchments. Flooding in towns and cities may be the result of local excess runoff, but it is very often the result of the runoff from the rural catchment upstream of the town. Indeed, urban flooding can often be mitigated by changes in land use in the rural catchment, such as by the provision of temporary flood storage or aforestation. There is also a growing interest in Sustainable Urban Drainage Systems (SUDS), which are based on the principle of using natural hydrological processes in urban drainage as far as possible. Another example of the interaction between urban and rural catchments is in the effect of urban runoff on the water quality of the receiving river, which is usually greatest after short, intense storms where the urban runoff occurs before the flow in the river has started to increase. These are just some examples which underline the need for connections to be made between rural and urban hydrology — hence the title of the book.

As well as emphasising these links, there were other reasons for producing this book. The opportunity has been taken to include a description of the various techniques in the *Flood Estimation Handbook*, which has now superseded the *Flood Studies Report* as the standard method of estimating flood discharges in UK rivers. Likewise, the method of estimating low flows recommended for the United Kingdom is described in some detail, as is the method of estimating potential evapotranspiration recommended by the Food and Agriculture Organization of the United Nations (FAO) (the FAO Penman-Monteith Method). It is hoped, therefore, that this book will be a useful reference for young practising engineers as well as for undergraduate students.

Another factor that is becoming increasingly important is climate change, and a chapter is devoted to describing the background to climate change, the latest predictions and the likely impacts, although it is recognised that the predictions may soon become out of date as new and improved modelling methods are developed.

Finally, spreadsheets are an invaluable tool in teaching hydrology and the author has developed the following spreadsheets to demonstrate various procedures (see below). These may be freely downloaded from http://hydrology.paisley.ac.uk. Readers may use or copy these spreadsheets as they wish, but neither the author nor the publisher can accept responsibility for their results.

Spreadsheets

Unit Hydrograph	Unit hydrograph method with *Flood Estimation Handbook* design rainfall
Reservoir	Flood routing through a reservoir
Channel	Flood routing through a river channel
Rational Method	Runoff calculation using the Rational Method
Evaporation	Evaporation using the FAO Penman-Monteith Method
Soil Water Balance	Soil water balance model using the Thornthwaite model
S-curve	Estimation of unit hydrograph using the S-curve technique
TOPMODEL	Rainfall–runoff modelling using the *TOPMODEL*
Depth–Duration	Calculation of depth–duration curves from rainfall records

Acknowledgements

The author is very grateful for the permission to reproduce material from various sources, in particular:

Flood Estimation Handbook	Centre for Ecology and Hydrology, Wallingford
The Wallingford Procedure	HR Wallingford
The HOST Soil Classification	The Macaulay Land Use Research Institute
UKCIP Maps	Tyndall Centre for Climate Change Research, University of East Anglia
Climate change	W. J. Burrows (2001), Cambridge University Press (Figures 1.3 and 1.5)
Understanding the Earth	G. C. Brown *et al.* (1992), Cambridge University Press (Figures 2.1 and 2.2)

It should be added that any discussion and interpretation of the material from these sources reflects the author's own opinions and are not endorsed by those bodies.

The author is also very grateful for the help received from colleagues at the University of Paisley, in particular from the library staff. Finally, the patient encouragement of my wife made this project much less daunting than it would otherwise have been.

Dr Martin Mansell started his engineering career in the drawing office of a local authority before studying civil engineering. After graduating, he worked in motorway construction and spent two years on secondment to the government of Papua New Guinea through Voluntary Service Overseas, working on small road and water-supply schemes. He undertook post-graduate studies in Hydraulics, Hydrology and Coastal Dynamics at the University of Strathclyde and obtained his PhD in 1982. After a period working on the UK wave-energy programme, he became a lecturer in the Department of Civil Engineering in the University of Zimbabwe and moved to his present post in the School of Engineering and Science at the University of Paisley in 1992. His research interests include urban drainage, the effect of climate change, small-scale hydropower and water resources in developing countries.

Contents

Chapter 1

Introduction

Water is the most abundant material on the Earth's surface. It covers over 70% of the surface of the planet and also makes up more than 70% of our bodies. Besides being necessary to sustain human and other life directly, it is essential in the production of food and is involved in transport, power generation, recreation and many other aspects of life.

The vital role of water in human development was recognised many centuries ago by major civilisations in the Middle East, China, South America and Europe. There is evidence of a dam being constructed by the Egyptians over 5000 years ago and wells being dug in Mesopotamia 3000 years ago. However, it was the Greek philosophers who carried out the first scientific study of hydrology, and it was not until the seventeenth century that the essential link between rainfall and river flow was demonstrated. In more recent years, there have been increasing pressures on water resources from the increasing population of the world, together with the prospect of substantial changes to the world's climate. At the same time, the tools available to the hydrologist have become more sophisticated with the development of automated data-recording systems and powerful numerical models.

1.1 The hydrological cycle

The hydrological cycle is the framework which links all the processes described in this book. It comprises the passage of water from the Earth's surface into the atmosphere by evaporation from both the land surface and the oceans and the return of the water via precipitation as well as the

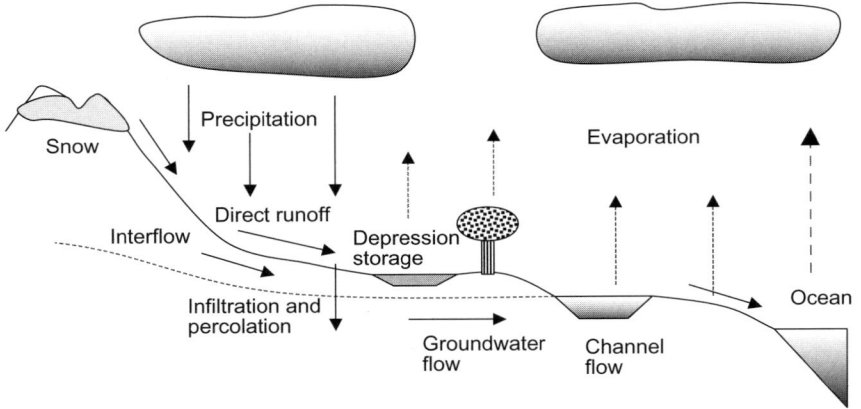

Figure 1.1   Movement of water on the Earth's surface

runoff from the land surface into the oceans. Figure 1.1 illustrates the main pathways of the hydrological cycle

The total rate of movement (flux) of water around the hydrological cycle is about 600 000 km^3/year. Of the flux into the atmosphere, by far the greatest proportion (84%) consists of evaporation from the oceans, the balance being evaporation from the land surface. The runoff from the land surface into the ocean is about 50 000 km^3/year.

Figure 1.2 summarises the global water balance. The width of the lines is proportional to the water fluxes expressed as a percentage of mean annual global precipitation. The areas of the squares are proportional to the storage on the Earth and in the oceans, although the volume of storage in the atmosphere is too small to be represented in this way.

Figure 1.2 demonstrates the enormous volume of water in the oceans relative to the volume stored on or in the Earth and in the atmosphere. Table 1.1 shows the relative volumes of water storage within different elements of the hydrological cycle as well as the replacement period (which can be considered as the average time a water particle will spend in each component of the cycle). Although the oceans cover about 70% of the Earth's surface, they contain over 97% of the water volume. The remaining 3% is freshwater, of which more than two-thirds is contained in glaciers and snowfields with a replacement time of several thousand years. Less than a third of the total freshwater resource (or less than 1% of the total volume of water) is readily available as groundwater or surface water. Because of the very small volume of water stored in the

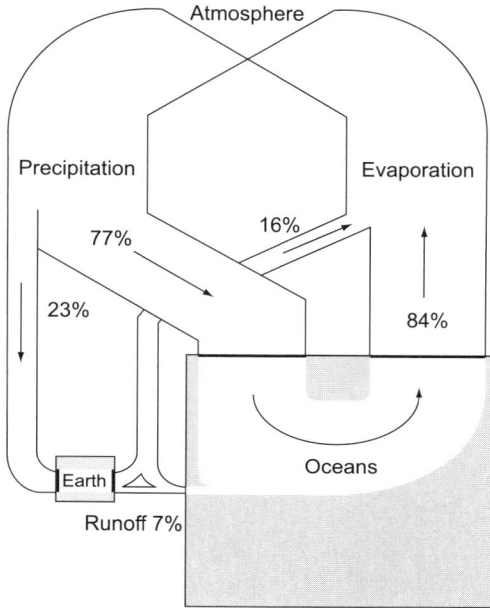

Figure 1.2   Global water balance

atmosphere (equivalent to about 25 mm water depth), the average residence time of water there is only about nine days: in other words, water cycles through the atmosphere some 40 times a year.

1.2 Radiation balance

The hydrological cycle is ultimately powered by the sun: about 20% of the solar radiation received by the Earth is consumed in driving the hydrological cycle. Therefore, some consideration of the Earth's radiation balance is useful in understanding the nature of the hydrological cycle.

The total net incoming solar radiation averaged over the Earth's surface is about 324 W/m^2, which is balanced by the outgoing radiation. Little more than half of this radiation flux reaches the Earth's surface (Figure 1.3); the balance is mostly either reflected by clouds or absorbed by the atmosphere. Clouds play a crucial role in the distribution of solar

Table 1.1   Replacement of water resources (Chorley, 1969)

Category	Total volume: km³ × 10³	Percentage of total	Percentage of freshwater	Replacement period
Oceans	1 350 000	97·6	—	2600 years
Glaciers/snow	26 000	1·9	77·8	2000 years +
Groundwater	7000	0·5	21·0	1400 years
Soil moisture	150	0·01	0·4	
Lake and rivers, etc.	125	0·01	0·4	Lakes 17 years Rivers 16 days
Inland seas	100	0·01	—	
Atmosphere	13	—	0·04	9 days
Total water	1 383 400	~100		
Total freshwater	33 400		~100	

energy of the Earth's surface. At any one time about half of the Earth's surface is covered in cloud but the actual distribution of clouds is very variable. The effect of clouds on the energy balance also depends on their thickness and the temperature of their tops, and predicting the behaviour of clouds is one of the major uncertainties in climate modelling. Another uncertain factor affecting the energy balance is the role of particulates, which are tiny particles, including fine dust and sulphate particles from the burning of fossil fuels. Unlike water vapour, these particles reflect sunlight more readily than they absorb terrestrial radiation and therefore they tend to reduce the net solar radiation received.

Of the proportion of energy that reaches the Earth's surface, only a small amount is reflected; most is absorbed and re-radiated. However, since the temperature of the Earth's surface is only about 300°K, the radiation from the Earth is at a much longer wavelength than the incoming radiation from the sun, which has a temperature of about 6000°K. The atmosphere is less transparent to longer wavelengths, and so much of the energy that is radiated from the Earth is absorbed by the

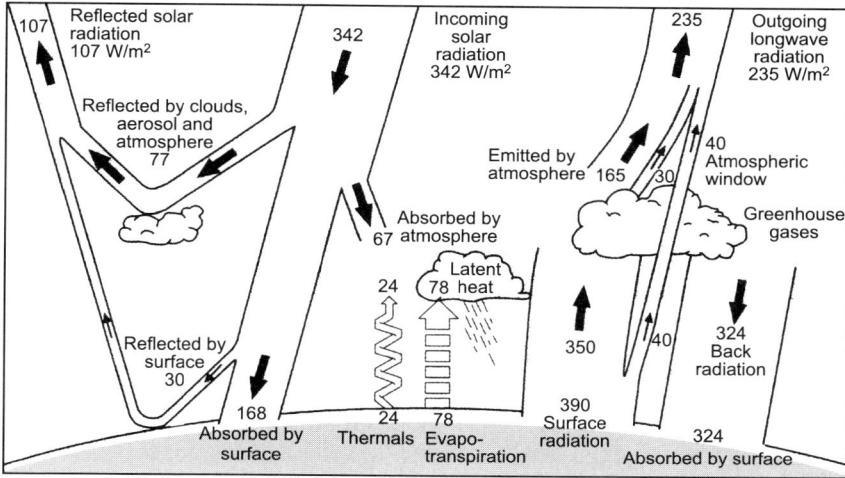

Figure 1.3   Radiation balance at Earth's surface (IPCC, 1995)

atmosphere and radiated back to the Earth, a process known as the greenhouse effect. In fact, the flux between the atmosphere and the Earth is about the same as the incoming solar radiation and if this back radiation did not occur, the average temperature of the Earth would be about 33°C lower than it is.

The radiation characteristics of land masses differs considerably from those of the oceans. The land areas heat relatively quickly compared to the oceans, as the heat is only stored in the surface layers, whereas the energy incident on the ocean penetrates to a much greater depth and, therefore, more energy is stored rather than re-radiated. The greater specific heat of water also accounts for the greater heat storage of the oceans.

The input of solar energy into the Earth's atmosphere system is not uniform, and the difference between the solar input and the net loss of energy from the Earth results in a surplus in low latitudes and a deficit in high latitudes, with a balance at about latitude 35°. In addition to acting as a great reservoir of heat, the oceans, together with the atmosphere, have a fundamental role in distributing the solar energy from the equatorial regions to the polar regions. Overall, the atmosphere and the oceans transport about the same amount of energy polewards and, in fact, there is a continuous interaction between them. There is a transfer of momentum from winds to the sea through waves and currents, as well as flows of heat and moisture in both directions. The Atlantic Ocean carries heat northwards and likewise in the Pacific Ocean there is a poleward movement of heat in the southern hemisphere. The Indian Ocean also transfers a significant amount of energy southwards.

1.3 Characteristics of the water, air and land phases of the hydrological cycle

1.3.1 Characteristics of water and the oceans

As mentioned above, the oceans play a vital role in determining climatic patterns over the world. 84% of evaporation into the atmosphere consists of evaporation from the oceans: nearly 2 m depth of water is evaporated annually in the sub-tropical areas of the North Atlantic over the Gulf Stream. In effect, evaporation from the oceans is a way of turning saltwater into freshwater. As has been noted, the oceans also distribute enormous amounts of energy over the Earth's surface and modify the climate of maritime areas.

These processes are possible because of several unique properties of water (Davis and Day, 1961). The structure of the water molecule, consisting of two hydrogen atoms and one oxygen atom, is similar to that of other substances which combine two hydrogen atoms with a single atom of another element, such as tellurium (H_2Te), selenium (H_2Se) and sulphur (H_2S). If the boiling and freezing points of these substances are plotted against their molecular weights, there is a clear and understandable trend, with the heavier materials having the higher boiling and freezing points. However, the actual boiling and freezing points of water are much greater than the trend would predict (Figure 1.4). The reason for this discrepancy lies in the peculiar structure of the water molecule. The bond between the hydrogen and oxygen atoms results from the sharing of electrons. The hydrogen atom consists of a nucleus and single electron, and, since the nucleus is the major component, there is a net positive charge. The oxygen atom has two 'shells' of electrons and the deficiency of the two electrons in the outer shell gives rise to a negative charge. Because of the uneven distribution of charges, the water molecule is, in effect, a polar molecule and therefore a bonding, known as hydrogen bonding, exists between the water molecules. More energy is therefore required to separate the molecules than would otherwise be needed, thus explaining the higher boiling and freezing points.

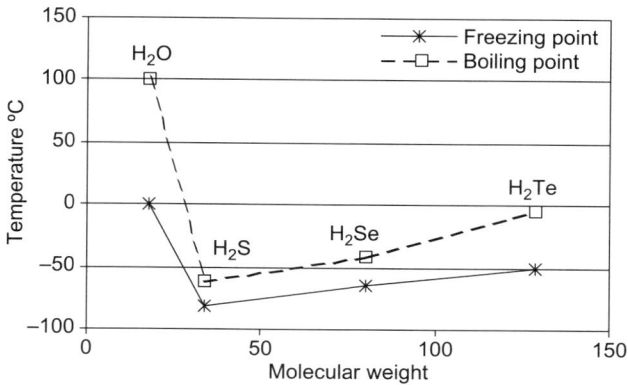

Figure 1.4   Freezing and boiling point of water and similar compounds (Davis and Day, 1961)

The unique structure of the water molecule also explains its exceptionally high specific heat compared with other common materials (Table 1.2). The high specific heat means that enormous amounts of heat can be transferred in the ocean currents with relatively small changes in temperature. Some surface ocean currents result from the prevailing winds and pressure systems. However, there is a system of *thermohaline*

Table 1.2   The physical properties of water in comparison with other common materials

Substance	Specific heat: kWhr/t/°C	Latent heat of fusion: kWhr/t	Latent heat of vaporisation: kWhr/t
Water	1·16	93·0	628
Alchohol	0·62	29·0	237
Benzene	0·45	3·5	109
Mercury	0·03	0·3	83
Sand	0·23	—	—
Iron	0·13	—	—

circulation due to variations in temperature and salinity. The temperature depends on where the surface water comes from and the heat balance between the ocean and the atmosphere. The salinity depends on the balance between the losses due to evaporation and the gains due to precipitation, river inflow and melting of ice sheets. In areas of high density, the water tends to sink and conversely where the density is low. The pattern of thermohaline circulation in the world's oceans has become known as the Great Ocean Conveyor (GOC) (Figure 1.5) and has a major effect on the world's climate. The warm surface currents, such as the Gulf Stream, which flows north-eastwards across the North Atlantic, move from warmer areas to cooler areas, and the cooler water then descends and returns southwards to Antarctica. The Gulf Stream carries a volume of about 400 km^3/h and for a difference in temperature of just 1°C the amount of heat transferred at that flow rate is about 460×10^6 MW. The result of this particular current is that countries in north-west Europe, such as the United Kingdom and those of Scandinavia, have temperatures some 20–25°C greater than would otherwise be the case.

As well as a high specific heat, both the latent heat of fusion (the energy required to convert a solid into a liquid) and the latent heat of vaporisation (the energy required to convert a liquid into a gas) are much

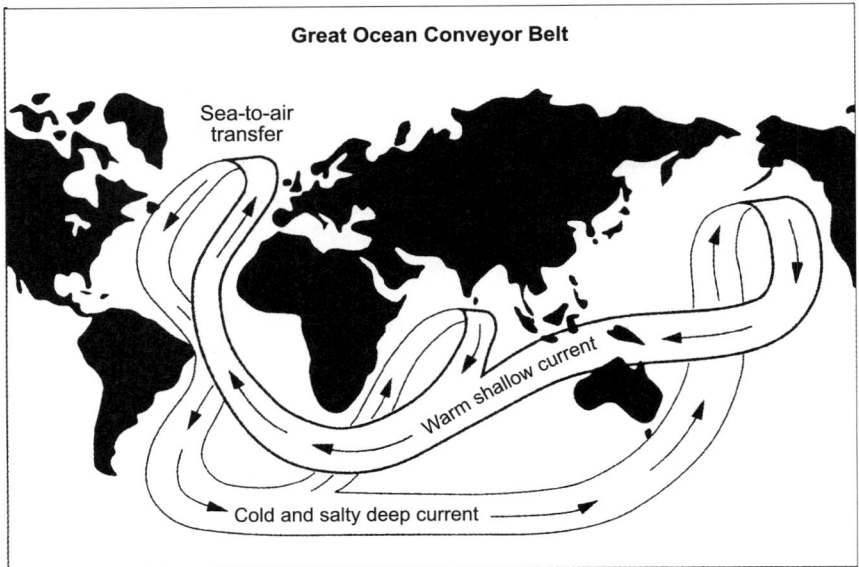

Figure 1.5   Major ocean currents (from Trenberth, 1992)

higher for water than for similar materials (Table 1.2). This also has an enormous effect on the distribution of solar energy over the Earth's surface. The relatively high solar energy in the tropical regions allows the air to absorb large amounts of water vapour which is then transported north and south to the temperate regions. In these regions this air comes into contact with colder, drier polar air and much of the moisture condenses and falls as precipitation. The latent heat which is released in the condensation, powers the cyclones and storms that occur in these latitudes.

Another unique feature of water also has profound consequences for human life. As the temperature of water is reduced from boiling point, it contracts, as do most materials. However, when the temperature falls below 4°C, water begins to expand due to the rearrangement of the molecules. As it freezes, close to 0°C, the ice has a density about 90% of that of the surrounding water. The significance of this simple property is quite profound. If ice were heavier than water, many of the lakes and oceans would gradually freeze from the bottom as their temperature fell below zero. In temperate and polar areas they would remain frozen permanently, apart from a small surface layer. Thus, they would be unable to support fish or other life, and their moderating effect on the world's weather described above would be severely limited.

The molecular bonding of water also explains other unique features of water. For example, the high surface tension of water is due to the cohesive effect of the bonding between the molecules on the surface of water. The concave meniscus of water in a thin glass tube is due to the bonding between the hydrogen atoms in the water and the oxygen atoms of the silica in the glass. The same effect occurs in the pores of clays and other soils, resulting in a capillary force, which causes water to rise above the water table and also partly explains how water is able to pass upward through plants.

1.3.2   *Characteristics of the atmosphere*

The atmosphere extends for about 100 km above the Earth's surface and, although the atmospheric pressure decreases rapidly and continuously with height, the variation of temperature is more irregular. The atmosphere is divided into four zones or 'spheres' based on the temperature variation. The most important of these regions is the *troposphere,* which extends to about 16 km above the Earth's surface and contains about 75% of the mass of the atmosphere and almost all of its moisture. It is a region of decreasing temperature, with an average

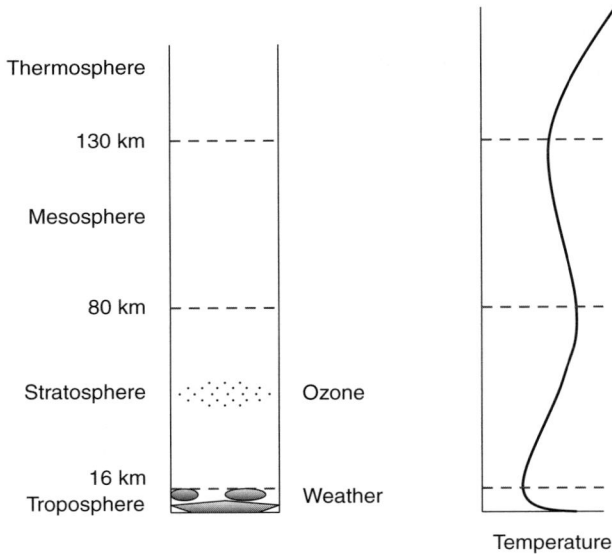

Figure 1.6   *The structure of the atmosphere*

decrease of between 5°C and 10°C per km. Above the troposphere the temperature increases again in the *stratosphere*, which extends up to about 80 km. This region contains ozone which absorbs short-wave solar radiation, releasing some of the energy as heat. Above the stratosphere are the *mesosphere* and the *thermosphere*, which are areas of decreasing and increasing temperature respectively (Figure 1.6).

Near the land surface the air is heated by radiation from the ground below rather than from direct solar radiation, and the actual temperature gradient over the first 100 m or so of the atmosphere varies continually and this has an important effect on the weather pattern. In particular, the difference between the actual temperature gradient and the *adiabatic lapse rate* determines the *stability* of the atmosphere (Figure 1.7). The adiabatic lapse rate is the rate at which the temperature of a parcel of air decreases as it rises, assuming that there is no heat exchange with the surrounding atmosphere. It typically varies between 0·5°C and 1°C per 100 m depending on the moisture content of the atmosphere. Where the actual temperature gradient is less than the adiabatic lapse rate, a small parcel of warm air (A) released at ground level will tend to remain at the same position since, by rising to say B, its temperature would be less (and its

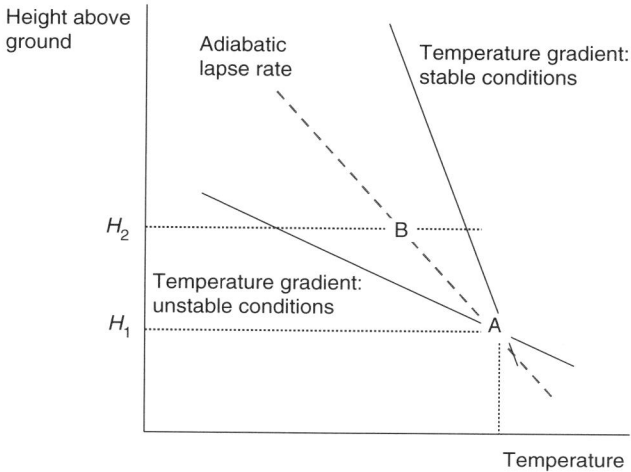

Figure 1.7   Stability of air masses

density greater) than the surrounding air. Such conditions are described as *stable*. Conversely, *unstable* conditions occur when the temperature gradient of the surrounding air is greater than the adiabatic lapse rate. In this case, a parcel of warm air would continue to rise at an increasing rate since, as it rose, it would still be warmer (and lighter) than the surrounding air. Stable air masses are normally associated with high-pressure systems with light winds, and often lead to fog and increased air pollution due to the trapping of warm air discharges. Where there is an increase in temperature with height, it is known as an *inversion* and extremely stable conditions result.

The basic circulation patterns of the atmosphere have a major effect on the patterns of the world's climate. At the equator, air tends to rise because of the higher temperatures in relation to higher latitudes. If the Earth were a simple stationary sphere, the rising air would be replaced by the cooler sinking air from the polar regions leading to a single-cell circulation pattern in each hemisphere (Figure 1.8(a)). In practice, the rotation of the Earth causes a deflection of the air masses, known as the *coriolis* effect, and, as a consequence, there is broadly a three-cell circulation pattern, a simplified version of which is shown in Figure 1.8(b). The zone near the equator, where the air masses converge, is known as the *Inter-Tropical Convergence Zone* (ITCZ) and is an important source of precipitation in those areas. Within a few degrees of the equator,

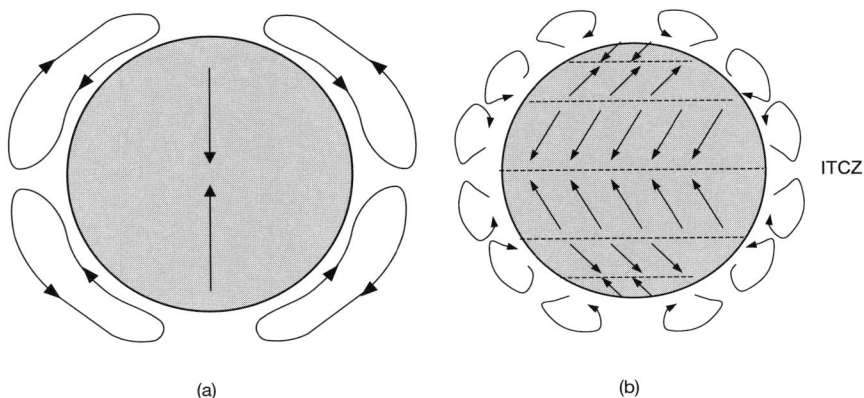

(a)   (b)

Figure 1.8 (a) Global air circulation without the coriolis effect; and (b) global air circulation with the coriolis effect

there is precipitation over the whole year sometimes with maxima at two seasons. Further north and south there are two distinct summer maxima with little winter rainfall and the summer rainfall becomes a single maximum with a longer dry period further from the equator. Between 20° and 30° north and south, in the region of descending dry air, there are deserts with very little rainfall, especially on the west side of continents. Further north and south is the converging zone between the polar air and the temperate air, which is the source of most of the depressions bringing precipitation over throughout the year. The polar areas tend to be relatively dry over most of the year.

The concept of *air masses*, which are homogeneous bodies of air normally associated with stagnant high-pressure areas, is important in the study of the world's climate as it is the boundaries between these air masses (known as *fronts*), which are a main source of precipitation. The air masses, in general, are classified as either *polar*, which are typically cold and stable air masses, or *tropical*, which are warm and unstable. There is a further classification into *continental* and *maritime* according to whether they were formed over a land mass or the oceans. Air masses move in response to differences in pressure and may be modified as they pass through different environmental conditions.

The atmospheric circulation patterns described above also act to distribute the incoming solar radiation from the equatorial regions to the polar areas. In part, this is through the sensible heat of the moving air, but

Table 1.3 Lamb weather types

Direction	Anticyclonic		Cyclonic
North-east	ANE	NE	CNE
East	AE	E	CE
South-east	ASE	SE	CSE
South	AS	S	CS
South-west	ASW	SW	CSW
West	AW	W	CW
North-west	ANW	NW	CNW
North	AN	N	CN

it is also due to the high latent heat of the moisture contained in the air, as mentioned above. The relatively high solar energy in the tropical regions allows the air to absorb large amounts of water vapour which is then transported north and south to the temperate regions. In these regions this air comes into contact with colder, drier polar air and much of the moisture condenses and falls as precipitation, releasing the latent heat as condensation.

It is recognised that precipitation and other meteorological characteristics are linked to weather systems, which may last for several days. Such weather systems broadly fall into one of a number of classes. A general classification of daily weather types has been developed, based on the wind direction and barometric pressure (Lamb, 1972) (Table 1.3).

1.3.3 Characteristics of the land phase of the hydrological cycle

Along with the atmosphere and the oceans, the land mass of the Earth represents the third major element in the hydrological cycle. The land phase of the hydrological cycle, in general, consists of the transformation of intermittent scattered pulses of precipitation into a continuous hydrograph of runoff which can be represented by two basic processes: *attenuation* and *loss*. Attenuation refers to the decrease and delay in peak flow resulting from storage in various components, such as groundwater, lakes, etc. Loss refers to the diversion of water back into the atmosphere through the evaporation of water directly from the land surface and water bodies, as well as transpiration by plants and trees.

On average, about 70% of precipitation over the land surface returns as local evaporation or transpiration. Significant evaporation occurs directly from the land surface and from the surfaces of lakes and rivers. Some of the precipitation will be used by vegetation and will pass back into the atmosphere by transpiration. Some will be intercepted by the leaves of vegetation and will be evaporated back into the atmosphere.

The remaining 30% of precipitation over land, flows over or under the Earth's surface and is termed *runoff*, and it is this fraction which constitutes the potentially available freshwater supply. Runoff can generally be equated with streamflow, although the former is normally in the same units as precipitation and evaporation (i.e. depth), whereas streamflow is usually measured in terms in cubic metres per second. Some of the runoff flows directly over the surface into rivers and streams; the rest will infiltrate into the ground. Some of the surface flow will collect in depressions in the ground and will later infiltrate into the ground or evaporate (Figure 1.1). Water that infiltrates into the ground may flow

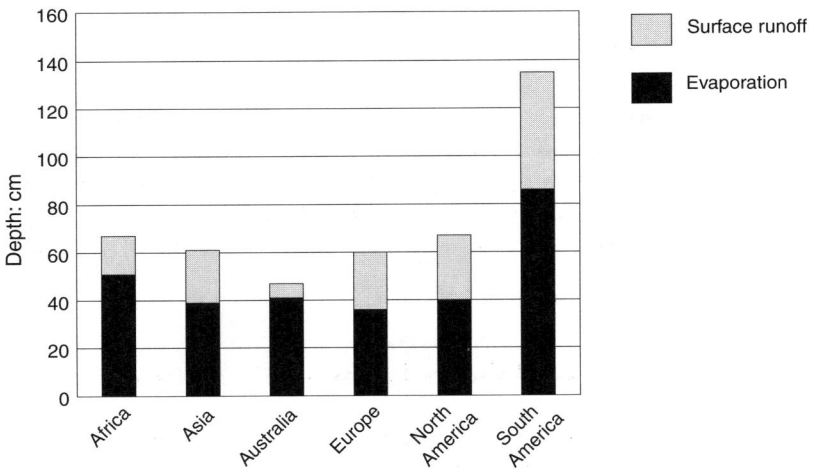

Figure 1.9   Average water balance for different continents (Budyko, 1974)

horizontally just below the surface (interflow) or may percolate downwards to the water table below which the soil is completely saturated. The surface of the water table in general slopes gradually towards the nearest channel, since there is a slow movement of water within the water table towards the channel. Whereas surface or direct runoff may reach a channel within a matter of minutes or hours, the pathway through the water table may take days, months or even years. Virtually all of the precipitation that does not evaporate eventually becomes streamflow, but the term runoff is sometimes restricted to the direct or surface runoff and the infiltration is then regarded as a loss. The exact balance between evaporation and runoff varies according to local conditions. Figure 1.9 shows the average water balance for different continents.

1.4   Water requirements

As has been shown in Table 1.4, the total circulation of water in the hydrological cycle is about 600 000 km^3/year, or about 270 m^3/person per day, which is over 1000 times the typical consumption in developed countries. However, only a small proportion of this amount falls as precipitation over land and, of this, a significant fraction is lost by evaporation, as described above. The amount of available water is further

Table 1.4   Total water flux

	Total flow: km^3/year	Per capita: m^3/person per day
Total circulation	600 000	270
Precipitation over land	95 000	41
Effective precipitation	66 000	30
Available runoff	8000	3·7
Water withdrawal	4000	1·8

reduced by the spatial and temporal variability of runoff: the supply of fresh water is not evenly distributed over the Earth's surface. In temperate areas most runoff occurs in winter and flows to the oceans without being used. Even in arid areas much runoff occurs after short heavy storms and is not available without extensive storage.

At present, the use of water amounts to about 4000 km^3/year (1·8 m^3 per day). However, the term *use* can include many processes. It is sometimes taken to mean any water which is temporarily withdrawn from the natural hydrological system. Such water may be returned to the system almost immediately, such as the water used for generating hydro-electric power or cooling thermal power stations. The term *consumptive use* is sometimes used to describe the use of water in such a manner as to prevent its immediate re-use because of contamination, evaporation or transpiration. In some cases, the actual water use may be greater than the consumptive use, for example, where water is recirculated or re-used many times and this offers a potential approach to satisfying the increasing demand for water. Figure 1.10 shows how the various uses of water contribute to the aggregate withdrawal and the consumptive use.

It can be seen that one of the greatest uses is for electricity supply, although not all the water used for electricity supply is lost in the sense that it is evaporated directly into the atmosphere. Most of the water is used for cooling and similar purposes, and can therefore be used again almost immediately. Also, much of the water used by industry and agriculture

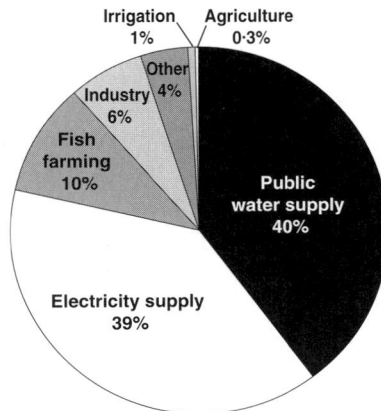

Figure 1.10   Distribution of water use in the UK (from Arnell, 1996)

ends up either in the groundwater or in rivers and is therefore potentially available for re-use. Although the requirement for public water-supplies is quite large, human beings require a daily intake of only about 2·5 l of water in order to survive. A further 20 l per person per day is required to provide the water necessary for proper sanitation and an additional 25 l per person per day for washing, bathing and food preparation, giving an aggregate basic water requirement of about 50 l per person per day. In most developed countries the actual water consumption is between 150 and 250 l per person per day, the balance being used for non-essential purposes, such as garden watering and car washing, etc. Although most municipal water ends up as wastewater, if it is treated locally to a high enough standard, it can be regarded as a temporary withdrawal.

There are 55 countries in the world where the average consumption is less than 50 l per person per day, resulting in widespread health problems. The World Health Organization estimates that there are 250 million cases of water-related diseases annually resulting in the deaths of over 5 million people. Only a small proportion of these deaths are due to lack of adequate amounts of drinking water: most cases are due to inadequate supplies of water for washing and personal hygiene, and the lack of proper water treatment or poor drainage leading to insect vector diseases such as malaria.

1.5   The effect of human activity on hydrological processes

In the past, human beings have adapted to spatial and seasonal variations in water availability by large-scale migration, and many animal and bird species still migrate annually over long distances. However, the introduction of political borders has made seasonal human migration almost impossible. Therefore, mankind has interfered deliberately with the natural hydrological processes in order to enhance the availability of water for direct consumption or for agricultural use. Sometimes the intervention has been to mitigate the effects of too much water in the form of soil erosion or flooding or, in other cases, to harness the potential energy of rivers to generate electrical power. In addition, many other human activities have had unintentional impacts on natural hydrological processes. The general effect of these activities is to reduce the attenuation of runoff and increase its volume. In the past centuries, this human intervention has tended to be on a small scale, but more recently the scale of human intervention has increased.

For example, the scale of some major dam projects is such that they may have significant impacts outside the country in which they are sited.

The control of water resources is becoming a major source of conflict among many countries where water is scarce. Political borders rarely coincide with hydrological divides and so the sharing of water resources is essential. It has been estimated (Gleick, 1996) that 220 river basins are shared between at least two countries. The construction of a large dam by one country could be seen as a threat by a neighbouring country and such a dam would be an obvious target for militant elements. Enormous potential destruction from a dam failure could be caused by a relatively small device. Tensions also exist within countries about rights to water. Many African countries, for example, have a system of Water Courts, which are specifically to resolve conflicts over water rights.

The following three sections describe some of the specific human activities that have an impact on water resources.

1.5.1   Forestation and deforestation

The hydrological effects of forestation and deforestation are sometimes contradictory. When land is cleared of trees there is an increase in water yield due to the reduction in losses from interception and transpiration as well as due to the reduced infiltration resulting from the compaction of the soil. Peak flows will therefore increase and peak stage (water level) may increase even more because of aggradation (increase in the bed level) in the channels resulting from increased sediment runoff. However, there is an argument that planting forests can increase precipitation due to the supply of moisture from transpiration.

The effect of forestry on low flows is even more contentious. It is normally held that forests tend to enhance low flows because of the greater infiltration and storage of forest soils, although increases in low flows have been reported as a result of forest clearance. This could be due to a combination of decreased evapotranspiration resulting from the removal of vegetation and the lowering of the water table together with the increased depth of soil contributing to runoff and therefore greater groundwater recharge.

1.5.2   Grazing and agriculture

The effect of grazing is generally to reduce infiltration because of compaction and the removal of vegetation. As before, this results in increased runoff and sediment yield and consequent changes in the channel morphology. In addition, grazing near the channel can cause bank erosion. The introduction of exotic vegetation can increase bank roughness and reduce the channel cross-section as well as reduce evaporation from lakes. Some agricultural practices, such as terracing and multi-cropping, may preserve the natural storage by reducing the speed of runoff. However, ploughing will tend to enhance the speed and volume of runoff.

1.5.3   Urbanisation

The general effect of urbanisation on the hydrological characteristics of a catchment is to reduce the amount of infiltration into the ground and to increase the speed of runoff. The lag time between precipitation peak and discharge peak can be reduced by a factor of up to 8 in urban areas (Wohl, 2000). The effect of urbanisation on the overall response of a catchment tends to be greater for small, frequent floods rather than more extreme events. In the latter case, the catchment is saturated and the extra runoff from the paved areas will be marginal.

It is also commonly assumed that rainfall losses in urban areas are less than those from adjacent rural areas because of the lower infiltration. However, if the significant areas of urban parks and gardens are taken into account, as well as the infiltration capacity of many paved surfaces, the differences may not be significant. Nevertheless, it does appear that losses due to evaporation are less for impermeable surfaces.

The other important effect of urbanisation is that it increases the exposure to the hazard of flooding. Almost all of the ancient towns and cities in the United Kingdom were built close to rivers. In some cases, the original towns were built on nearby high ground but more recent expansion has occurred on the level, fertile floodplains adjacent to the river and in recent years there has been considerable damage to these areas from flooding (Figure 1.11).

There is also some evidence that urban areas can actually increase the amount of precipitation. It is possible that the higher temperatures in urban areas lead to increased convection and that the roughness of the land surface also enhances the upward movement of air. Urban areas also provide additional sources of water vapour from combustion and industrial processes, as well as greater concentrations of particles that act as nuclei for the formation of raindrops. The increase in precipitation in, and downwind of, urban areas can be up to 15%.

Figure 1.11   Flooding in Paisley (courtesy of Meiklewall Ltd)

Urbanisation also has an effect on the water balance because of changes in the drainage patterns resulting from the provision of artificial stormwater drainage as well as mains water supply and foul drainage systems. An extensive stormwater drainage system will direct water into channels and rivers rather than allow it to infiltrate to the groundwater. Another feature of urban areas is a well developed mains water supply, which means that flows are introduced from outside the urban limits. Since the direct consumption of this water is generally less than 10%, most of this water will end up either being removed by the foulwater system or recharging the groundwater through leaking pipes or outflows from septic tanks or from irrigation of parks and gardens (Foster *et al.*, 1994).

1.6   Hydrological management

In addition to the incidental impact of various human activities, the hydrological cycle has been modified by specific management measures such as the following.

1.6.1   Irrigation

Irrigation is required where the total annual rainfall is less than that required by the crop or where the distribution of rainfall does not coincide with the crop requirements. More than half of the world's agriculture depends on some form of irrigation. In most cases, supplies need to be provided from some form of storage, which might be the natural storage provided by a groundwater aquifer or the storage of surface water in specially constructed reservoirs. Where there is adequate dry season flow in a nearby river, flow can be diverted into irrigation channels by constructing small weirs in the river channel or by pumping directly from the channel.

The use of groundwater storage has certain advantages over surface storage. There is very little evaporation and a lower risk of contamination by humans or animals, and the resource can be exploited close to the point of use, reducing the cost of conveyance. However, there is a problem of dissolved salts being retained in the upper layers of the soil, leading to a progressive loss of water quality.

It has been seen that agricultural irrigation is by far the largest single component of water use and it is sometimes argued that there is considerable wastage of water in agricultural use. However, much of the water that is not used by the crops is returned to the groundwater and is therefore, in theory, available for re-use. Only the evaporation from surface storage and conveyance channels can really be considered as a loss. Nevertheless, there is a trend to more efficient, intensive irrigation, which has led to developments such as micro-drip irrigation.

1.6.2   Dams and reservoirs

One of the most obvious examples of large-scale anthropogenic interference in the natural hydrological processes is the construction of large reservoirs. Such reservoirs are constructed for a number of reasons, including flood mitigation, power generation and irrigation. About 6000 km^3 of water is stored in reservoirs worldwide and in some rivers the volume of reservoir storage is two or three times the mean annual flow. Hydropower generates around 20% of the world's electrical energy consumption, amounting to about 640 GW. In 63 countries, hydropower produces more than 50% of total electricity supply. The Three Gorges Project in China, for example, will have a total storage capacity of 40 km^3 and will generate about 18 GW of electricity. However, fewer large dams are now being constructed: of the ten largest dams in the

USA, only one was completed after 1970 (Gleick, 1996). This is partly because many of the most suitable sites have now been exploited and partly because of increased opposition to large dam projects.

The major hydrological effect of reservoirs is to increase the storage of water and attenuate the peak flows in the river. However, they also have many other hydrological consequences that may be less desirable (Adams, 1992). For example, the operation of hydro-electric dams can impose unnatural flow patterns on rivers in which the flow rate is mainly controlled by the demands of power generation. This can result in rapid changes in water level, which disrupts spawning fish and nesting birds as well as river-bank agriculture. Reservoirs trap a large amount of the sediment carried by rivers that previously enriched the soil of the floodplain (through periodic flooding which also recharged the aquifers). An extreme example of this effect is the Aswan High Dam, which traps 98% of the silt that formerly would have been spread over the floodplain every year. The removal of the sediment from the river flow, apart from reducing the efficiency of the reservoir, alters the regime of the river downstream, which leads to adjustments in the channel profile. The width of the channel may decrease, reflecting lower peak-flows and since the sediment capacity of the river will be greater than its load, there will be more erosion, leading to incision of the channel. Further downstream, aggradation may occur where the load exceeds the capacity.

River flows are reduced by the high rate of evaporation from reservoirs, which also tends to increase the salinity of the water. The quality of the water stored in a large dam is likely to be reduced by high-nutrient levels leading to algal growth and by depleted oxygen levels due to the decomposition of vegetable material. The temperature of the stored water when released is also often quite different from that of the natural river-water, which can have serious effects on fish and other fauna.

Large dams also bring many ecological problems. Apart from the considerable loss of agricultural land in the flooding of the reservoir, one of the most significant environmental impacts of large dams is the disruption of the natural ecosystem of the river. Dams are a physical barrier to longitudinal movement of not only sediment, but also fish and other species. Far fewer fish are able to survive in the still waters of reservoirs compared to the fast flowing waters of the original river. Many species are lost because of the loss of habitat when the reservoir is flooded and when the displaced farmers clear other land. The large areas of still water provide a breeding ground for mosquitos and snails, which spread malaria, bilharzias and other diseases. The risk of flooding and drought is replaced by the risk of waterborne disease. It also appears that the effect of the methane produced by the decomposition of vegetation in a large hydro-electric reservoir can exceed the greenhouse gases produced by an equivalent fossil-fuel power station.

The social consequences of large dams are well known. Many people are often displaced from relatively fertile agricultural land to much poorer soils with no irrigation capacity and, generally, these people do not share the benefits of the dam in terms of electricity and flood alleviation, etc.

1.6.3   Small-scale catchment management

A significant reduction in the risk of flooding can be achieved by enhancing the natural storage in a catchment rather than building large artificial reservoirs. Likewise, the irrigation benefits of large dams may also be achieved by small-scale traditional measures. Runoff harvesting, for example, is the channelling of water running off slopes to lower fields. In some regions, large areas of bare rock can be used to collect substantial amounts of water. The terracing of small seasonal creek beds slows the flow and spreads it over wide areas of soil. The heaping of weeds between plants also reduces runoff. Where there is seasonal flooding of a floodplain, the judicious construction of small dams can retain the floodwater in pools for agriculture. The general feature of these techniques is that they encourage replenishment of groundwater storage, which reduces the losses due to evaporation.

In the case of perennial rivers, which dry up for part of the year, a significant supply of water can often be obtained from shallow wells on the river banks that use the water within the bed of the river. This supply may be enhanced by the construction of small rock dams within the sediment of the river bed.

1.6.4   Channelisation and levees

Channelisation is the removal of obstructions in the channel and the artificial widening, straightening or deepening of the channel to increase its capacity, usually with an artificial lining. Channelisation has been practised for centuries and almost 40% of urban rivers in the UK have been modified in this way. Levees are linear banks constructed on the river banks parallel to the flow but usually set back at some distance from the main channel in order to contain the overbank flow. Although channelisation and levees are a common traditional form of flood protection, they do present a number of problems. Their main disadvantage is that they merely pass the danger of flooding further downstream and may in fact increase the risk. They are also often difficult to construct in urban areas without seriously reducing the amenity value of the area. Temporary or demountable flood barriers are now being developed to overcome this problem.

1.6.5   Land drainage

The installation of land drains can have a significant effect on the hydrological characteristics of a catchment. They are generally designed to remove water from the upper saturated zone through gravity. The main hydrological impact of land drainage relates to the increase in drainage density and soil-water capacity. In general, higher drainage densities result in increased peak runoff flows. This is because channel flows are much faster than interflow or overland flow. However, it has been argued that closely spaced drains can reduce peak runoff rates as they reduce or eliminate surface runoff. The effect of the increased storage depends on the storm characteristics and antecedent conditions. Where the rainfall is not sufficient to produce saturated conditions, the drainage tends to reduce peak flows: for higher rainfall levels it will tend to increase peak flows. Robinson (1990) found that field drainage increased peak flows for permeable soils but not from clay soils. This was thought to be because the permeable soils already drained by subsurface flow and this was encouraged by the land drainage. By contrast, clay soils tend to be waterlogged for long periods and storm period flows are predominantly on the surface. Artificial drainage largely eliminates the surface flow and therefore reduces peak flows.

1.6.6   Urban drainage

Urban drainage systems are designed to remove the rainfall which accumulates on relatively impermeable surfaces in towns and cities. In doing so, they further increase the speed of runoff and reduce the natural attenuation of the land surface. The runoff from an urban area may have a significant effect on the hydrograph of a natural watercourse which is receiving the runoff, since the latter may not have responded to the rainfall. More attention is now being paid to increasing the attenuation of urban drainage systems by the use of tanks and reducing the flows by promoting infiltration.

1.7 Recent developments in enhancing water resources

1.7.1 Wastewater re-use

The major part of a municipal water-supply ends up in an urban drainage system being discharged to a watercourse. Although in some cases, the water is re-used as part of an abstraction further down the river, the possibility of re-using wastewater for specific applications is now receiving increasing attention. The particular applications include: the recharge of groundwater aquifers, the supply for industrial processes, irrigation for agriculture and urban parks and gardens, as well as potable water-supply. It is estimated that the annual potential for wastewater re-use is between 2 and 3 km^3 per year, which is small fraction of the total municipal consumption of some 50 km^3 per year. However, in some areas it is already a valuable enhancement to the water resource. In Israel, 70% of wastewater is treated and re-used for irrigation (Gleick, 1996) and in Windhoek, Namibia, up to 30% of drinking water comes from treated wastewater. Strict regulations are normally required about the standard of treatment needed for different uses, which range from primary treatment for irrigated grass and non-edible foliage to tertiary disinfection for edible food crops.

1.7.2 Desalination

Although nearly 97% of the water on the Earth is too saline to drink, there is a natural tendency to regard the oceans as an infinite source of water for drinking or irrigation. The process of desalination is technologically well developed but the high cost and energy requirement of desalination plants have severely limited their contribution to freshwater resources. At present, the total production from desalination plants is about 7 km^3 per year, of which the majority is in the Middle East and the USA. The high cost of transporting water also limits the use of desalination to coastal areas.

1.7.3 Cloud seeding

Precipitation normally results from the freezing of super-cooled water onto small atmospheric particles (nuclei). It was discovered that certain salts, notably silver iodide, can induce precipitation by acting as additional nuclei. Cloud seeding is the introduction of the salt crystals, usually from an aircraft, into existing clouds so that immediate freezing occurs. Seeding is unlikely to be effective in clear skies or where the cloud temperature is above –5°C because of the absence of super-cooled water droplets. Raindrops are only formed in such conditions after the collision and coalescence of liquid water droplets, which requires much larger nuclei. Cloud seeding is particularly effective with convective cumulus clouds and is less effective with the stratus type of clouds. The overall effectiveness of cloud seeding over a large area is difficult to evaluate. It can certainly produce local increases in rainfall but this may be the result of redistributing the existing moisture in the cloud over different areas.

1.8   Summary

This chapter has shown that the movement of water in the hydrological cycle depends partly on the complex interaction between the ocean-atmospheric systems and the incoming solar energy, and partly on the way man has, deliberately or otherwise, altered the nature of the land-based phase of the hydrological cycle. The next chapter explores the natural and human-induced changes in the Earth's climate systems which have occurred, or are likely to occur in the future.

1.9   References

Adams, W. N. (1992). *Wasting the Rain*. Earthscan Publications, London.

Budyko, M. I. (1974). *Climate and Life*. Academic Press, London.

Chorley, R. J. (1969). *Water, Earth and Man*. Methuen, London.

Davis, K. S. and Day, J. A. (1961). *Water The Mirror of Science*. Heinemann, London.

Foster, S. S. D., Morris, B. L. *et al*. (1994). *Groundwater problems in urban areas*. Thomas Telford, London.

Gleick, P. H. (1996). *The World's Water*. Island Press, Washington DC.

IPCC (1995). *Climate Change 1995: The Science of Climate Change*. Cambridge University Press, Cambridge.

Lamb, H. H. (1972). *Climate: Present, Past and Future*. Methuen, London.

Robinson, M. (1990). *Impact of Improved Land Drainage on River Flows*. Institute of Hydrology, Wallingford.

Wohl, E. E. (2000). *Inland Flood Hazards*. Cambridge University Press, Cambridge.

Chapter 2

Climate change

This chapter considers the question of whether there is a long-term trend in the pattern of the Earth's climate and, if so, what the consequences are likely to be. It sets the current debate about climate change in the context of changes that have occurred in our climate both on a geological time-scale and in more recent times. The various causal mechanisms which might initiate a change are considered, and the chapter then looks at how the nature of the Earth's climate determines the scale and nature of the resulting change. The role of numerical modelling in predicting climate change is described, together with the limitations of the modelling approach, and the chapter concludes by considering the possible effects and impacts of climate change.

2.1 Historical perspectives

The climate of the Earth is continually changing in time-scales that vary from seconds in the case of wind speed to the daily variations in temperature and seasonal variations of weather patterns, and to the variations in climate extending over many decades. The present debate on climate change is concerned with whether there is a gradual progressive trend in various climate parameters. The essential problem is to detect such a trend among the natural wide variations in these parameters.

One of the difficulties in determining whether a trend exists in a particular record is deciding on the time-scale. What might appear to be a linear trend when viewed in a time-series of, say, 100 years (which is roughly the order of the longest directly recorded rainfall series), might be part of a cyclical variation with a period of several thousand years. One must therefore set the discussion of climate change in the context of the natural variations in the Earth's climate, which have occurred over the thousands of millions of years of its history.

2.1.1 *Measurement of climate over geological time-scale*

There are various techniques, some of which have been developed relatively recently, which can help to infer the Earth's climate from geological records. Useful information about temperature and snowfall can be obtained from ice sheets and glaciers that were formed from snow accumulated over many thousands of years. It has been found, for example, that the ratio of two isotopes of oxygen (^{16}O and ^{18}O) in the air trapped in bubbles within the ice, gives a good indication of the temperature at the time the snow fell, and the technique has been used on ice cores from Greenland up to 15 000 years old (Burroughs, 2001). Inferences can also be made about the amount of dust in the atmosphere due to volcanoes and other sources, from the composition of air trapped in bubbles within the ice. It has been established that the dust content of the ice is related to changes in temperature, indicating abrupt shifts in atmospheric circulation patterns, and the analysis of the gas bubbles has been useful in providing a record of variations of trace gases, such as carbon dioxide and methane and how they relate to past climate changes.

Ocean sediments also provide a valuable source of information on past climate change. In particular, the types of species found in the sediment and their relative abundance are good guides to surface temperatures, and it is possible to build up a spatial picture of temperature and its variation with time. For the period within the last 40 000 years or so, this can be done by carbon dating, which relies on the predictable decay of the carbon isotope ^{14}C, but dating of older sediments relies on a knowledge of sedimentation rates, which increases the uncertainty. On longer time-scales, some evidence of dating can be found from the residual magnetism in the rocks. It has been established that the Earth's magnetic field undergoes periodic reversals, on average about every 700 000 years, and this is imprinted in the magnetic materials of the rocks as they are formed.

The stratigraphy of oceanic sediments can also give information about climate change through changes in sea level. Where the sediments are deposited on the edge of a continental shelf, the highest point of the layer will represent the highest sea level when it was formed, and the sequence of layers can indicate if the sea level was rising, falling or stationary at that time. However, considerable caution is needed in interpreting these records, since the shape and position of the continental land-masses have altered dramatically over the geological time-scale, and it is only in the last 600 000 years that continents have been in their present positions.

Finally, tree rings can provide data on the variations in climate over the life of a tree, as well as providing a means of dating the record. The width of the annual rings in a cross-section of the trunk of a tree is proportional to the amount of growth which occurred in that year. This is a composite effect of the mean temperature, sunshine and rainfall, and isolating each individual element can be difficult. The normal approach is to correlate the ring width in more recent years with appropriate local measurements of rainfall, temperature, etc., and then to extend this relationship back to the early life of the tree. In some cases, more detailed information can be obtained from the analysis of the structure of the wood within each ring. This might include the length and conditions within each season and the presence of frost damage in early spring growth. Care has to be taken in this analysis to allow for the normal decrease in the ring width as the tree matures.

It can be seen from the above that proxy measurements may be useful in providing a picture of long-term climate changes and, indeed, these are the only way of obtaining such a record. However, the limitations of proxy measurement need to be recognised. Firstly, in most cases the measurement is not a direct measurement of a meteorological characteristic. It has been seen that tree rings, for example, are related to both temperature and rainfall; the depth of ice only indirectly measures precipitation, and the use of fossil and pollen records depends on complex relationships with climatic conditions. In the second place, it may not always be certain that the geological record being used has not been disturbed in some way, leading, for example, to fossils in one layer becoming mixed with another layer. Thirdly, the records are not usually continuous, either spacially or temporally. Some fossils may be the product of exceptional conditions, and as a result the record may not be representative of the general conditions at the time.

2.1.2   *Variations in climate over geological time-scale*

Very little is known of the Earth's climate more than 1000 million years ago, although it was thought that a general warming in the Cambrian period was followed by two ice ages between 900 and 600 million years ago, in which most of the Earth's surface was frozen (Figure 2.1). Since that time there have been several dramatic shifts in the fossil record, indicating global cataclysmic events such as asteroid impacts or volcanic

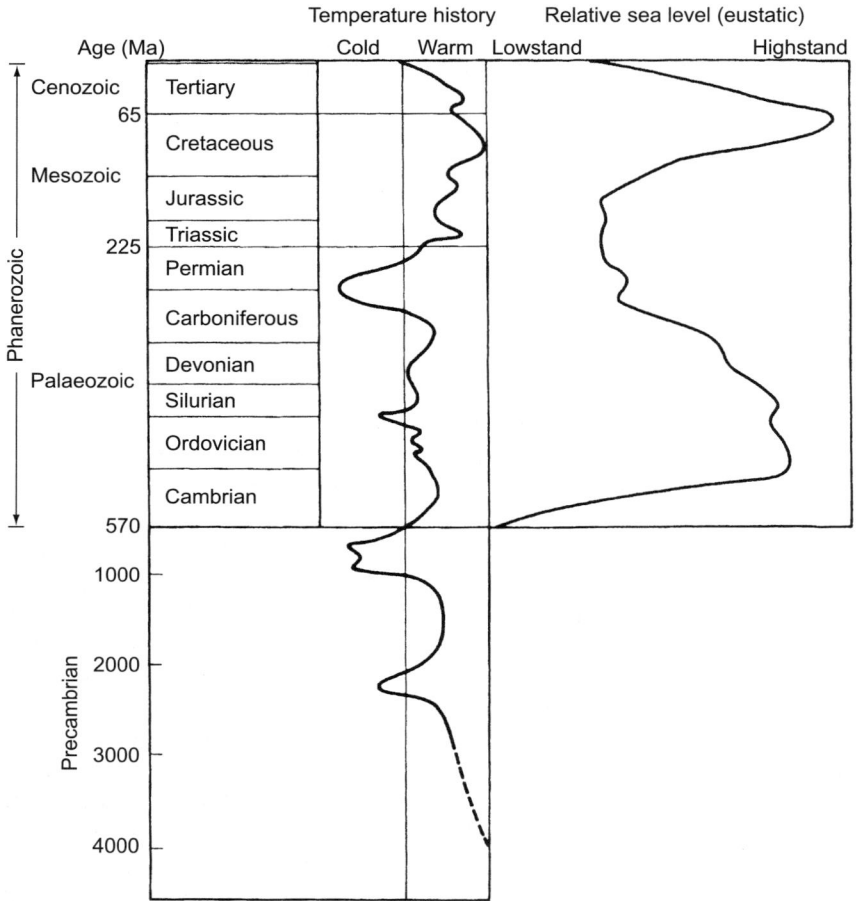

Figure 2.1   Variation of temperature and sea level over geological time (Brown et al., 1992)

eruptions. The overall impact of these events is such that 99·9% of all the species that have ever lived have become extinct over the last 600 million years (Burroughs, 2001).

Over the same period of 600 million years or so, there have also been major changes in the mean sea level, which has varied between about 250 m above the present level and 200 m below. Part of this variation may be due to changes in the volume of ice stored in the ice caps, and part to changes in the volume of water due to thermal expansion.

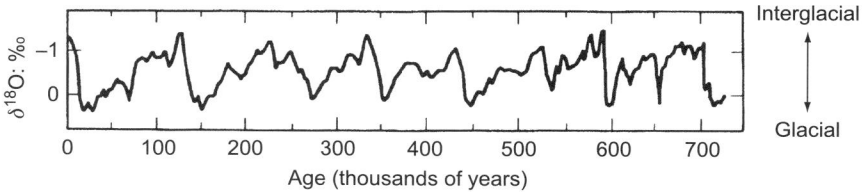

Figure 2.2   Variation of temperature from ice cores (Brown et al., 1992)

Although there have been several periods when a large part of the Earth has been covered in ice, most interest has centred on the ice ages which have occurred in the last million years. The concept of ice ages only became widely accepted in the 1950s and there has been some dispute about the exact number of such periods. It is now generally agreed that there have been four such ice ages in the last 800 000 years, occurring on average about every 200 000 years, with warmer interglacial periods. From ice records, it appears that the onset of the ice ages was usually fairly gradual, taking place over 70–80 000 years, whereas the end of the ice ages tended to be quite sudden, occurring in less than 10 000 years and giving rise to a 'saw tooth' pattern in the temperature record (Figure 2.2).

The last ice age occurred between 75 000 and 25 000 years ago, and we are currently near the peak of an interglacial period. During the most recent ice age, global temperatures were about 5°C lower than present values, but there were interludes of relatively warmer climate with time-scales of years to millennia. After the end of the last ice age there was a period of very rapid change between about 15 000 years ago and 6000 years ago (Figure 2.3), during which time the global sea level rose by about 90 m, equivalent to nearly 1 m per century. About 6000 years ago, temperatures in the northern hemisphere reached a maximum which was about 2–3°C warmer than present temperatures, with a predominance of westerly type weather and higher precipitation, and this is thought to have increased the spread of agriculture. From about 5000 years ago, the climate has been gradually cooling and becoming drier. Glaciers in Europe began to advance, the rise in mean sea level reduced to about 50 mm per century, and there was a noted decline in rainfall in the Middle East and North Africa. However, there have been considerable fluctuations in this trend especially in the last 2000 years. For example, there was a warmer period in the ninth and tenth centuries AD but the length of this period varied considerably. There was also a 'mini ice age'

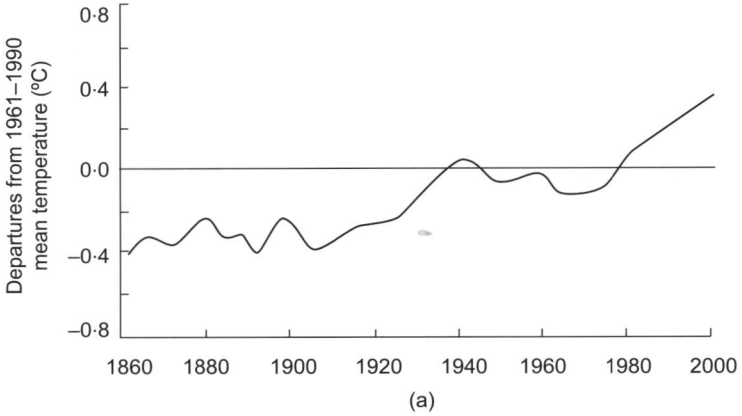

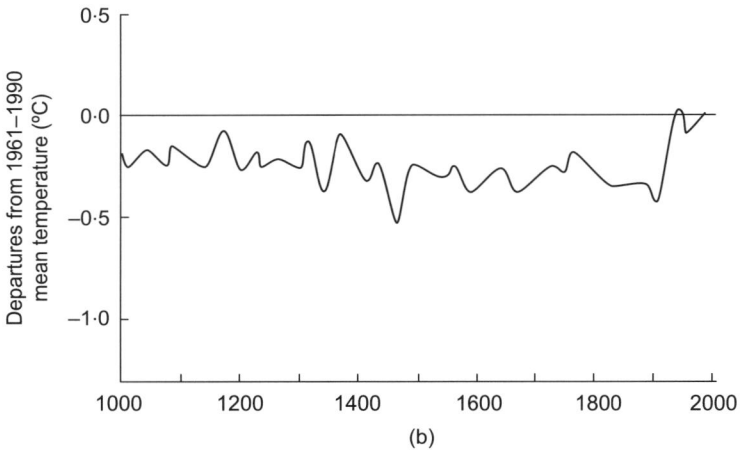

Figure 2.3 Variation in the Earth's surface temperature: (a) global; and (b) northern hemisphere (from IPCC, 2001a)

in the seventeenth and eighteenth centuries, with frequent freezing of the River Thames in London and an expansion of the alpine glaciers.

2.1.3 Climate variations in recorded history

From 1690 there is a record of temperature for central England and this provides detail of the more short-term fluctuations in temperature. From the beginning of the eighteenth century, temperatures changed from some of the coldest to some of the warmest on record within 40 years. There was a sudden return to colder conditions in 1740, which lasted until the end of the nineteenth century. Another feature which is demonstrated by this record, is the variability of the temperatures over periods of years or decades. Elsewhere in the world there was similar variability, but the colder periods did not always coincide with those of Europe.

In the twentieth century there was a gradual warming that has resulted in an increase in global average temperatures of about 0·06°C per decade. However, this increase has not been uniform; in two 20-year periods, 1925–44 and 1978–97, the temperature increase in the northern hemisphere has been over three times this rate. Moreover, the more recent increases have mainly been restricted to winter temperatures, especially night-time temperatures with a general reduction in the diurnal range.

The 1990s was almost certainly the warmest decade since 1861 and probably for more than 1000 years. Five of the six warmest years on record have occurred in the 1990s. Satellite images show that in the last 30 years of the twentieth century, snow cover reduced by about 10% with a similar reduction in the extent of sea ice. The average sea level has risen by between 0·1 and 0·2 m over the past 100 years. Overall, precipitation has increased by up to 1% per decade in the mid and high latitudes of the northern hemisphere, with a lower increase in equatorial areas and a decrease in sub-tropical areas of the northern hemisphere. However, the

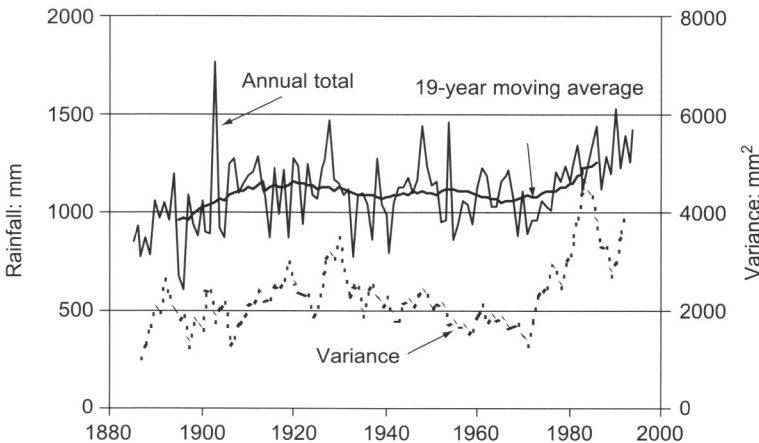

Figure 2.4   Annual rainfall in Paisley 1885–1995

changes in the last 30 years have been much more than this. Figure 2.4 (Mansell, 1997) shows the annual rainfall for Paisley, in the west of Scotland, between 1885 and 1995. The average increase is about 5% per decade, but in the last 30 years it has been about double this. The graph also shows how the annual variance has increased significantly in the last 30 years.

2.2   Causes of climate change

The climate is a complex interaction of many processes. Although climate change may be initiated by external factors, such as variability in solar radiation or changes in human activities, the effects of such factors may be magnified, reduced or distorted by the response of the Earth's climate system. In this section, the primary external destabilising effects are considered first, followed by a discussion of the ways in which the Earth's climate system could respond to these stimuli.

2.2.1   Solar variability

It has been known for several hundred years that there are variations in the sun's activity. Sunspots are relatively cool areas of the sun's surface, which fluctuate in number in a regular pattern, with a mean period of about 11 years. Although the sunspots are cooler than the rest of the sun's surface, it has been found that the sun's output actually increases slightly during sunspot activity. It is thought that the increased convection associated with sunspots brings hotter gases from within the sun to the surface. Although the change in the total energy emitted by the sun over an 11-year sunspot cycle is only of the order of 0·1%, most of this change occurs in the ultraviolet (UV) part of the spectrum, which is preferentially absorbed in the higher atmosphere. It is thought that the increased absorption of radiation in the upper atmosphere causes photochemical changes which reduce the amount of solar radiation entering the lower atmosphere. It is also possible that increased UV radiation leads to an increase in the concentration of condensation nuclei, which results in more cloud formation. Recent research has suggested that there is a significant correlation between increases in UV radiation and temperatures in the northern hemisphere, which accounts for about half of the warming between 1860 and 1970 and about a third of the warming since then.

Another mechanism by which changes in sunspot activity causes changes in climate is through the magnetic fields of sunspots. Sunspots alternate in polarity in successive 11-year cycles and it has been suggested that this could influence the occurrence of thunderstorm activity. A periodicity of 20–22 years has been observed in many climate records (for example, Figure 2.3) and is in fact more prevalent than the 11-year sunspot cycle.

Although the mean period of sunspot activity is about 11 years, a more detailed spectral analysis indicates that there are two peaks at periods of 11·1 and 10·0 years. The interaction of these two signals generates a cycle with a frequency equal to the difference between the frequency of these components, which is about 100 years and, in fact, cycles of sunspot activity with such periods have been noted in observed records. The analysis of tree-ring data indicates there may also be periods of 350, 500 and 2300 years.

2.2.2 Orbital variations

The orbit of the Earth about the sun is in the form of an ellipse, with the sun at one focus of the ellipse. The distance of the sun from the centre of the ellipse and the eccentricity of the orbit change with time, as does the time of year when the Earth is farthest from the sun and the inclination of the Earth's axis to the plane of its orbit (obliquity). The periods of these variations range from about 20 000 years to over 400 000 years. However, if the effects are integrated over the whole of the Earth's surface for a complete year, the energy flux depends only on the eccentricity. The current value of the eccentricity is 0·017, although it has varied between 0·001 and 0·054 with a period of about 100 000 years. The other factors, nevertheless, have a major influence on the seasonal and spatial distribution of solar radiation over the Earth's surface. For example, it is the obliquity which controls the seasonality of climate: if it were zero there would be no seasons.

The 100 000-year periodicity of the eccentricity matches the periods observed in the ice ages, although it is known that the eccentricity is the weakest of all the orbital effects. Recent modelling suggests that ice ages are triggered by the shorter period variations in the obliquity and other effects, and that there are time constants in the growth and decay of the ice sheets which explain the longer periods. The 100 000-year period may also be the result of differences between the shorter period perturbations.

2.2.3 Volcanic activity and major impacts

Sudden volcanic explosions or impacts of major extra-terrestrial objects can inject vast amounts of dust into the atmosphere. If this reaches the upper atmosphere 15 to 30 km above the surface, it can remain suspended for several months or years. Sulphur dioxide in volcanic emissions is even more important, as it is converted into sulphuric acid aerosols. The combined effect of these particles is a heating of the stratosphere and a cooling of the lower atmosphere.

Such effects have been observed following the eruption of Krakatau in 1883, Mount Pinatubo in the Philippines in 1991, and other volcanoes. Where the eruption does not reach the upper atmosphere, such as the Mount St Helens eruption, the effect is limited. However, where the emission contains large amounts of sulphur, such as the eruption of El Chichon in 1982, even a modest event can have a significant effect.

There is considerable evidence that the Earth has been struck by large stellar objects, such as meteorites or asteroids, in the past, the most recent example being a meteorite which burnt out over Siberia in 1908. Although this was a relatively small object, some 50 m across, it caused a seismic impact equivalent to five on the Richter scale. There is evidence that a much larger object, some 10 km across, formed a crater in Mexico. This would certainly have caused a massive earthquake, tidal wave and widespread fires, and would have ejected enormous amounts of dust into the atmosphere. It is estimated that such an impact would have resulted in a reduction in temperatures of between 20°C and 40°C, which might have lasted for up to a year.

Although the results of a volcanic eruption or a meteorite impact would be catastrophic in the short term, it is not certain that they would result in any long-term climate change. However, there is some evidence that major volcanic eruptions which coincide with a long-term but unconnected cooling can be enough to trigger the start of an ice age. There is also a suggestion that volcanic activity itself may be triggered by changes in the stresses in the Earth's crust following changes in sea level as a result of glaciation or sustained changes in atmospheric pressure.

2.2.4 Human activities

Human activity has been shown to influence the climate both on a local scale and a global scale. The aggregation of populations in urban areas leads to local increases in temperatures, as well as reduced wind speeds and increased particulate concentrations, etc. However, the effect of urbanisation on a global scale is not very significant, since urban areas occupy such a small proportion of the Earth's surface.

One major potential anthropogenic threat to global climate systems, at least in the short term, is the possibility of the detonation of one or more nuclear weapons. The effect of such an event would be similar to that of a volcanic eruption or a meteorite impact and has been referred to as a 'Nuclear Winter'.

However, the major concern regarding human activities on a global scale is the effect on the Earth's radiation balance of certain gases which are emitted from the burning of fossil fuels. The energy from the sun is concentrated in wavelengths between 0·2 and 4 μm. Part of this radiation is reflected by the atmosphere and the remainder heats the Earth's oceans and land surface. The warm surfaces re-radiate the heat at longer wavelengths (7–25 mm) and certain gases in the atmosphere absorb energy at these longer wavelengths. For example, carbon dioxide (CO_2) absorbs energy with wavelengths of around 15 μm and methane (CH_4) at around 8 μm. These gases are known as *greenhouse gases*, since the process is the same as the trapping of heat by the panes of glass in a greenhouse.

The result of this 'greenhouse effect' is that the atmosphere becomes warmer. In fact, if this process did not operate, the mean temperature of the Earth would be 33°C lower than it is at present. The principal gases involved in the greenhouse effect are shown in Figure 2.5. The gas which has received most attention is CO_2, the concentration of which has increased by about 30% since 1750 (Figure 2.6) and which is now at its highest level for over 400 000 years (IPCC, 2001b). It is estimated that a

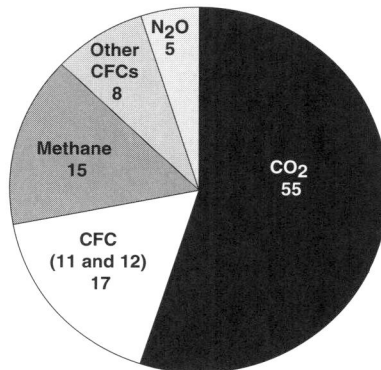

Figure 2.5   Principal greenhouse gases

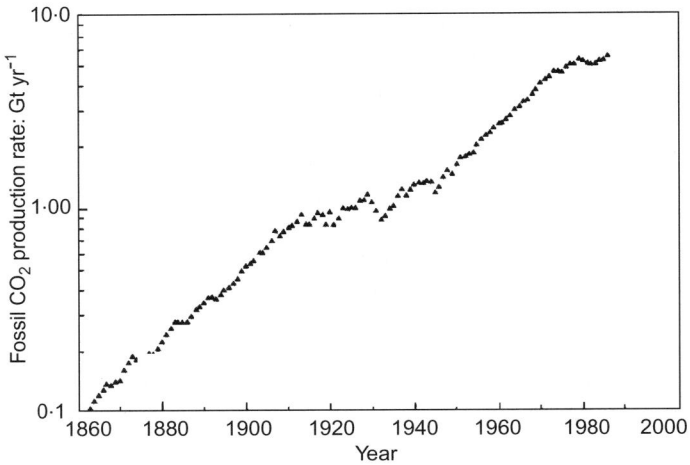

Figure 2.6   The production of CO_2

doubling of the concentration of CO_2 will be equivalent to an increase of 4 W/m^2, which is about 0·9% of the average radiation received from the sun.

Carbon circulates in the Earth's biosphere and atmosphere in a similar way to water in the hydrological cycle. It is absorbed from the CO_2 in the atmosphere by plants, especially trees, which can store the carbon for up to several hundred years. If the wood is used for combustion, the carbon re-enters the atmosphere as CO_2. If the vegetation is not removed it may be stored as coal, oil or gas for millions of years before being released as CO_2 when the fossil fuels are burnt. About 75% of the increase in CO_2 can be attributed to the burning of fossil fuels and the remainder is due to the removal of large areas of forest which act as 'carbon sinks'.

2.3   The response of the Earth's climate system

Although the driving forces behind changes in climate are now fairly well established, the way in which climate responds to these changes is complex and less well understood. In the short term, the world's weather can be considered as chaotic in the sense that it is highly sensitive to initial conditions, so that any prediction for more than about 10 days becomes almost impossible. In the longer term, the Earth's climate may be considered as a non-linear system whose response is not directly proportional to a given change.

One of the characteristics of such a system is that when it is acted on by an input which is varying at a given frequency, the system may oscillate at frequencies which are multiples (harmonics) of the exciting frequency. If the input is oscillating at two frequencies (f_1 and f_2), the system may also respond at sub-harmonics such as f_1-f_2, $2(f_1-f_2)$, etc. In practical terms, if there were variations in solar radiation with periods of say 10 and 11 years, there might be a response with a frequency of $(1/10 - 1/11)$ years^{-1}, corresponding to a period of 110 years. Thus, what may appear to be a long-term variation may in fact be the result of a particular short-period fluctuation.

Another particular problem with complex systems is *feedback*. When a climate variable alters, it may alter another variable in a way which affects the variable which triggered the change. If this circular process leads to a reinforcement of the original impact, it could lead to an escalating response out of proportion to the original stimulus and this is referred to as positive feedback. An example of this could be where the effect of a general warming is to reduce the area covered by snow, which would reduce the amount of reflected radiation, causing a further increase in temperature. Where the process leads to a dampening of the response, it is known as negative feedback. This might occur, for example, where an increase in temperature results in more water vapour entering the atmosphere, leading to more cloud cover and a reduction in temperature.

2.3.1 Atmospheric–ocean interaction

It has been noted in Chapter 1 that both the oceans and the atmosphere act to redistribute the incoming solar radiation from the equatorial to the polar regions. However, the oceans and the atmosphere interact in various complex ways, which have a considerable significance in relation to climatic trends. For example, the salinity of the oceans is partly controlled by the balance between the evaporation to the atmosphere and the input of freshwater from rainfall and runoff. Also, the heating of the oceans depends on the amount of cloud cover. Conversely, the air temperature and movement around land masses depend on the ocean temperature. It has been mentioned above that a small change in one input can cause complex changes in the patterns of climate, with magnitudes and periods quite different from those of the input, depending on the nature of the feedback and inertia of the system. The inertia characteristics of the atmosphere and the ocean, that is their speed of response to change, are quite different. The oceans respond very slowly to change, usually over a time-scale of months or years, whereas the atmosphere responds in days or even hours. The land surface responds on a time-scale somewhere in-between the two extremes.

One example of the atmosphere–ocean interaction is the so-called *Southern Oscillation*. It was noticed in the 1920s that, when the atmospheric pressure was high in the Pacific Ocean, it tended to be low in the Indian Ocean and vice versa. In effect, this represents a transfer of air mass around the circumference of the Earth at low latitudes. It has since been quantified as the *Southern Oscillation Index* (SOI), in terms of the difference in surface pressure anomalies between Tahiti and Darwin. The Southern Oscillation is linked to fluctuations in the temperature of the Pacific Ocean. Normally, trade winds cause a general westward movement of relatively cool water across the Pacific. Occasionally, these winds diminish and even reverse, allowing warmer water to spread from the western Pacific. This disruption usually begins as a reduction in the normal up-welling of cold currents off the Pacific coast of South America between January and March, which has become known as *El Nino* (Christ's Child), and gradually spreads as a narrow tongue of warm surface water across the Pacific and reaching Papua New Guinea by late summer. The anomaly in sea surface temperatures is only of the order of 1–2°C and is therefore difficult to detect using remote sensing, although the reduced density of the seawater leads to an increase in mean sea level of 5 to 15 cm, which can be more readily detected (Figure 2.7). It generally retreats eastwards and, by the end of the year, temperatures have returned to normal. Six months later the sea temperatures in the eastern Pacific are often below normal and the usual large area of warmer water develops in the western Pacific. Sometimes the El Nino can last two or three seasons or more, and it is particularly important for the fishermen in the west coast of South America, as it severely disrupts their fishing activities by reducing the plankton on which the fish feed.

The combination of the presence of El Nino and a negative SOI is referred to as ENSO, and in the last half of the twentieth century there has been, on average, about one ENSO event every five years. An ENSO event usually starts to develop in October and November, and usually causes widespread disruption to weather patterns over many areas of the world. For example, it affects the *Inter-Tropical Convergence Zone* (ITCZ), which is the area of converging air which produces precipitation around the equator. Normally the ITCZ moves from 10°N in August/September to

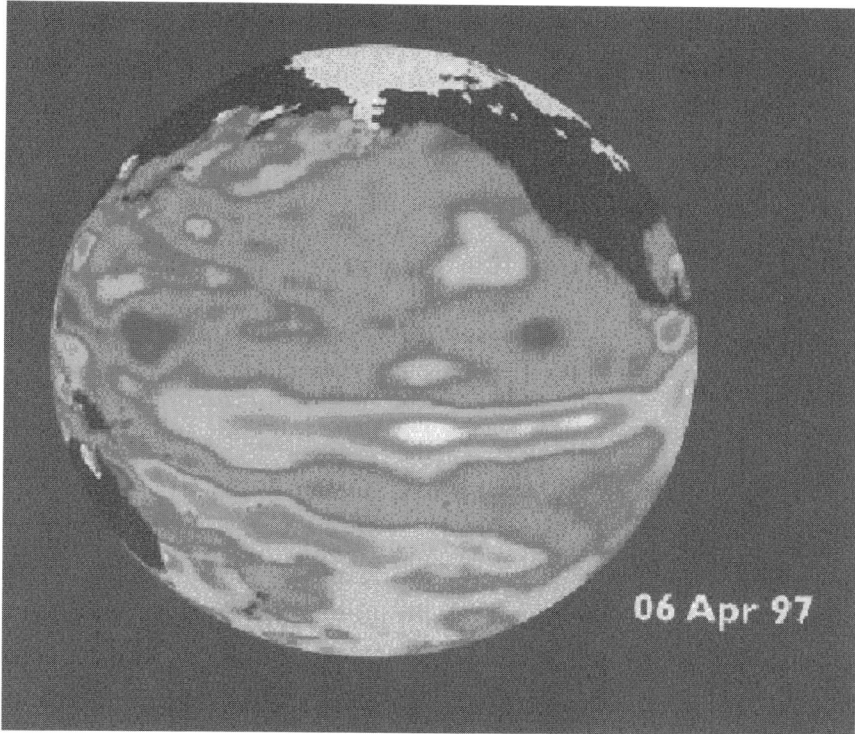

Figure 2.7 Sea surface anomalies in the Pacific Ocean. The light areas show positive height anomalies indicating higher temperatures (image courtesy of NASA)

about 3°N in February/March, but during an ENSO event it moves further south, even crossing the equator. As the warm water spreads across the Pacific, the ITCZ causes very high precipitation in the central and western Pacific. Many countries in Africa and Asia also face disruption to their rainfall patterns and, in some cases, severe drought. In particular, the ENSO causes fluctuations in the amount of monsoon rainfall on the Indian subcontinent. In some cases, the effect of an ENSO event may be combined with a shorter-term climatic disturbance known as the *Madden-Julian Oscillation*, which consists of pulses of strong winds and rains that travel eastwards around the equator in 30 to 60 days, pumping huge amounts of heat into the atmosphere. It appears that if one of these perturbations occurs as an ENSO event is about to break, it can stimulate its rapid development (Burroughs, 2001).

It has also been established that there is similar oscillation in atmospheric pressures in the northern hemisphere known as the *North Atlantic Oscillation* (NAO), which is measured in terms of the pressure difference between Iceland and Lisbon or the Azores. A high NAO Index occurs when pressure is low over Iceland and high over the Azores, which produces strong westerly winds and mild air over Europe, together with cold polar air over Greenland. The opposite situation brings cold air down over Europe and produces much greater snow cover, which tends to reinforce the low temperatures. For example, the cold winter of 1995–96 coincided with an extremely low NAO Index (Figure 2.8). The NAO fluctuates on a time-scale of several years to a few decades and seems to correlate well with changes in sea ice-cover and the frequency of depressions. There has been a gradual increase in the NAO Index from the 1960s to the 1990s (Alcock and Rickards, 2001). However, the 1990s have been characterised by extreme variations in the NAO, with both the highest and the lowest values on record occurring in that decade, as well as a sudden 'flip' between 1994–95 to 1995–96. Recent research has also shown that the NAO is linked to changes in the circulation patterns in the North Atlantic Ocean and that sudden changes in the circulation patterns have occurred in the past. It seems that changes in precipitation from low-pressure systems and runoff from continents could be sufficient, for example, to trigger relatively sudden changes in the extent

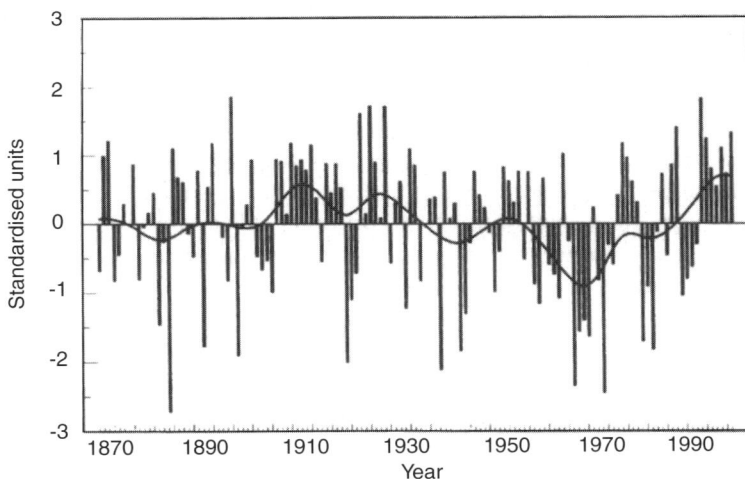

Figure 2.8   Variation in the North Atlantic Oscillation Index (IPCC, 1995)

and position of the Gulf Stream, which would have major climatic impacts over periods of several decades. It is also suggested that there is a feedback mechanism by which the changes in the position of the Gulf Stream itself influence the fluctuations in the NAO.

There is evidence that changes in ocean currents have generally coincided with the end of ice ages. However, it is not clear whether the shifts in the ocean currents are caused by climate change or vice versa. It also seems that relatively small changes in energy flux, for example, can trigger significant changes in the distribution of ocean currents which result in major climatic changes. Further back in geological history there is the added complication of continental drift, which profoundly affected ocean circulation by opening and closing certain seaways at different times. What is of current concern is whether the recent, relatively sudden increase in global temperatures resulting from human activity is sufficient to cause the ocean currents to 'switch' into a different mode.

2.3.2   *General conclusions about the causes of climate change*

It has been seen that changes in climatic patterns may be due to either the natural variability of the Earth's climate system or changes in external forcing systems, such as the sun's radiation or due to the results of human activities (anthropogenic forcing). Some of the observed changes may also be the result of complex non-linear interactions between the ocean and atmospheric systems, which can produce relatively sudden step-changes in climate following relatively minor changes in natural or anthropogenic-forcing mechanisms. In other words, different climate regimes may be able to exist without any great changes in the global energy balance.

There is still much debate about the relative importance of these factors. However, it is now generally accepted that, while much of the changes in the early part of the twentieth century might have been the result of natural variations, there is strong evidence for human influences in the records for the second half of the century.

2.4   Modelling of climate

Predictions of future patterns of climate are normally made using numerical models of the Earth's climate, which have developed rapidly with the advent of relatively cheap, powerful computers. Until recently, the main driving force for developing these models has been the provision of detailed short-term weather forecasts (i.e. for periods of up to five or six days). The basic process used for such models is to divide the atmosphere into a number of layers vertically and, then, at discrete grid points in each layer, to solve sets of differential equations based on the principles of the conservation of mass, momentum, energy and water (in all its phases), as well as the Newtonian equations of motion and the various laws of thermodynamics. The results of the models consist of values of pressure, temperature, wind speed, humidity and precipitation at the grid points. The total number of equations which need to be solved obviously depends on the number of layers and the density of grid points, as well as the time-step used in the calculations. Current weather-forecasting models use up to 30 layers with a horizontal grid spacing of 50 km, giving some four million points. With a time-step of 15 minutes, a typical forecast can involve up to 20 trillion calculations.

The general circulation models used to investigate long-term changes in climate are based on a similar approach, but there are significant differences. Firstly, the limitations of even the most powerful computers mean that the models cannot use the same spatial resolution, especially the horizontal grid-spacing. The spacing used in climate-change models is typically two or three degrees of latitude or longitude (3–400 km). The vertical resolution is also reduced slightly to about 20 layers. The second important difference is that in the short-term (five to six days) the state of the ocean can be considered as being constant and forecasting models can use current values of surface temperature. Over the period for which climate-change models are run (typically several decades) this is obviously not the case and climate-change models also model the oceans in layers (typically also about 20 layers). This presents a number of problems because of the different characteristics of the oceans and the atmosphere. Many of the ocean currents and other features are on a scale that is too small to model accurately with the grid spacing which is necessary in climate models. It is also difficult for models to deal with the short-term fluctuations in the atmosphere and the long-term, but much greater, fluxes which occur in the oceans. These factors often cause problems in the running of climate models, such as a general drift in the output over a period of time, which may require empirical adjustments.

2.4.1   Uncertainties in climate models

There are now over a dozen global climate models (GCMs) in operation and various studies have been carried out to evaluate their performance.

The variations between the different models show that there are still significant uncertainties which need to be understood more fully. One of the major areas of uncertainty is in predicting the amount of cloud cover. Clouds have a major impact in the radiation balance in terms of the amount of energy they absorb or reflect. Another area which requires more study is the effect of land use on the climate. Different types of land surface reflect and absorb energy in different ways, which are not always well represented in models. The high albedo (reflectance characteristic) of deserts tends to reinforce dry conditions and satellite data suggest that, where large-scale weather patterns increase precipitation, the increased vegetation reduces the albedo and reinforces the wet conditions. There is, likewise, a complex relationship between the soil moisture and the amount of evaporation and precipitation. It has also been shown that, in very dry conditions or where there has been considerable removal of vegetation, there is a reduction in surface radiation because of the absorption of radiation by the dust in the atmosphere. A similar effect occurs with sulphur dioxide aerosols, which are small particles formed from emissions of sulphur dioxide (SO_2) from power stations, and there is considerable uncertainty about the extent to which this effect is significant in mitigating the effects of global warming. Predicting the future emissions of SO_2 require assumptions about the regional patterns of industrial activity in the world, which are difficult to make.

2.4.2   The use of global climate models

Global climate models (GCMs) allow the effects of a given change in the levels of greenhouse gases to be assessed in terms of mean temperature, precipitation, etc. The first step in the use of climate models is to select a given scenario. This is a set of assumptions about population growth, economic growth, the use of energy, land use, etc., which results in a specific increase in the level of greenhouse gases, normally represented by CO_2. Table 2.1 shows examples of four scenarios which were used by the UK Climate Impacts Programme (UKCIP) in its 2002 report (UKCIP, 2002).

It is important to remember the difference between emissions and the atmospheric concentration of greenhouse gases. CO_2 has an effective lifetime in the atmosphere of about 100 years, so it will take several decades for any reduction in global emissions to have any effect on the rise in global atmospheric concentrations. To stabilise concentrations of CO_2 it would be necessary for global emissions to fall by over 60%. Furthermore, even if CO_2 levels were stabilised, the inertia of the world's

Table 2.1   UKCIP climate change scenarios (UKCIP, 2002)

Climate change scenario	Increase in global temperature: °C	Atmospheric CO_2 concentration: ppm
Low emissions	2·0	525
Medium-low emissions	2·3	562
Medium-high emissions	3·3	715
High emissions	3·9	810

climate system is such that it would be many decades before temperatures would stop rising.

For a given scenario, the model is usually run using historical input-values over a given period of time, and the results are compared with the output using the same input time-series but with increased levels of greenhouse gases. A common baseline is the period 1961 to 1990, which is defined by the World Meteorological Organization as the '*Climatic Normal*'. However, this includes periods of exceptionally high temperatures, such as the 1980s, and so some researchers use the period 1951 to 1980.

The early climate models compared current climatic conditions with equilibrium conditions when the concentration of CO_2 or other greenhouse gas had been increased. Later models simulate a gradual change in the concentration of CO_2 over a period of time, i.e. a compound annual percentage increase. These so-called *transient models* bring out the effect of the inertia of the ocean systems on the pattern of climate change, which generally results in lower temperature increases on the margins of continents. They also may incorporate an element of feedback, such as the effect of changes in the area covered by snow. However, it is difficult to extract climate conditions for specific dates from a transient model because of the non-linear response of the climate system, and transient models also tend to underestimate change in the first few years of simulation because of problems of initialisation.

As has been mentioned, several different climate models are being used at different research institutions. Bearing in mind the variability between the different models, some consideration needs to be given about which model or combination of models to use. One option is to use the separate outputs from a number of models to give a range of possibilities. Alternatively, the output from several models can be averaged or weighted to reflect their ability to simulate current climate in the region of interest.

As has been mentioned, GCMs use a relatively coarse grid, which means that the results are averaged over a large area (typically 80 000 km^2). The models therefore cannot take into account local variations in relief and land use, etc., and their output needs to be interpolated or 'downscaled' to fit the typical scales used for hydrological models, which are of the order of 10 000 km^2. The interpolation may be a simple weighted average of adjacent grid-points or may involve the application of additional meteorological or topographical knowledge, such as the length of the rainy season (Arnell, 1996). An alternative approach to interpolation from a large grid is to 'nest' a higher resolution regional climate model within the GCM. The nested model uses the same algorithms as the global model and uses the values for adjacent grid-points as boundary conditions. The extent and scale of the regional model needs to be sufficiently fine to represent regional variations, but not too fine to generate large-scale atmospheric features that are incompatible with the larger model.

The results from the global models are used to drive the hydrological models in various ways. One approach is to use an existing observed time-series of say, precipitation or evapotranspiration, and to perturb it (i.e. scale it by a factor) corresponding to the increase in the parameter found from the global model. However, this practice has the disadvantage of preserving the existing distribution of rainfall, including the periods of drought, etc. (Wardlaw *et al.*, 1996). It is more realistic to assume that, in a period of changing levels of rainfall, the pattern of rainfall would also change. Furthermore, rainfall is not uniform across catchments and therefore increases are not likely to be uniform.

An alternative approach is to generate a synthetic time-series with parameters based on the existing time-series but modified to allow for the trends produced by the climate model. Some researchers have used spatial or temporal analogues, which are records from a different geographical area or time period and which show the characteristics predicted by the climate model. The argument is that changes in the distribution as well as levels of precipitation will be included, but it is difficult to remove geographical or other differences, which might make the data unreliable.

The final stage of the modelling process is to establish the impacts of the climate change. These impacts will depend on the vulnerability of particular communities, as discussed later in this chapter, and may be

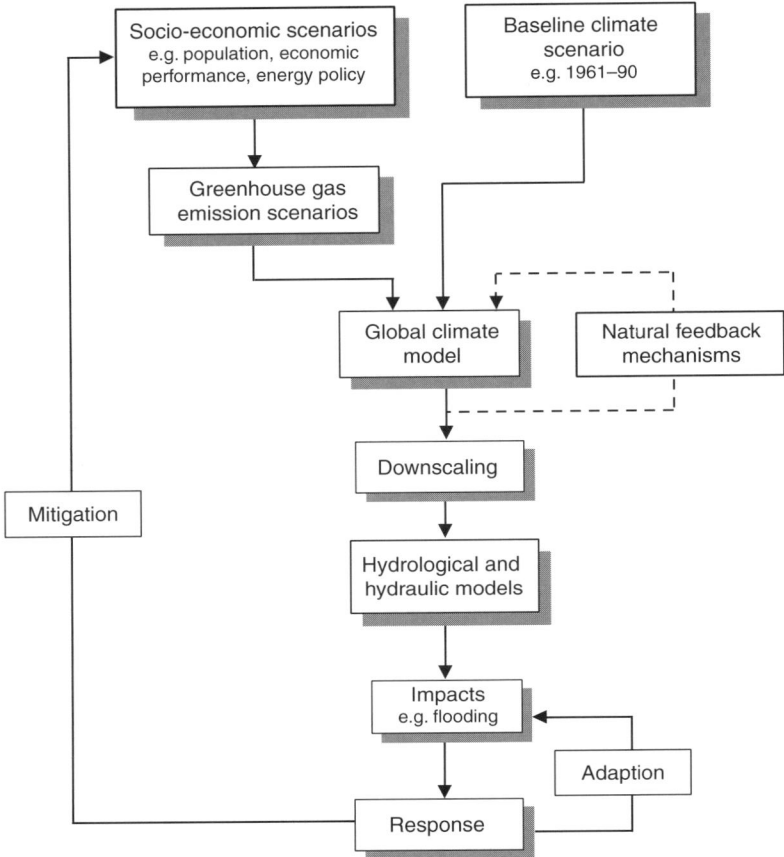

Fig 2.9   Climate modelling procedure

offset either by adaptive measures or by mitigation in the form of, for example, changes in energy policy. These factors should ideally be built into an iterative modelling procedure, for example as shown in Figure 2.9.

2.4.3   Confidence in climate models

For climate models to be of real benefit to decision-makers, the levels of uncertainty associated with predictions must be understood and, if possible, quantified. As has been indicated above, there are many uncertainties associated with forecasting climate change. Some can be quantified relatively easily, while others can be expressed in terms of probabilities and others only in qualitative terms.

The main sources of uncertainty in climate prediction can be summarised as (IPCC, 2001):

- data errors:
  - missing data
  - noise
  - sampling errors
- modelling problems:
  - lack of knowledge of processes
  - lack of knowledge of functional relationships
  - lack of knowledge of parameter values
  - approximation techniques for equation solving
  - the effect of 'trigger' mechanisms, such as changes in the Gulf Stream
- physical problems:
  - long time-lags between forcing and response
  - low-frequency variability with periods greater than the record length
- human problems:
  - projecting human behaviour
  - projecting technological change.

As one moves from the chosen emission scenario to the global climate response and the regional changes in climate parameters, the uncertainty increases rapidly. To a certain extent, models can be validated against historic records, although the extent of such records is often less than the period for which the forecasts are being made. Models can also be validated against themselves, as there is now a dozen or more independent models at different institutions. As more runs of these models are carried out, the database of projections becomes larger and it becomes increasingly possible to ascribe probabilities to certain outcomes.

2.5   Climate change predictions

The main changes in climate that are predicted are:

- increases in global temperatures

- increases and decreases in precipitation
- changes in patterns of runoff
- increases in mean sea level.

Most models output air and sea surface temperatures and precipitation directly, and these can be compared with other models or historic records. The prediction of runoff is more difficult as it involves a complex interaction between temperature, evaporation, vegetation and precipitation, and therefore the predictions are less certain. Likewise, the prediction of changes in mean sea level involve assumptions about the rate of melting of ice caps and glaciers, as well as the expansion of seawater.

An indication of the likely trends in the main hydrological parameters is given below.

2.5.1 Temperature

Although there is considerable variability between the different models in terms of the numeric values, there is a consistent trend in their predictions of temperature change. This is that the predicted increase in the levels of greenhouse gases will cause a significant increase in mean global temperatures. The general consensus figure is that a doubling of the concentration of CO_2 will lead to an increase in mean global temperatures of between 1·5°C and 4·5°C by 2100, with a most likely estimate of 2·5°C. However this could be reduced by about 0·8°C if the effect of sulphur aerosols is taken into account.

The warming is predicted to be greater in higher latitudes than in the tropics. The general forecast for the United Kingdom is a warming of between 2°C and 3·5°C by 2080 for the UKCIP02 low- and high-emission scenarios respectively (see Table 2.1) (UKCIP, 2002). The predicted seasonal temperatures for the medium-high emission scenario are given in Figure 2.10. There is expected to be greater warming in summer and autumn than in winter and spring, and greater warming in the daytime in summer and at night in winter, with an increasing trend towards the south-east. The diurnal temperature range is expected to decrease. The mean sea temperatures are also expected to increase, as is the average length of the growing season.

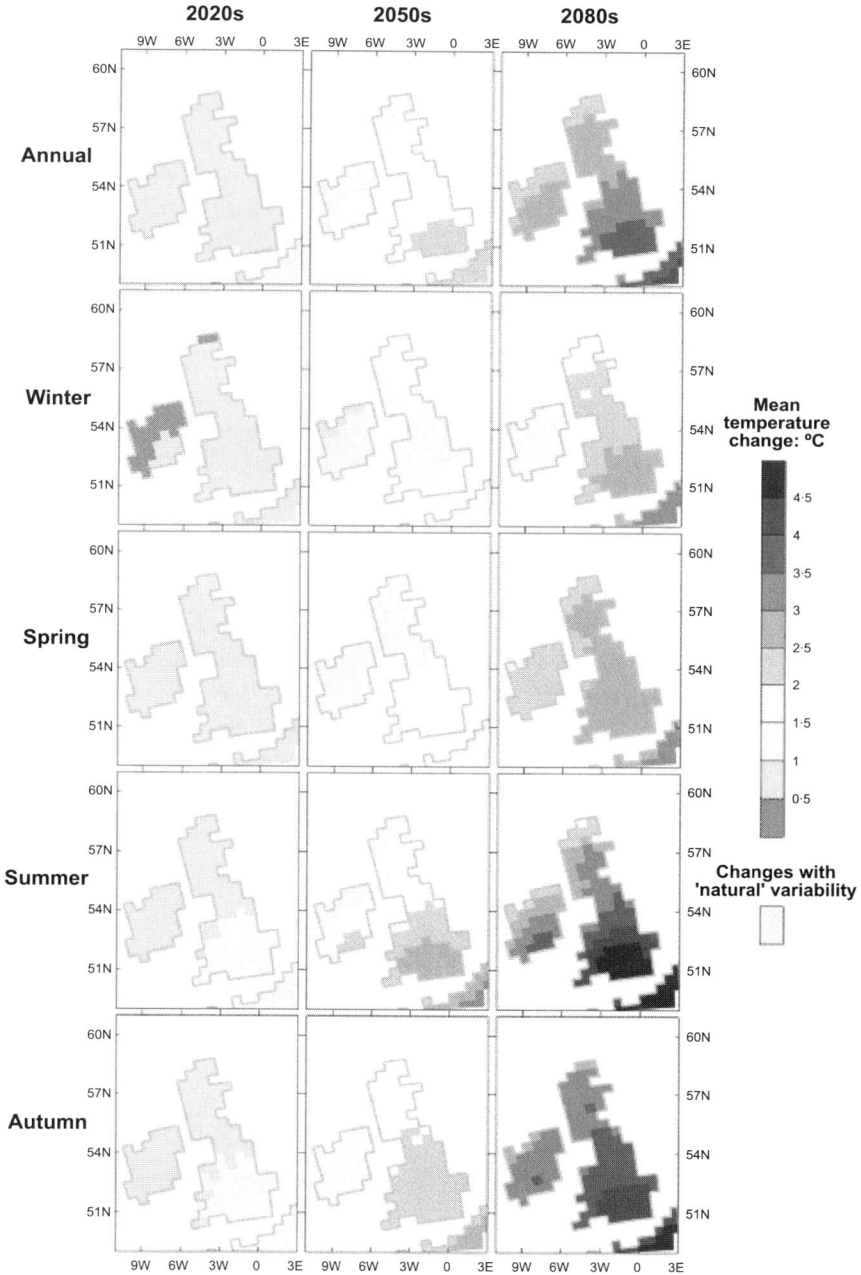

Figure 2.10   Mean changes in UK temperature under the medium-high emission scenario (UKCIP, 2002)

2.5.2 Precipitation

The predicted trends in precipitation vary in different parts of the world. There is expected to be a general increase in mid and high latitudes and in equatorial regions with a decrease in sub-tropical areas (IPCC, 2001b), although in many areas these changes will be small compared with the natural variability. There is also likely to be an increase in the frequency and intensity of tropical cyclones.

The general trend in precipitation for Britain is for increases in winter of up to 30% and decreases in summer of up to 50% by the 2080s, with the greatest changes occurring in the south-east of Britain and generally little change in the north-west (Figure 2.11) (UKCIP, 2002). Annual rainfall is likely to decrease by up to 20% over most of Britain. There is evidence (Marsh, 1996; Mansell, 1997) that the ratio of winter to summer rainfall has already shown a significant increase in the last 30 years. It is also expected that the intensity of winter rainfall will increase but the amount of precipitation falling as snow is likely to decrease. The variability of precipitation is likely to increase and there is also likely to be an overall increase in the frequency of intense rainfall especially in western Scotland.

Although GCMs forecast increased convective activity, they do not predict the preponderance of weather types, such as westerly or anticyclonic, which is also a major factor in determining the amount of precipitation.

2.5.3 Evapotranspiration

Evapotranspiration is a complex function of incoming radiation (which can vary due to changes in cloud cover), temperature, relative humidity and wind speed. The actual evapotranspiration from vegetated surfaces will further depend on plant properties. The physical properties of plants, such as the size and leaf area index, affect the rate at which water is removed, while the physiological properties of plants determine the rate at which water passes through the plant. The effect of climate change on evapotranspiration is therefore a composite result of several processes and, although many predicted climate trends would suggest an increase in evapotranspiration, the effects of other factors may oppose these trends.

Much of the uncertainty regarding future trends in evapotranspiration centres on the effect of changes in plant growth. Changes in temperature and precipitation will affect the seasonal growth of plants and, in the long

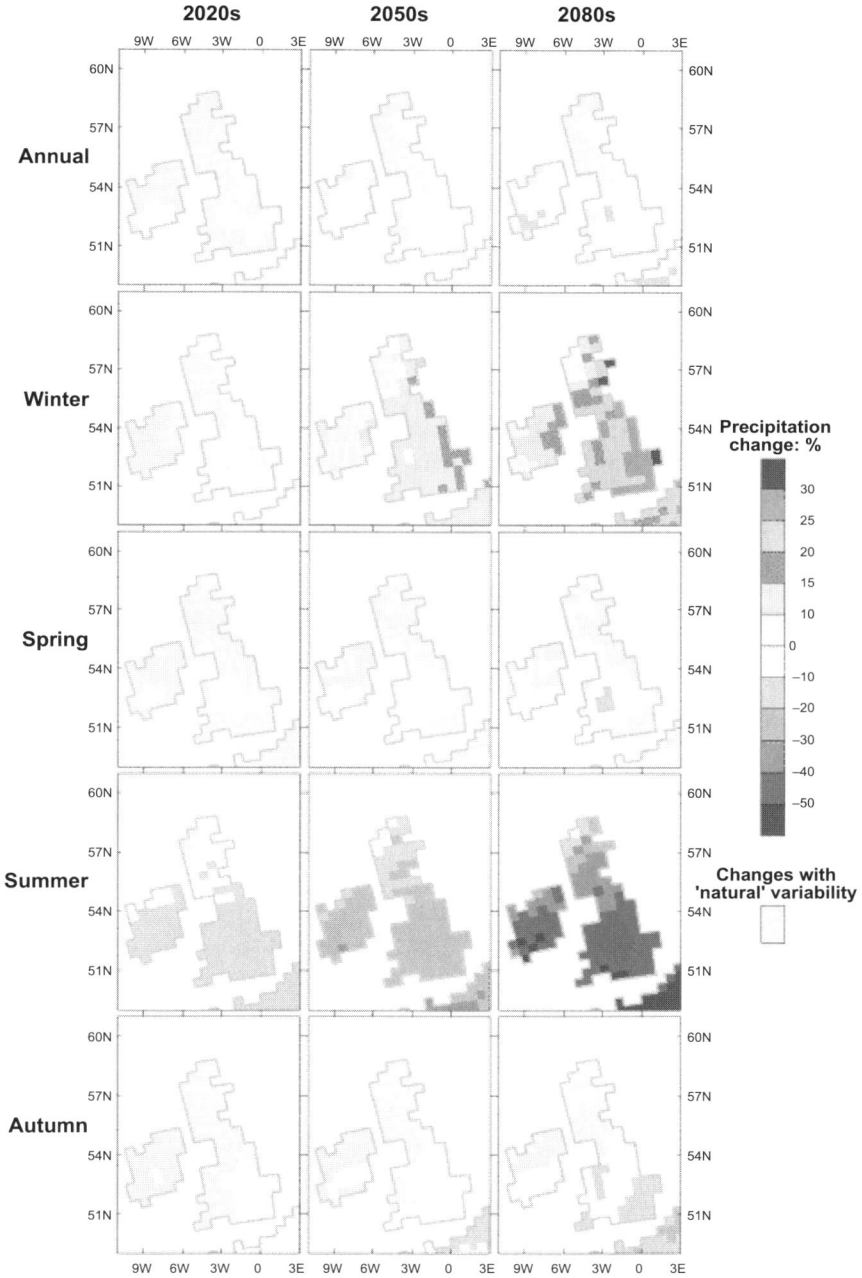

Figure 2.11   Changes in UK precipitation under the medium-high emission scenario (UKCIP, 2002)

term, could alter the geographical distribution of plant species. Plants will also respond directly to changes in the concentration of CO_2 in the atmosphere. CO_2 affects plant growth to different degrees, depending on the plant type. In some plants, such as trees and root crops, an increase in atmospheric CO_2 will stimulate growth, but grasses will be much less affected. The other mechanism by which CO_2 could influence plant activity is in the resistance of the stomata, which are the pores through which the water vapour passes from the plant to the atmosphere. When plants are exposed to increased levels of CO_2, their stomata close and, in general, transpiration is reduced. However, these conclusions are largely based on laboratory experiments, which are difficult to replicate in the field. In the long term, plants may adapt to changes in CO_2 concentration and this effect may become less significant.

Net radiation is predicted to increase generally in summer with little increase in winter (Arnell and Reynard, 1993). In the UK the major factor influencing potential evaporation is the relative humidity of the atmosphere. Although *absolute* humidity may increase, for most places *relative* humidity will decrease because the higher temperatures will mean an increase in the moisture capacity of the atmosphere. Only in northern Scotland will the temperature effect be insufficient to outweigh the increase in absolute humidity. Overall potential evaporation is expected to increase especially in the south-east of Britain and particularly in winter. The percentage increase in potential evaporation may exceed 100% in some places, but, because of the low levels of evaporation in winter, the absolute increases will not be very large.

2.5.4 Groundwater

The effect of climate change on groundwater resources is also uncertain. Aquifers are generally recharged by the net rainfall (i.e. after evapotranspiration and soil losses), and sometimes by rivers and lakes. Unconfined aquifers are recharged directly by local rainfall and rivers, etc., whereas confined aquifers are recharged by rainfall and rivers which may be several hundred kilometres away. In general, an increase in rainfall may be expected to increase groundwater recharge, but it may be offset by the large soil moisture deficit at the start of the wet season as a result of higher evaporation, which would reduce the effective recharge season. For the same reason, any reduction in rainfall, combined with a higher soil moisture deficit, could have a more than proportionate effect on groundwater levels. However, where the recharge is through large fissures or macro pores, the recharge will be less susceptible to changes in the soil moisture deficit. In arid and semi-arid floodplains, where the shallow alluvial aquifers are recharged from seasonal streamflow, changes in recharge will be determined by changes in the duration of flow of these watercourses and the permeability of the overlying beds. Shallow coastal aquifers are also particularly sensitive to increases in mean sea level, which can increase the extent of saline intrusion.

For confined aquifers which are fed a long distance away, it is more difficult to quantify recharge rates or to assess the possible impact of changes in climate, not least because of the lack of data on the properties of the aquifer through which the water passes.

2.5.5 Runoff

The expected changes in runoff resulting from climatic change are similar to the predicted changes in precipitation, with increases in the high latitudes and many equatorial regions and decreases in mid-latitudes. However, runoff is also influenced by the amount of evaporation, and some areas which experience increases in rainfall may have decreased runoff because of the increased evaporation.

In colder regions, some of the precipitation which currently falls as snow will fall as rain and the peak runoff will then tend to occur in winter rather than in the spring snowmelt. The runoff in arid regions is particularly sensitive to changes in precipitation. It has been estimated that a 1–2°C increase in mean annual temperature and a 10% decrease in precipitation could reduce annual runoff by 40–70%, although the reduction in transpiration resulting from an increase in CO_2 may mitigate this effect (van Dam, 1999). In mountainous areas, there is likely to be a reduction in snow cover, reducing the snowmelt flood but increasing the runoff due to rainfall.

The sensitivity of annual runoff to changes in precipitation and evaporation can be represented by (Wigley and Jones, 1985):

$$\frac{R_1}{R_0} = \frac{\alpha - (1 - \gamma_0)\beta}{\gamma_0} \tag{2.1}$$

where R is the annual runoff, α is the fractional change in annual precipitation, β is the fractional change in actual evaporation, γ is the runoff coefficient (the ratio of annual runoff to annual precipitation), and the subscripts 0 and 1 refer to current and changed conditions

Table 2.2   Sensitivity of four British catchments to changes in precipitation (Arnell, 1996)

Catchment	Runoff coefficient	Percentage change in precipitation				
		−20	−10	0	+10	+20
Don	0·56	−32	−19	−5	10	25
Greta	0·61	−32	−18	−4	10	25
Harper's Brook	0·31	−44	−27	−9	12	35
Medway	0·48	−38	−23	−6	11	29

respectively. The expression shows that the fractional increase in runoff (R_1/R_0) increases as the runoff coefficient (γ) decreases. Results from various hydrological model studies indicate that annual runoff is more sensitive to changes in precipitation than changes in evapotranspiration, and that, in most cases, the proportionate change in runoff is greater than the proportionate change in precipitation. For example, in the west of Scotland it has been found that rainfall from 1965 to 1995 increased by about 40%, while runoff increased by up to 60% (Mansell, 1997). Table 2.2 shows the sensitivity of four British catchments to changes in precipitation (Arnell, 1996).

The sensitivity of runoff to changes in precipitation also depends on how the changes in precipitation are distributed throughout the year. Where the existing (and increased rainfall) mostly occurs in winter, the sensitivity of runoff to rainfall will be greatest, since the effect of evapotranspiration will be a minimum.

In the UK, runoff is expected to increase by about 15% or more in Scotland, while in the south-east of Britain there will be decreases of up to 25%. In the north, most of the increase is likely to occur in the winter, with little change in the summer, while in the south a decrease in runoff is predicted in almost every month. In catchments underlain by large aquifers, the changes will probably be less than in the more responsive catchments (i.e. those with a higher proportion of direct runoff). In urban areas the increase in convective rainfall in summer could also increase the frequency of flooding, and in coastal areas there may be an additional risk of flooding due to increased storm activity as well as a rise in sea level.

Of particular interest to water-resource engineers is the change in the low-flow characteristics of rivers under climate-change scenarios. This can be described in terms of the flow which is exceeded for a given percentage of time: a typical low-flow parameter being the 95 percentile flow (Q_{95}). Low flows in Britain usually occur in late summer after a prolonged dry period. They are determined by a combination of the recession characteristics of the catchment, which depend on the geology and soil properties, and the length of the dry period and amount of rainfall before it. In the north of Britain, there is little change predicted for Q_{95} since the summer rainfall is not expected to change very much (Arnell, 1996). In the extreme south, where there are significant aquifers, the low flows are partly dependent on the previous winter's rainfall. The exceptional drought of 1996, for example, did not deplete the levels in the chalk aquifer of southern England much below their normal values because of the high rainfall of the preceding winter (Marsh, 1996). Little change is expected in low flows in the south of Britain, although there are likely to be significant decreases in Q_{95} in the central areas of the UK.

The probability of high flows can likewise be described by frequency parameter; in this case Q_5 is normally used, although it is normally calculated on the basis of average flows and therefore cannot be taken to represent the occurrence of flood peaks. The value of Q_5 is expected to increase by up to 20% in the north and decrease by a similar amount in parts of the south.

2.5.6   Sea level rise

Global mean sea level is expected to rise as a result of global warming, mainly due to the thermal expansion of the oceans and the increased melting of glaciers and ice sheets. The projected rise over the twenty-first century is between 0·09 and 0·88 m, with about 60% of the increase attributable to thermal expansion (UKCIP, 2002). In many places, such as the UK, allowance also has to be made for isostatic land adjustments, which are vertical land movements resulting from changes in crustal loading after the last ice age. The isostatic adjustment in the northern part of the UK is between 0·5 and 1 mm upwards per year, whereas the south-eastern part is sinking by between 1 and 1·5 mm per year. The net sea level rise predicted for the UK therefore varies from about 10 mm per year in the south to about 5 mm per year in the north.

2.5.7 Wind speed and storminess

There are expected to be modest increases in mean wind-speed over the UK by 2050 of about 6% in winter and 2% in summer. However, there do not appear to be any clear trends in storminess (Woolf, 2001). It seems that the incidence of storms is more related to variations in the NAO than in any human-induced climate change. There is, in any case, a high variability in wave heights which tends to mask any long-term trends.

2.6 Impact of climate change on water resources

2.6.1 The general impact on water resources

The impact of climate change on water resources depends not only on the physical changes in the climate described in the preceding section, but also on the nature of the water-resource system within a particular country or region. Some countries will be more sensitive to a given change than others. Regions that are likely to be particularly sensitive are those with (among other factors):

- a highly seasonal hydrological regime
- a high proportion of consumptive water use in relation to the resource
- a low reservoir storage
- a topography and land use pattern which is sensitive to flash flooding and soil erosion
- a low skills base for the effective management of water resources
- a high population density and lack of mobility.

The impact of a given climatic change depends on the extent to which institutions and individuals can adapt to the changes. The various adaptive techniques available can be broadly classified into 'supply-side' and 'demand-side' strategies. The former include structural strategies, such as new reservoirs, large-scale transfer by canals and artificial recharge of groundwater, etc., as well as non-structural strategies, such as changes in institutions, procedures and planning controls. Demand-side policies include water-demand management, leakage control and improved irrigation techniques.

It should also be noted that non-climatic factors may compensate for the effects of climate change or may exacerbate them. For example, in

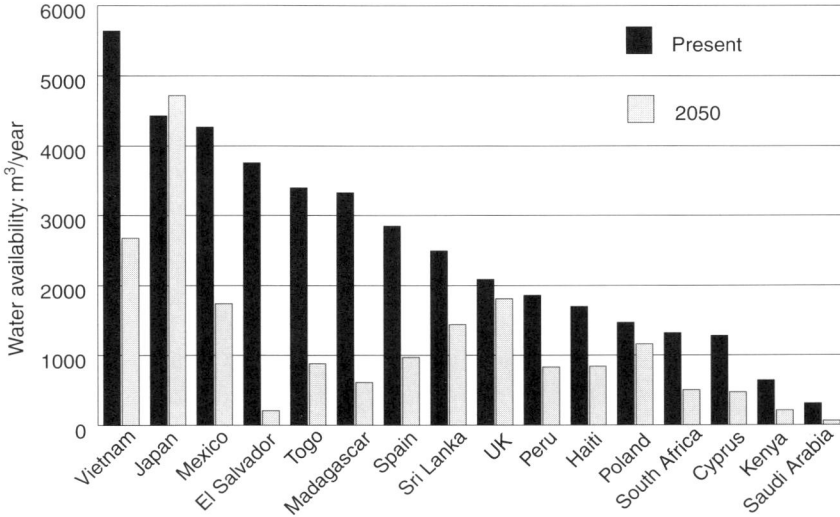

Figure 2.12 Present and future water availability by country (from Kaczmarek, 1996)

some developed countries, water consumption is falling as a result of more efficient water use, while, in many developing countries, per capita water consumption is relatively low and is therefore likely to increase (Figure 2.12).

2.6.2 The impact on water supply

The water stress of a country can be measured in terms of the volume of water consumed in relation to the total volume available. Where withdrawals are greater than about 20% of the resource, or where the resource is less than about 1700 m^3/year/person, water is generally the limiting factor on development (IPCC, 2001b). In 1990 about one-third of the world's population lived in such areas and this is expected to increase to about 60% by 2025, even without climate change. The effect of climate change is expected to increase this figure by less than 8%.

A water supply system can be characterised by its:

- storage capacity — a direct river abstraction has zero storage, while reservoirs and groundwater have storage capacities

- firm yield — this is the supply which can be safely abstracted, usually measured as a percentage of average inflow
- probability of failure — this is the probability of not meeting the specified yield.

For a given storage capacity, a change in the pattern of inflow can lead to either a change in the probability of failure or a change in the yield which can be met with a given degree of risk (Cole *et al.*, 1991). The change may be caused either by a change in the mean flows or by a change in the distribution of flows, for example reduced summer flows relative to winter flows. The notional effect of climate change on water supply is illustrated in Figure 2.13. Assuming that the effect of climate change is to reduce mean flows or to redistribute the flows such that more storage is required to achieve a given yield, the yield–storage curve under current conditions (A) would move to say curve B. Thus, in order to meet the existing yield (Y_1), storage would need to be increased from S_1 to S_2. If the storage were not increased, the yield would be reduced to Y_2. Alternatively, if the storage were increased only to S_3, the probability of not meeting the yield would be increased (curve C).

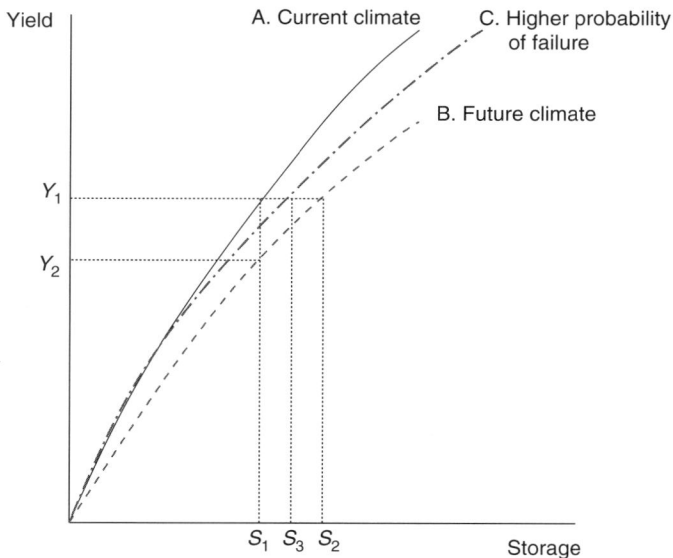

Figure 2.13   The effect of climate change on water supply

Water supply systems may be highly sensitive to climate change in that the changes in yield or storage requirements may be considerably greater (in relative terms) than the changes in climatic inputs (Arnell, 1996). Kirshen and Fennessey (1995) showed that under one climate-change scenario, a 1·6% decrease in precipitation together with a 20% increase in evaporation resulted in a 23% decrease in the safe yield.

2.6.3   Domestic water demand

Domestic water demand comprises many elements which have varying sensitivities to climate change. Table 2.3 (Herrington, 1996) shows the estimated increase in domestic demand in the south-east of England with and without climate change.

Table 2.3   Estimated increase in domestic demand in the south-east of England with and without climate change (Herrington, 1996)

Component	1991	2021 No climate change	2021 +1·1°C warming
WC	35·5	33·6	33·6
Showering	5·3	24·0	26·8
Other personal washing	41·2	37·6	37·6
Clothes washing	21·7	22·0	22·0
Dish washing	11·8	11·0	11·0
Waste-disposal unit	0·4	1·5	1·5
Car washing	0·9	1·5	1·5
Lawn sprinkler	2·5	8·7	11·7
Other garden use	3·8	7·2	8·6
Miscellaneous use	23·9	31·3	31·3
Total domestic use	147·0	178·4	185·6

It can be seen from Table 2.3 that the major increase is in garden use, but most of the increase is due to an increase in horticultural activity, which is expected to occur regardless of any climate change. Non-domestic public water demand is generally less sensitive to climate change except for water used for irrigating parks, etc. A considerable amount of water in public distribution systems is lost through leakage and it is possible that leakage rates might fall if there are significantly fewer frosts.

2.6.4 Agriculture

The effect of climate change on the demand for irrigation water is rather mixed. Where temperatures and evapotranspiration rates increase there will obviously be an increase in the demand for irrigation water, especially if the growing season also lengthens. In other areas, the increase in precipitation will outweigh the increases in evapotranspiration and demand may actually fall. There will also be an increased demand for non-irrigation water, for example for watering stock. One of the most serious consequences of climate change for agriculture is in the greater variability which is likely to occur. Where the climate changes only gradually, farmers can adjust their crop selection and farming practices to cope with changes, but the predicted increase in the inter-annual variability of the climate is likely to cause a reduction in productivity.

2.6.5 Industrial demand

The great majority of industrial demand for water is for cooling, especially for power stations. It is possible that the demand for cooling water will reduce as a result of the reduced demand for electricity in the milder winters and the increased efficiency of modern generators. However, increased temperatures in rivers will tend to reduce the efficiency of cooling systems.

2.6.6 Hydro-electric demand

Although the demand for electricity may be reduced in winter, it may well increase in summer as a result of the increased use of air conditioning and refrigeration. The impact of climate change on hydro-electric generation depends on the amount of storage included in the scheme. Run of river

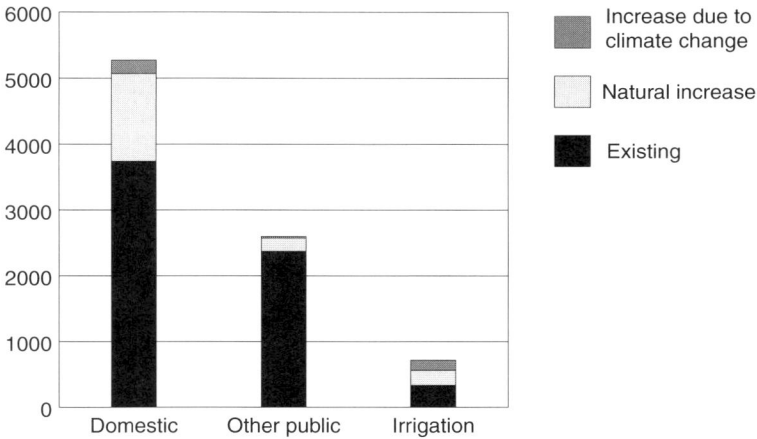

Figure 2.14   Forecast change in water consumption

schemes, where there is no storage, will be susceptible to changes in the patterns of flow in the river. For example, lower snowmelt will tend to reduce spring flows and increase winter flows. Where there is significant reservoir storage, there could be increased losses due to evaporation from the water surface. In some cases, hydro-electric reservoirs also serve to mitigate floods, and more storage may need to be set aside for this purpose in a future climate scenario. The general impact of climate change on different types of consumption is summarised in Figure 2.14.

2.7   General implications of climate change

The use of GCMs has indicated the likely direction of the world's climate even if there is still uncertainty about the quantities. There are going to be some direct effects, such as increases in temperature and precipitation, and some more indirect effects, including changes in potential evaporation, runoff and sea level. The latter are more difficult to quantify because they involve assumptions about the response of the climate system to changes. In addition to the progressive effects, there will be problems associated with the increased variability of the climate, both within years and between years. The actual impacts which can be measured in terms of the probability of flooding, storm damage and water shortage, etc., are still more difficult to estimate because they depend on the vulnerability of our infrastructure to specific types of damage, as well as the policies adopted to mitigate the damage.

2.8 References

Alcock, G. and Rickards, L. (2001). *Climate of UK Waters at the Millennium*. Inter Agency Committee on Maritime Science and Technology.

Arnell, N. (1996). *Global warming, river flows and water resources*. Wiley, Chichester.

Arnell, N. and Reynard, N. S. (1993). *Impact of Climate Change on River Flow Regimes in the United Kingdom*. Institute of Hydrology, Wallingford.

Brown, G. C., Hawkesworth, C. J. and Wilson, R. C. L. (1992). *Understanding the Earth: A New Synthesis*. Cambridge University Press, Cambridge.

Burroughs, W. J. (2001). *Climate Change: A Multidisciplinary Approach*. Cambridge University Press, Cambridge.

Cole, J. A., Slade, S., Jones, P. D. and Gregory, J. M. (1991). Reliable Yield of Reservoirs and Possible Effects of Climate Change. *Hydrological Science Journal*, **36**, 579–597.

Herrington, P. (1996). *Climate Change and the Demand for Water*. HMSO, London.

IPCC (1995). *Climate Change 1995: The Science of Climate Change*. Cambridge University Press, Cambridge.

IPCC (2001a). *The Third Assessment Report*. Intergovernmental Panel on Climate Change.

IPCC (2001b). *Climate Change 2001: Impacts, Adaption and Vulnerability*. Cambridge University Press, Cambridge.

Kaczmarek, Z., Arnell, N. W., Stakhiv, E. Z., Hanaki, K., Mailu, G. M., Samlyody, L. and Strzpek, K. (1996). *Water Resource Management in Climate Change 1995 — Impacts, Adaptations and Mitigation of Climate Change*. Cambridge University Press, Cambridge, pp. 471–486.

Kirshen, P. H. and Fennessey, N. M. (1995). Possible Climate Change Impacts on Water Supply of Metropolitan Boston. *Journal Water Resource Planning and Management*, **121**, 61–70.

Mansell, M. G. (1997). The Effect of Climate Change on Rainfall Trends and Flooding Risk in the West of Scotland. *Nordic Hydrology*, **28**, 37–50.

Marsh, T. J. (1996). The 1995 UK drought — a signal of climatic instability? *Proceedings of the Institution of Civil Engineers, Water Maritime & Energy*, **118**, 189–195.

UKCIP (2002). *Climate Change Scenarios for the United Kingdom*. UKCIP.

van Dam, J. C. (1999). *Impacts of Climate Change and Climate Variability on Hydrlogical Regimes*. Cambridge University Press, Cambridge.

Wardlaw, R. B., Hulme, M. *et al.* (1996). Modelling the Impacts of Climatic Change on Water Resources. *Journal of CIWEM*, **10**, 355–364.

Wigley, T. M. L. and Jones, P. D. (1985). Influences of Precipitation Changes and Direct CO_2 Effects on Streamflow. *Nature*, **314**, 149–152.

Woolf, D. (2001). *The Vulnerability of the UK Coastline, Storminess and Sea Level Rise*. Institution of Civil Engineers, London.

Chapter 3

Statistical tools

Hydrology is essentially concerned with observing and understanding the processes of water movement on land and in the atmosphere, and using the knowledge to estimate the probability of future hydrological events. This chapter outlines some of the statistical methods that are available to describe and quantify the processes.

The statistical aspects of hydrological studies generally fall into four groups as follows.

1. Estimating the probability of future hydrological events, e.g. flooding, by studying the statistical distribution of such events within an observed time-series (with the assumption that this distribution will remain unchanged in the future).
2. Investigating relationships between variables such as evaporation and wind speed.
3. Determining the presence of any trends in a hydrological time-series and extrapolating into the future.
4. Generating synthetic time-series by stochastic modelling in order to extend the length of a record or to fill gaps in the data.

3.1 Statistical distributions

3.1.1 *General properties of distributions*

If a large number of measurements of a single variable (say temperature) are taken at one time there will be some variation in the results. The spread of results reflects the errors in procedure and the limitations of the equipment used. If the observations are divided into classes according to their value, it will be found that most of the results will fall into a class with a central value, which can be demonstrated when the results are plotted as a frequency histogram showing the number of observations within each class. As the number of measurements is increased and the class interval is reduced, the histogram becomes a smooth curve. If the vertical scale is adjusted so the area under the curve is unity, the curve is known as the *probability density function* (PDF) and indicates the probability of a result occurring within a given range of values. The data can also be plotted as a *cumulative distribution function*, which indicates the probability of a sample being above or below a certain value. The cumulative distribution of flow data, for example, is referred to as the *flow–duration curve* and is a common technique to estimate flows with a given probability of exceedance.

The shape of any PDF is described by a number of parameters. The mean value is a parameter which defines the location of the distribution and, in the case of symmetrical distributions, is at the peak of the curve. Figure 3.1(a) shows similar distributions with different means. The mean of a sample $X_1, X_2, X_3 \ldots X_n$ can be considered as the first moment of the data, i.e.:

$$\mu = \sum_{i=1}^{N} \frac{X_i}{N} \tag{3.1}$$

The second moment of the data (the variance) measures the spread of the distribution. Since the area under the PDF remains the same (unity), increasing the variance reduces probability density around the mean value (Figure 3.1(b)).

$$\sigma^2 = \sum_{i=1}^{N} \frac{(X_i - \mu)^2}{N} \tag{3.2}$$

Generally, it is required to estimate the properties of an infinite population of values based on a limited sample of data. In such a case, the sample mean is taken as the best estimate of the population mean, but the variance of the sample will generally underestimate the variance of the population, since the sample data will tend to be closer to the sample mean than the population mean. The variance of the population should therefore be estimated using:

$$\sigma^2 = \sum_{i=1}^{N} \frac{(X_i - \mu)^2}{N-1} \tag{3.3}$$

although, for large samples, the difference is marginal.

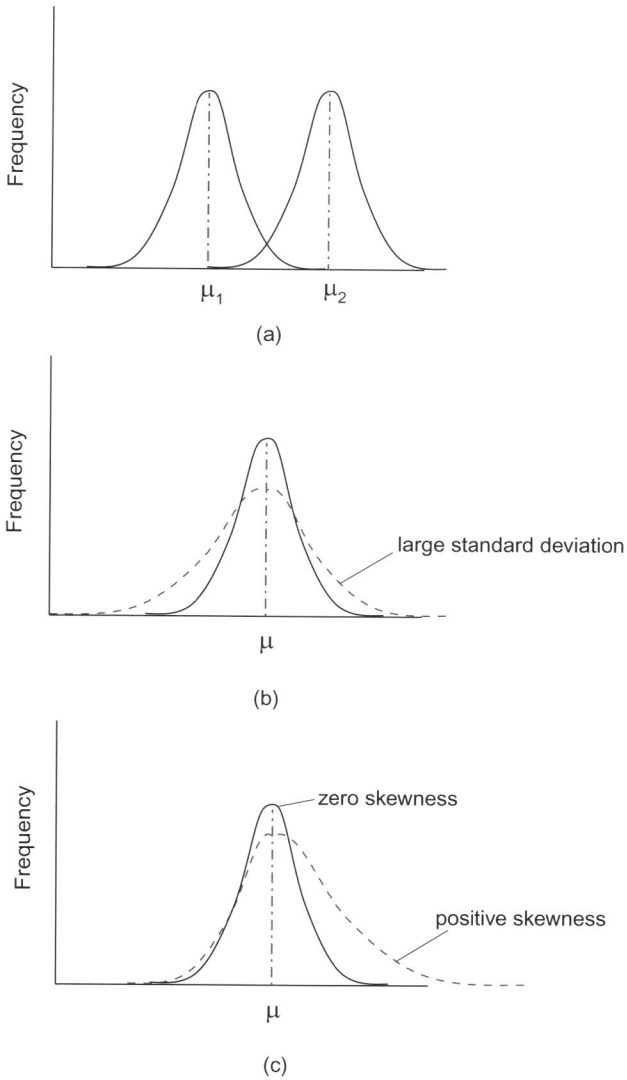

Figure 3.1   The effect of changes in: (a) the mean; (b) the standard deviation; and (c) the skewness on a frequency distribution

The third moment of the data measures the *skewness* or asymmetry of the data:

$$\beta = \sum_{i=1}^{N} \frac{(X - \mu)^3}{N} \qquad (3.4)$$

with the correction noted above being applied for small samples. A positive skewness indicates that the data are skewed to the left of the mean (Figure 3.1(c)) and are a common feature of the distribution of stream flows, where low flows predominate and high flows occur only intermittently (Example 3.1).

The fourth moment (*kurtosis*) measures the peakedness of the distribution, but is not normally used in distribution fitting.

Example 3.1

The following values of mean daily flow (m³/s) were recorded in a river:

0·444, 2·685, 1·982, 0·309, 0·945, 0·846, 2·380, 0·162, 0·871, 1·405

Calculate the mean, variance, standard deviation and skewness of the data.

Calculation of mean

0·444
2·685
1·982
0·309
0·945
0·846
2·380
0·162
0·871
1·405
Total = 12·029

Mean = 12·029/10 = 1·203

Variance

$(0{\cdot}444 - 1{\cdot}203)^2 = 0{\cdot}576$
$(2{\cdot}685 - 1{\cdot}203)^2 = 2{\cdot}196$
$(1{\cdot}982 - 1{\cdot}203)^2 = 0{\cdot}607$
$(0{\cdot}309 - 1{\cdot}203)^2 = 0{\cdot}799$
$(0{\cdot}945 - 1{\cdot}203)^2 = 0{\cdot}067$
$(0{\cdot}846 - 1{\cdot}203)^2 = 0{\cdot}127$
$(2{\cdot}380 - 1{\cdot}203)^2 = 1{\cdot}386$
$(0{\cdot}162 - 1{\cdot}203)^2 = 1{\cdot}083$
$(0{\cdot}871 - 1{\cdot}203)^2 = 0{\cdot}110$
$(1{\cdot}405 - 1{\cdot}203)^2 = 0{\cdot}041$
Total = 6·993

Sample variance = 6·993/9 = 0·777
Sample standard deviation = $\sqrt{0{\cdot}777}$ = 0·881

Skewness

$(0{\cdot}444 - 1{\cdot}203)^3 = -0{\cdot}437$
$(2{\cdot}685 - 1{\cdot}203)^3 = 3{\cdot}255$
$(1{\cdot}982 - 1{\cdot}203)^3 = 0{\cdot}473$
$(0{\cdot}309 - 1{\cdot}203)^3 = -0{\cdot}714$
$(0{\cdot}945 - 1{\cdot}203)^3 = -0{\cdot}017$
$(0{\cdot}846 - 1{\cdot}203)^3 = -0{\cdot}045$
$(2{\cdot}380 - 1{\cdot}203)^3 = 1{\cdot}631$
$(0{\cdot}162 - 1{\cdot}203)^3 = -1{\cdot}128$
$(0{\cdot}871 - 1{\cdot}203)^3 = -0{\cdot}037$
$(1{\cdot}405 - 1{\cdot}203)^3 = 0{\cdot}008$
Total = 2·989

Sample skewness = 2·989/9 = 0·332

One major difficulty with using a specific distribution for a time-series is that it assumes that the series is stationary, i.e. that the mean and variance do not vary with time. Where there is evidence of a significant trend, it may be appropriate to select only part of the record even though this will reduce the sample size.

3.1.2 Common statistical distributions

As has been noted above, the distribution of a sample will not exactly match the distribution of the whole population, depending on the size of the sample. In order to estimate the distribution of the parent population, one of a number of theoretical distribution curves is normally fitted to the sample data distribution, the parameters of the curve being determined by the mean, variance and other moments of the sample distribution.

The distribution of most hydrological data will tend to follow one of a number of recognised distributions depending on the nature of the process. Table 3.1 lists some of the most common distributions used in hydrological estimation.

Where several repeated measurements are taken of a single value, one would expect the results to be randomly distributed above and below a central value, with most of the results close to the central value. In such a case, the data will tend to follow a *normal distribution* with a classic symmetrical bell-shaped PDF, the equation for which is:

$$p(x) = \frac{1}{\sigma\sqrt{2\pi}} \exp\left[-\frac{(x-\mu)^2}{2\sigma^2}\right] \tag{3.5}$$

In most practical problems it is required to find the probability of a result occurring above or below a given value. To determine this, it is necessary to use the cumulative distribution function (CDF), which, as has been noted, is the integral of the PDF. Unfortunately, the mathematical expression for the PDF of the normal distribution does not generally allow integration, except numerically, and the most practical

Table 3.1 Common distributions used in hydrology

2 parameter	3 parameter
Normal	Pearson Type 3
Gumbel	Log-normal
Log-normal	Generalised extreme value
Exponential Weibull	Generalised logistic

solution is to refer to a published statistical table, for example as shown in Appendix 3.1.

The normal distribution requires that the results are equally distributed above and below the mean value, leading to a symmetrical distribution with zero skewness. As noted above, in many hydrological time-series, such as river flow, there are far more low values than high values and the resulting distribution is usually skewed to the left (positive skew). The normal distribution can be used to describe such data if the logarithms of the values are used. The distribution is then known as the *log-normal distribution* and the probability density function is given by:

$$p(x) = \frac{1}{\sigma_y e^y \sqrt{2\pi}} \exp{-\left[\frac{(x-\mu_y)^2}{2\sigma_y^2}\right]} \qquad (3.6)$$

where $y = \log(x)$ and μ_y and σ_y are the mean and standard deviation of y.

If the data to be analysed consist, not of observations of, say, daily flow but of the maximum daily flow occurring in a period of time (e.g. a year), then the results will generally follow a type of extreme value distribution, depending on the distribution of the original flow data. The *Gumbel Distribution* is the limiting form of the distribution of maxima from a large number of equally sized samples where the initial distribution is exponential. The probability of a flow (q) exceeding a given flow (Q) is given by the cumulative distribution:

$$P(q \geq Q) = 1 - \exp[-\exp\{-y\}] \qquad (3.7)$$

where:

$$a = 0.5772c - \mu$$

$$c = \frac{\sqrt{6}}{\pi}\sigma$$

$$y = \frac{a+Q}{c}$$

where μ and σ are the mean and standard deviation of the time-series respectively.

The Gumbel Distribution is commonly used to estimate the flows with a given probability of exceedance. To obtain the annual maximum flow (Q_T) with a given return period (T), the above equation can be rearranged to give:

$$Q_T = Q_m - 0.45\sigma + 0.78\sigma y \tag{3.8}$$

where:

$$y = -\ln\left[-\ln\left(1 - \frac{1}{T}\right)\right]$$

Q_m and σ are the mean and standard deviation, respectively.

The value of Q is often plotted against y, which is known as the reduced variate, in which case it can be seen that graph will be linear (Example 3.2).

Another of the distributions which may be used for extreme probabilities is the *Poisson Distribution*, which is an extension of the Binomial Distribution for small probabilities (p) and large sample size (N). The probability density function of the Poisson Distribution is:

$$p(x) = \frac{m^x e^{-m}}{x!} \tag{3.9}$$

where m is the product Np. The mean and standard deviation of the distribution are both equal to m.

Example 3.2

The following annual maximum flows (m³/s) were recorded at a gauging station:

36·6, 69·9, 99·0, 76·2, 62·6, 44·2, 49·2, 53·1, 58·8, 64·1, 77·8, 71·2, 59·6, 55·1, 49·6, 58·6, 39·7, 38·2, 103·0, 47·9

Using the Gumbel Distribution, estimate the return period of an annual maximum flow of 105 m³/s and the annual maximum flow with a return period of 100 years.

From flow data:
  Mean flow, $\mu = 60.72$ m³/s
  Standard deviation, $\sigma = 18.23$ m³/s

$$c = \frac{\sqrt{6}}{\pi} \times 18.23 = 14.21$$

$$a = 0.5772 \times 14.21 - 60.72 = -52.52$$

For a flow of 105 m³/s:

$$P(q \geq 105) = 1 - \exp\left[-\exp\left\{\frac{-(-52\cdot52 + 105)}{14\cdot21}\right\}\right]$$

$= 0\cdot025$

$$T = \frac{1}{0\cdot025}$$

$= 40\cdot6 \text{ years}$

For a return period of 100 years (from Equation (3.8)):

$$Q_{0\cdot01} = 60\cdot72 - 0\cdot45 \times 18\cdot23 + 0\cdot78 \times 18\cdot23\left[-\ln\left(-\ln\left(1 - \frac{1}{100}\right)\right)\right] = 118\cdot0 \text{ m}^3/s$$

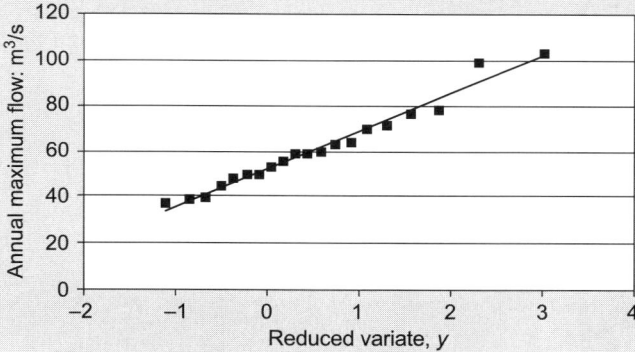

The *Weibull Distribution* is given by:

$$F(x) = 1 - \exp\left(\frac{x^{\gamma}}{\vartheta}\right)$$

(3.10)

where γ and ϑ are parameters which define, respectively, the scale and shape of the distribution, and also provides a good fit for extreme value data.

A *Rectangular Distribution* is an artificial distribution which, although not found in natural time-series, is useful in generating synthetic time-series, as it describes the distribution of numbers produced by pseudo-random number generators available in most spreadsheets and statistical packages. Such numbers are assumed to be uniformly distributed between zero and one, i.e. the probability density function is:

$$p(x) = 0 \quad \text{for} \quad x < 0 \quad \text{and} \quad x > 1$$
$$p(x) = 1 \quad \text{for} \quad 0 < x < 1 \tag{3.11}$$

3.1.3   The use of L-moments

Where the data are highly skewed, the method of fitting distributions by using the moments estimated from the samples often gives poor results because estimating the skewness from the sample data is not very reliable. It has been found that distributions based on linear combinations of data are more robust and this is the method recommended by the *Flood Estimation Handbook* (Robson and Reed, 1999). The moments associated with linear combinations of data are known as *L-moments* and the first three moments are calculated as below.

$$l_1 = \frac{\displaystyle\sum_{j=1}^{N} x_j}{N}$$

$$l_2 = 2b_1 - l_1$$

$$l_3 = 6b_2 - 6b_1 + l_1 \tag{3.12}$$

where:

$$b_1 = \sum_{j=2}^{N} \frac{(j-1)}{N(N-1)} x_j$$

$$b_2 = \sum_{j=3}^{N} \frac{(j-1)(j-2)}{N(N-1)(N-2)} x_j$$

x_j is the *j*th element in a sample size N which has been sorted into ascending order.

The moments all have the same units and are often normalised by reference to the first or second moments to give dimensionless L-moment ratios as below.

$$L\_CV = t_2 = \frac{l_2}{l_1}$$

$$L - skewness = t_3 = \frac{l_3}{l_2} \tag{3.13}$$

The distribution recommended by the *Flood Estimation Handbook* is the *Generalised Logistic Distribution*, which gives the flow (x) with a given probability of exceedence (P) as:

$$x(P) = \xi + \frac{\alpha}{k} \left\{ 1 - \left(\frac{(1-P)^k}{P} \right) \right\} \tag{3.14}$$

where:

$$\alpha = \frac{\xi t_2 k \sin(\pi k)}{k\pi(k + t_2) - t_2 \sin(\pi k)}$$

$$k = -t_3$$

ξ is the location factor and can be taken as the mean or median value, α is a scale factor and k is a shape factor. The parameters t_2 and t_3 are the second and third L-moment ratios as defined above.

3.1.4   Goodness of fit

Where it is not clear how well the data fit a given distribution, an objective statistical test can be used. The so-called $\aleph^2$ *Test* measures the difference between the observed frequencies (f_o) of the data within certain classes and the frequencies (f_e) which would be expected from a given distribution. The $\aleph^2$ statistic is given by:

$$\aleph^2 = \sum \frac{(f_o - f_e)^2}{f_e} \tag{3.15}$$

and the significance of the result can be assessed from a table of percentage points (Appendix 3.2) with the number of degrees of freedom equal to the number of classes minus 1 (Examples 3.3 and 3.4).

Example 3.3

Construct the frequency distribution for the data in Example 3.1 and find the goodness of fit with a normal distribution.

Normal distribution
Using an appropriate class interval, the frequency and probability of values within each class are calculated as below.

Class	Observed frequency, f_o
0–0·75	3
0·75–1·50	4
1·50–2·25	1
2·25–3·00	2
3·00–3·75	0

By reference to the table of the cumulative normal distribution, the expected probability of values within each class can be calculated. For example:

Probability of a value below 0·75:

$$z = \frac{(0\cdot75 - \mu)}{s} = \frac{(0\cdot75 - 1\cdot203)}{0\cdot881} = -0\cdot514$$

from tables, $P_{0\cdot75} = 1\cdot00 - 0\cdot696 = 0\cdot304$

Probability of value below 1·50:

$$z = \frac{(1\cdot50 - \mu)}{s} = \frac{(1\cdot50 - 1\cdot203)}{0\cdot881} = 0\cdot337$$

from tables, $P_{1\cdot50} = 0\cdot632$

Probability of value between 0·75 and 1·50 = 0·632 − 0·304 = 0·328

The goodness of fit is assessed using the $\aleph^2$ test, where:

$$\aleph^2 = \sum\left[\frac{(\text{expected frequency}-\text{observed frequency})^2}{\text{expected frequency}}\right]$$

Class		z	P	Expected probability	Expected frequency, f_e	$(f_o - f_e)^2/f_e$
0·00		−1·365	0·086			
0·75		−0·514	0·304	0·218	2·175	0·313
1·50		0·337	0·632	0·328	3·283	0·157
2·25		1·188	0·883	0·251	2·506	0·905
3·00		2·039	0·979	0·097	0·967	1·104
3·75	4·50	2·890	0·998	0·019	0·188	
					$\aleph^2$	**2·478**

From the table, $\aleph^2_{0·05,2} = 5·99$.
Therefore, the fit is reasonable.

Example 3.4

Construct the frequency distribution for the data in Example 3.1 and find the goodness of fit with a log-normal distribution.

Log-normal distribution
The logarithms of the flows are:

−0·353, 0·429, 0·297, −0·510, −0·025, −0·073, 0·377, −0·790, −0·060, 0·148

Mean = −0·056
Standard deviation = 0·398

Class		Observed frequency, f_o
−1·4	−1·0	0
−1·0	−0·6	1
−0·6	−0·2	2
−0·2	0·2	4
0·2	0·6	3
0·6	1·0	0

Class		z	P	Expected probability	Expected frequency, f_e	$(f_o - f_e)^2/f_e$
−1·4	−1·0	−2·374	0·009			
−1·0	−0·6	−1·368	0·086	0·077	0·769	0·070
−0·6	−0·2	−0·362	0·359	0·273	2·730	0·195
−0·2	0·2	0·644	0·740	0·381	3·815	0·009
0·2	0·6	1·650	0·950	0·210	2·104	0·382
					$\aleph^2$	**0·656**

From the table, $\aleph^2_{0·05,2} = 5·99$.
Therefore, the fit is better than for the normal distribution.

3.2   Bivariate and multivariate analysis

3.2.1   *Bivariate analysis*

In many hydrological studies there is a need to investigate if there is a causal relationship between two variables, for example, between temperature and evaporation. Assuming that simultaneous records of both parameters are available for the same location, plotting one variable against the other may show evidence of a relationship but with considerable variability or scatter.

The mean of each parameter can be calculated (μ_x and μ_y) and the variance of each parameter is defined as the mean of the squares of the deviation of each data point from the mean value (Figure 3.2(a)). If the variance of the dependant variable y is calculated with reference to a sloping line through the data (instead of the mean), it will be considerably reduced and the line which gives the minimum variance is known as a regression line of y on x (Figure 3.2(b)). If the data were all located on a straight line the variance will have been reduced to zero. (There will be a corresponding different regression line for x on y.)

The degree of scatter of the data about the regression line is given by the correlation coefficient which measures the covariance of x and y in relation to the variance of x and y, and is defined as:

$$r = \frac{1}{N} \sum_{i=1}^{N} \left(\frac{x_i - \mu_x}{\sigma_x} \right) \left(\frac{y_i - \mu_y}{\sigma_y} \right) \tag{3.16}$$

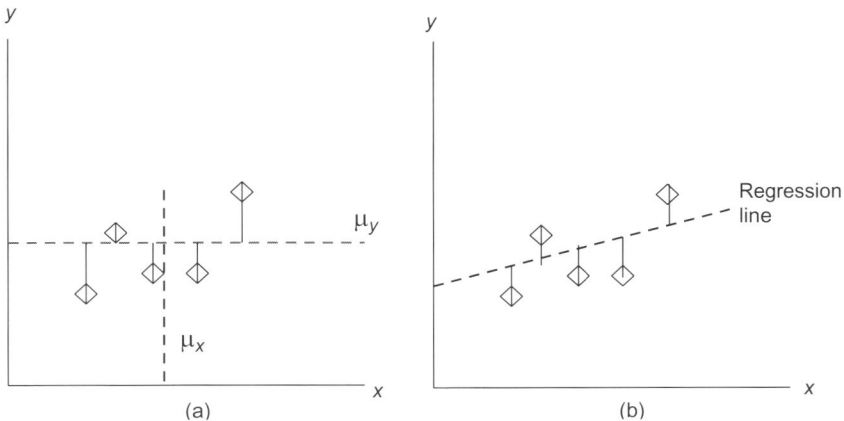

Figure 3.2   Variance and regression: (a) variance about mean (μy); and (b) variance about regression line

where μ and σ^2 are the mean and variance of x and y respectively. The value of the correlation coefficient will vary between 1, indicating an exact positive linear relationship and –1, which indicates an exact negative relationship. A value close to zero indicates a very weak correlation.

The total variance of, say, y, can therefore be considered to comprise the variance due to its relationship with x plus the variance not explained by the regression, i.e.:

$$\sum_{i=1}^{N}\left(y_i - \mu_y\right)^2 = \sum_{i=1}^{N}\left(y_i - \hat{y}_i\right)^2 + \sum_{i=1}^{N}\left(\hat{y}_i - \mu_y\right)^2 \tag{3.17}$$

where $\hat{y}_i = mx_i + c$. This can be written as:

$$\sigma_y^2 = \sigma_y^2 r^2 + \sigma_y^2 (1 - r^2) \tag{3.18}$$

The slope and intercept of the regression line $y = mx + c$ can be estimated from the expression for the mean squared deviation:

$$S^2 = \sum_{i=1}^{N} \frac{(y_i - c - mx_i)^2}{N-2} \tag{3.19}$$

Since, by definition, the slope and intercept of the regression line are such that the mean squared deviation of the data from the line is minimised, the partial differentials $\partial S / \partial c$ and $\partial S / \partial m$ are set equal to zero, giving:

$$m = \frac{\sum_{i=1}^{N}(x_i - \mu_x)(y_i - \mu_y)}{\sum_{i=1}^{N}(x_i - \mu_x)^2} \tag{3.20}$$

and:

$$c = \mu_y - m\mu_x \tag{3.21}$$

It should be noted that the regression line only provides the best estimate of the relationship between x and y and will still involve some uncertainty. The degree of uncertainty can be assessed from the standard error given by:

$$S_y = \sigma_y \sqrt{(1 - r^2)} \tag{3.22}$$

The practical significance of the standard error is that there is about a 95% probability that values of y will lie within plus or minus two standard errors from the regression line.

Although the values of two variables may show a good correlation, there is the possibility that this relationship could have happened by chance because of the particular samples of data used. The probability of a given correlation coefficient occurring by chance can be estimated by a test using the *t statistic* given by:

$$t = \frac{r\sqrt{N-2}}{\sqrt{1-r^2}}$$ (3.23)

The probability of a given value of t being exceeded can be looked up in standard tables (Appendix 3.3) using $N-2$ degrees of freedom. It is generally accepted that if the probability of t being greater than the calculated value is less than 5%, the correlation is statistically significant.

It should also be remembered that the correlation coefficient only measures the degree of *linear* correlation. There may be a very strong relationship between two variables but if this relationship is, say, logarithmic or exponential, it may not be reflected in the correlation coefficient. However, it is possible to transform a non-linear relationship into a linear form in order to assess the degree of correlation. For example:

$y = ax^b$ becomes $\log(y) = \log(a) + b \times \log(x)$

$y = ae^x$ becomes $\log_e(y) = \log_e(a) + x$ (3.24)

etc.

It is also important to realise that a high correlation coefficient may not always indicate a causal relationship. For example, a time-series of evaporation may show a high correlation with atmospheric pressure, when the real causal variable may be temperature or radiation. In that case, the problem could arise because warm clear days are often associated with high atmospheric pressure.

Another frequent problem is to determine whether the slope of the regression line is significantly different from zero (Example 3.5). Again the t test is used, the t-statistic on this occasion being given by:

$$t = \frac{m\sqrt{\sum(x-\mu_x)^2}}{S_y}$$ (3.25)

Example 3.5

For the following readings of mean daily wind speed and evaporation, determine:

(a) the total variance of the wind speed and evaporation data
(b) the best fit linear regression line of evaporation on wind speed
(c) the correlation coefficient between the variables
(d) the residual variance of the evaporation data
(e) the level of significance of the correlation

Day	Wind speed (x): m/s	Evaporation (y): mm/day
1	1	10
2	9	72
3	3·5	36
4	6	50
5	8	70
6	2	14

	Wind speed (x) m/s	Evap. (y): mm/day	$x - \mu_x$ (1)	$y - \mu_y$ (2)	(1) $\times$ (2)	(1)^2
	1	10	−3·92	−32	125·33	15·34
	9	72	4·08	30	122·50	16·67
	3·5	36	−1·42	−6	8·50	2·01
	6	50	1·08	8	8·67	1·17
	8	70	3·08	28	86·33	9·51
	2	14	−2·92	−28	81·67	8·51
Mean	4·92	42·00	0·00	0	433·00	53·21
Sample std. dev.	3·26	26·80				
Variance	10·64	718·40				
Pop. std. dev.	2·98	24·47				
Variance	8·87	598·67				

Slope of regression line $m = \dfrac{\sum xy}{\sum x^2} = \dfrac{433 \cdot 0}{53 \cdot 21} = 8 \cdot 14$

Intercept of regression line $c = y_m - m \times x_m = 42 \cdot 0 - 8 \cdot 14 \times 4 \cdot 92 = 1 \cdot 95$

$(x - \mu_x)/S_x$ (1)	$(y - \mu_y)/S_y$ (2)	(1) × (2)
−1·315	−1·308	1·720
1·371	1·226	1·681
−0·476	−0·245	0·117
0·364	0·327	0·119
1·035	1·144	1·185
−0·979	−1·144	1·121
	Σ	5·943
	r	0·990
	r^2	0·981

Variance explained by regression $= \sigma^2 r^2 = 718 \cdot 4 \times 0 \cdot 981 = 704 \cdot 8$

Residual variance $\sigma^2 \times (1 - r^2) = 718 \cdot 4 \times 0 \cdot 019 = 13 \cdot 6$

3.2.2 Multiple linear regression

The above analysis can be extended to cases where the dependant variable (y) is affected by more than one independent variables (say x,u,v). The form of a multiple linear regression equation would be:

$$y = c + m_x x + m_u u + m_v v \tag{3.26}$$

The mean squared deviation of the data points from the regression line (S) is then calculated as before and the partial differentials $\partial S/\partial c$, $\partial S/\partial m_x$, $\partial S/\partial m_u$ and $\partial S/\partial m_v$ are set to zero, giving four simultaneous equations which can be solved for c, m_x, m_u and m_v. Several statistical software packages for multiple regression are available.

3.3   Time-series analysis

Most hydrological data are in the form of a time-series in which a particular parameter, such as river flow or temperature, is sampled at a regular time interval. Hydrologists need to analyse time-series for a number of reasons. For example, it is frequently required to estimate the probability of a flow exceeding (or otherwise) a given level. It may also be important to establish whether there is a long-term trend in a time-series or it may be necessary to generate a synthetic time-series which has similar properties to the observed time-series.

The probability of exceedence of a given flow can be estimated from the cumulative distribution of flows, commonly referred to as a *flow–duration curve*. However, statistical distributions are of limited use in describing a time-series as they give no information on the sequence of the data. In other words, it is assumed that each value in a time-series is independent of the previous value. A brief glance at a flow hydrograph will show that there is a high degree of correlation of an individual value with the preceding value. Time-series analysis is the study of the patterns and sequences of values in a time-series.

3.3.1   Components of a time-series

Any given time-series can be considered to be composed of a number of elements which might include (Figure 3.3):

- a linear or other type of trend
- a cyclical variation
- a sudden impulse
- an irregular or random component.

A trend may be defined as a long-term change in the mean. However, the difficulty is in the definition of 'long-term'. What might appear to be a trend in a 30-year record could be part of a cyclical variation with a period of 100 years. In effect, a trend can be considered as any variation with a period greater than the length of the time-series.

Cyclical variations are very common in hydrological time-series. These range from diurnal variations of temperature to seasonal changes in climate and may include variations with periods of several years or decades. In most cases they are relatively predictable and, like the trend, can be removed from the time-series to isolate other components.

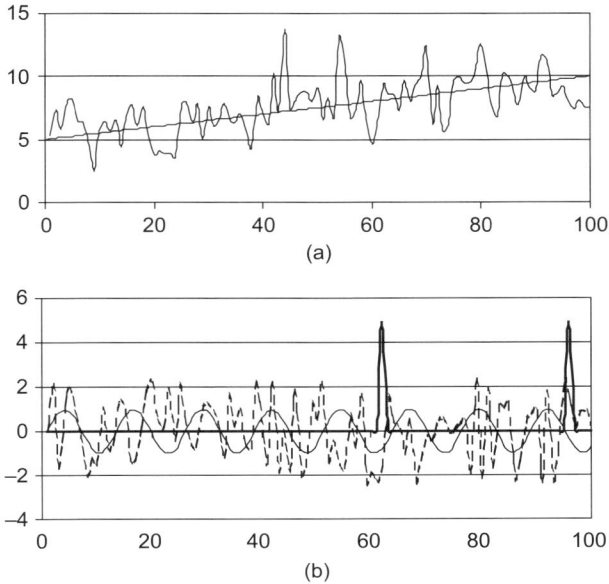

Figure 3.3 Components of a time-series: (a) trend component; and (b) cyclical, impulse and random components

While trends and cyclical components of a time-series are generally deterministic in the sense that they are the result of specific processes which may or may not be identifiable, impulses are sudden changes in a time-series as result of random events, such as volcanic eruptions or earthquakes. Although the event itself may be short lived, the effect of the event may last for some time and it may generate various periodic signals, due to complex feedback mechanisms within the ocean and atmosphere systems.

If all the deterministic and impulse components are removed from the time-series, a series of random variations, which are often termed *noise*, will remain. Noise is characteristic of all hydrological time-series and is the result of several different factors. All instruments used for recording data have a certain level of inaccuracy or uncertainty. For example, a simple daily total rain gauge, consisting of a container placed on the ground, will give readings which have inherent errors. There may be losses due to evaporation and gains due to splashing, as well as errors due to shielding from buildings, etc.; not to mention errors in recording the data. Many of the errors can be minimised by good design and careful siting of the instrument, but they cannot be completely eliminated. For instrument records there will also be errors and inaccuracies due to the response of the transducer or the analysis of the data or other factors. In some measurements, such as flow recording, errors also arise from the limited number of the velocity readings. In records based on proxy measurements, such as tree-ring width, there will be errors associated with the conversion from the raw data to an appropriate climatic parameter. Finally, even if instruments could perfectly record the instantaneous variations in a parameter, there would still be apparently random fluctuations in the record as a result of the chaotic nature of the weather systems.

In principle, the random component of a time-series can be represented by a random variable that has a normal distribution with a zero mean and a variance which represents the precision of the instrument and the degree of variability of the parameter. In practice, this component is not truly random in the sense that each value is independent of the preceding value. Part of this signal may be due, for example, to fluctuations in sea-surface temperatures, which occur relatively slowly and would therefore correspond more to a low-frequency cyclical variation.

If the trend component of a time-series is required, it is firstly necessary to identify and eliminate any cyclical and impulse component as far as possible. The remaining components will therefore be the trend and the residual noise. If the nature of the trend can be determined, the data can be linearised if necessary and a regression analysis carried out to estimate the parameters of the trend in a similar way to regression between two variables. This exercise, known as *curve fitting*, may involve linear, polynomial, logarithmic or other types of curve. The square of the correlation coefficient (r^2) then provides a good indication of how much of the variance of the data can be explained by the trend and how much is due to the residual noise.

3.3.2   Tests for randomness

In some cases it is obvious whether a time-series is random or whether there is a linear or cyclic component, but in others it may be necessary to carry out a statistical test. One of the simplest tests is a count of the number of turning points, a turning point being defined in terms of successive values where:

$$x_{i-1} < x_i > x_{i+1}$$

or:

$$x_{i-1} > x_i < x_{i+1} \tag{3.27}$$

For any sequence of three values there are six possible orders, but there will be turning points in only four of these, i.e. where the central value is either greater or less than both of the adjacent values. Neglecting the first and last items in a series of n values, the expected number of turning points is therefore:

$$E\{N_{tp}\} = \frac{2}{3}(n-2) \tag{3.28}$$

It can be shown (Kendall, 1976) that the values of N_{tp} will be normally distributed with a variance given by:

$$\text{var}\{N_{tp}\} = \frac{(16n-29)}{90} \tag{3.29}$$

Thus, it is possible to estimate the probability of a value of N_{tp} occurring by chance.

An extension to the comparison of adjacent values in a series is to compare all the values in pairs. For a series, the number of cases of $x_j < x_i$ for $j < i$ are counted. The total number of such pairs in a series of n values is $\frac{1}{2}n(n-1)$ and therefore the probability of $x_j < x_i$ in a random series is $\frac{1}{4}n(n-1)$. A greater number than this would indicate a rising trend and, conversely, a smaller number would suggest a falling trend. Kendall (1976) proposed a rank coefficient given by:

$$\tau = \frac{4P}{n(n-1)} - 1 \tag{3.30}$$

where P is the number of cases of $x_j < x_i$ in a series of n values. The coefficient varies from +1 to –1, and for a random series the expected value would be zero (Example 3.6). The variance of τ is given by:

$$\text{var}\{\tau\} = \frac{2(2n+5)}{9n(n-1)} \tag{3.31}$$

The level of significance of a given value of τ can therefore be estimated by reference to normal distribution tables.

A similar test for the possible trend in a series is the *Spearman Rank Test*. In this test the rank (y_i) of each element in the series is established by arranging the series in ascending order and calculating the correlation coefficient between the order (i) of x_i and the rank, which is given by (Sneyers, 1990):

Example 3.6

Test the randomness of the flow data from Example 3.1.

For the each values of x, the number of preceding values which are less than x are counted and summed.

x	No. of preceding values $<x$
0·444	0
2·685	1
1·982	1
0·309	0
0·945	2
0·846	2
2·380	5
0·162	0
0·871	4
1·405	6
Total	21

Number of pairs for $x_j > x_i$

$$P = 21$$

$$\tau = \frac{(4 \times P)}{[n(n-1)] - 1} = \frac{4 \times 21}{10 \times 9} - 1 = -0.067$$

Variance of $\tau = \frac{[2(2n+5)]}{[9n(n-1)]} = \frac{[2(20+5)]}{90 \times 9} = 0.062$

Standard deviation $= 0.248$

Probability $P < \left| \frac{0.067}{0.248} \right| = 0.106 \times 2 = 0.212$

Therefore there is a 21% chance of τ being less than $|0.067|$ and the result is not significant.

$$r_s = 1 - \frac{6}{n(n^2 - 1)} \sum_{i=1}^{n} (y_i - i)^2 \tag{3.32}$$

For a series where there is a significant increasing trend, there will be a high positive correlation between the order and the rank, while a low absolute correlation will indicate no significant trend and a high negative correlation coefficient will indicate a strong negative trend. The correlation coefficient can be considered as being normally distributed with a mean of zero and a variance given by:

$$\sigma^2 = \frac{1}{(n-1)} \tag{3.33}$$

The level of significance of a given value of $|r_s|$ is also estimated by reference to normal distribution tables (Example 3.7).

Example 3.7

Using the Spearman Rank Test, investigate the randomness of (a) the series from Example 3.1 and (b) the re-ordered series below:

0·162, 0·309, 0·846, 0·444, 0·945, 1·982, 1·405, 0·871, 2·38, 2·685

The correlation coefficient between the series (i) and the rank (y_i) is given by:

$$r_s = 1 - \frac{6}{n(n^2 - 1)} \sum_{i=1}^{n} (y_i - i)^2$$

This can be considered as being normally distributed with a mean of zero and a variance of:

$$\frac{1}{n-1} = 0{\cdot}11$$

Hence the probability of $r < |r_s|$ can be found from normal distribution tables and the probability of $|r| > |r_s|$ is:

$$2 \times \left(1 - P\left(r < |r_s|\right)\right)$$

The results of the calculations for the two series are tabulated below.

	(a)				(b)		
i	x	y	$(y - i)^2$	i	x	y	$(y - i)^2$
1	0·444	3	4	1	0·162	1	0
2	2·685	10	64	2	0·309	2	0
3	1·982	8	25	3	0·846	4	1
4	0·309	2	4	4	0·444	3	1
5	0·945	6	1	5	0·945	6	1
6	0·846	4	4	6	1·982	8	4
7	2·38	9	4	7	1·405	7	0
8	0·162	1	49	8	0·871	5	9
9	0·871	5	16	9	2·38	9	0
10	1·405	7	9	10	2·685	10	0

$\sum(y - 1)^2 = 180$ $\sum(y - 1)^2 = 16$

$r_s = 0·09$ $r_s = 0·90$

Variance = 0·11   Variance = 0·11

Standard deviation = 0·33   Standard deviation = 0·33

$P(r < |r_s|) = 0·61$ $P(r < |r_s|) = 1·00$

$P(|r| > |r_s|) = 0·785$ $P(|r| > |r_s|) = 0·007$

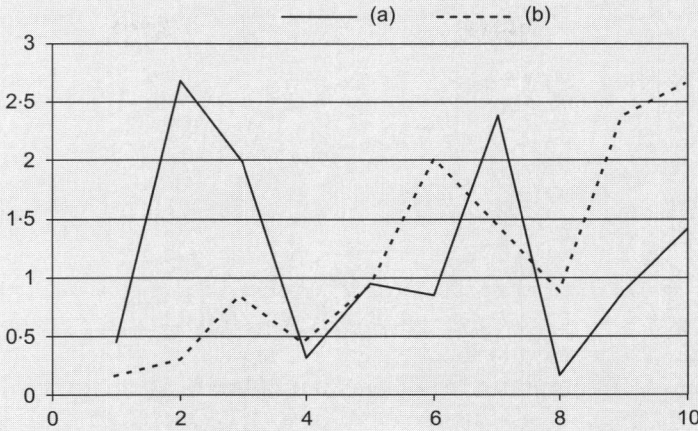

The trend in series (b) is significant at the 0·8% level.

3.3.3 Frequency analysis

The French mathematician Fourier showed that any time-series consisting of 2N equally spaced points may be expressed as the sum of N harmonic components of different amplitudes. Thus, a series of 100 daily values could be represented by the sum of 50 cosine components. The first component would have a period of 100 days, the second 50 days, the next 33·3 days and so on to the fiftieth component with a period of two days. The amplitude of each component can be found by a process known as *Fourier Analysis* and the square of the amplitude represents the variance contained in that component.

$$x_t = a_0 + \sum_{p=1}^{\frac{N}{2}-1}\left[a_p\,\cos\left(\frac{2\pi pt}{N}\right)+b_p\,\sin\left(\frac{2\pi pt}{N}\right)\right]+a_{n/2}\,\cos(\pi t) \qquad (3.34)$$

where:

$$a_0 = \mu$$

$$a_{n/2} = \sum \frac{(-1)^t x_t}{N}$$

$$a_p = \frac{2}{N} \left[\sum x_t \cos\left(\frac{2\pi pt}{N} \right) \right]$$

$$b_p = \frac{2}{N} \left[\sum x_t \sin\left(\frac{2\pi pt}{N} \right) \right]$$

The plot of the amplitude squared against frequency is known as the *power spectrum* of a time-series and is a useful way of indicating the principle frequency components of a time-series.

Where each component has a similar amplitude, i.e. there is a uniform mixture of frequencies, the signal is termed *white noise*, using the analogy of white light. As mentioned in the previous section, the random noise component found in most time-series should, in theory, be white noise. However, because of the inertia of the response of many hydrological variables, such as sea temperature, there is often a low frequency cyclic component, which is often called *red noise*, using the same analogy with light. The combination of this red noise with the white noise from instrumental errors, etc., form what is called *pink noise*.

3.3.4 Filtering

Much of the random and cyclical fluctuations in a hydrological time-series can be removed by a process of filtering or smoothing. The simplest and most common method of smoothing is using a *moving average*, by which a time-series is formed from the mean of a given number of successive values in the original time-series. For example, a moving average of size three would be:

$$a_i = \frac{(y_{i-1} + y_i + y_{i+1})}{3} \qquad (i = 2, 3, \ldots, N-1) \qquad (3.35)$$

The object of taking a moving average is to eliminate or filter out cyclical components with periods less than that used for the moving average. It can be seen that, for example, a nine-year moving average will eliminate a cyclical component with a period of exactly nine years since the values over any period of nine years will cancel each other.

Fluctuations with periods which have an exact integer number of cycles in nine years (e.g. periods of four and a half years, three years, etc.) will also be eliminated. However, components with other periods less than nine years (e.g. six years) will not be eliminated and may turn up as spurious signals out of phase with the original harmonic. This difficulty arises fundamentally because of the equal weight given to each of the values in the moving average 'box' and the zero weight given to values outside the 'box'. The problem can be reduced if the weighting within the moving average box is progressively varied from a low value at the outside to a maximum at the centre (bearing in mind that the sum of the weights should be equal to 1). A simple approach is to use a triangular weighting which, for a moving average of five, might be:

$$a_i = 0{\cdot}12y_{i-2} + 0{\cdot}22y_{i-1} + 0{\cdot}32y_i + 0{\cdot}22y_{i+1} + 0{\cdot}12y_{i+2} \qquad (3.36)$$

The effect of using a triangular weighting instead of the equal weighting is shown in Figure 3.4, which shows the amplitude of components of different frequencies after the application of the rectangular (uniform weight) filter and the triangular filter. The 'ideal' filter is one which removes all components with frequencies greater than the 1/period of the filter and leaves longer period components unaffected. This is virtually impossible to achieve and, in practice, there must be a compromise between the performance of the filter and the complexity and the amount of calculation.

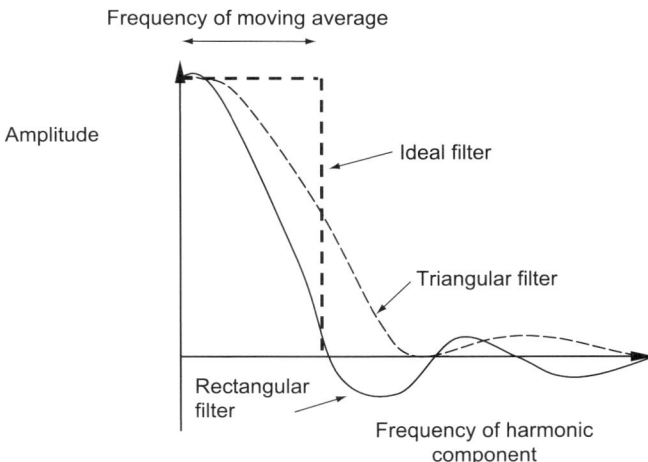

Figure 3.4   Performance of moving average filters

The moving average is an example of a *'low pass'* filter in which low-frequency components are allowed to pass and high-frequency components are removed. Corresponding *'high pass'* filters can be used to remove low-frequency components and, likewise, *'narrow band'* filters can remove frequencies outside a specific range. In all cases, the performance of the filter is never ideal and depends on the complexity and the number of terms in the filter function.

3.3.5 Autocorrelation

It has been seen that the relationship between two variables *x* and *y* can be described by their covariance which indicates the strength of the relationship between the variables. The same analysis can be applied to the relationship between successive values in a time-series. For a given time-series $x_1, x_2, x_3, \ldots , x_N$, pairs of values can be formed, i.e. x_1 and x_2, x_2 and x_3, $\ldots , x_{N-1}, x_N$. The pairs of values can be regarded as consisting of a dependent and an independent variable, and the covariance between the two variables is known as the *autocovariance* and is given by:

$$c_1 = \frac{1}{N}\sum_{i-1}^{N-1}(x_i - \mu)(x_{i+1} - \mu) \tag{3.37}$$

where μ is the mean of the series. (The above equation assumes that the mean of the first $N-1$ points in the series is the same as the mean of the last $N-1$ points and that $N-1 \approx N$). The suffix 1 indicates that the series is being compared with the same series displaced by one time-interval (lag 1).

The relationship can also be expressed in the form of the *autocorrelation coefficient* which, for a lag *k*, is given by:

$$r_k = \frac{\displaystyle\sum_{i=k}^{N-k}(x_i - \mu)(x_{i+k} - \mu)}{\displaystyle\sum_{i=1}^{N}(x_i - \mu)^2} \tag{3.38}$$

i.e.:

$$r_k = \frac{c_k}{c_0} \tag{3.39}$$

Values of the autocorrelation coefficient for lags of 2, 3, $\ldots$, $N-1$ can be calculated in the same way although there is usually little point in

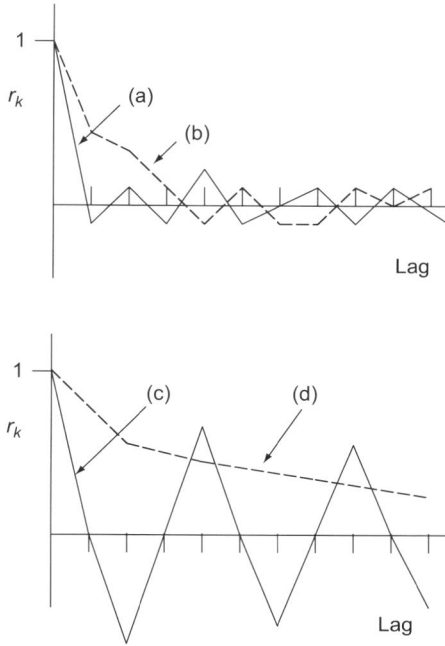

Figure 3.5   Examples of correlograms: (a) random time-series; (b) short-term correlation; (c) periodic series; and (d) times-series with a positive trend

calculating r for lags greater than about $N/4$. It is often useful to plot the autocorrelation coefficient against the lag k and the resulting *correlogram* provides information about the nature of the time-series. For a random time-series, r_k will generally be close to zero (apart from r_0) (Figure 3.5(a)). In fact r_k tends to be normally distributed with a zero mean and a standard deviation of $1/N$. In other words, 95% of the values of r_k for a random series would be expected to lie between $\pm 2/\sqrt{N}$. However, it is also possible to have an occasional relatively high value of r_k, even when the series is completely random.

Many hydrological records show some correlation in the short term. In other words, any particular observation is likely to be quite similar to the preceding observation and to one or two before that. The correlogram for such a series will show relatively high values of r_k, for the first few lags and values close to zero for longer lags (Figure 3.5(b)). Where the series contains a periodic element (for example, seasonal variations), the correlogram will also show an oscillation at the same frequency (Figure 3.5(c)). Usually, the seasonal component dominates the correlogram and it is normal practice to remove any seasonal component before constructing the correlogram, so that other information about the time-series can be found. If there is a significant trend in the time-series, the correlogram will not reach zero because of the strong correlation at all lags (Figure 3.5(d)). Again, the trend is normally removed from the time-series before obtaining the correlogram so that other useful information is not masked by the trend. If the time-series contains outliers (exceptionally high or low values) it is also advisable to remove these from the series, as they can have a disproportionately large effect on the correlogram.

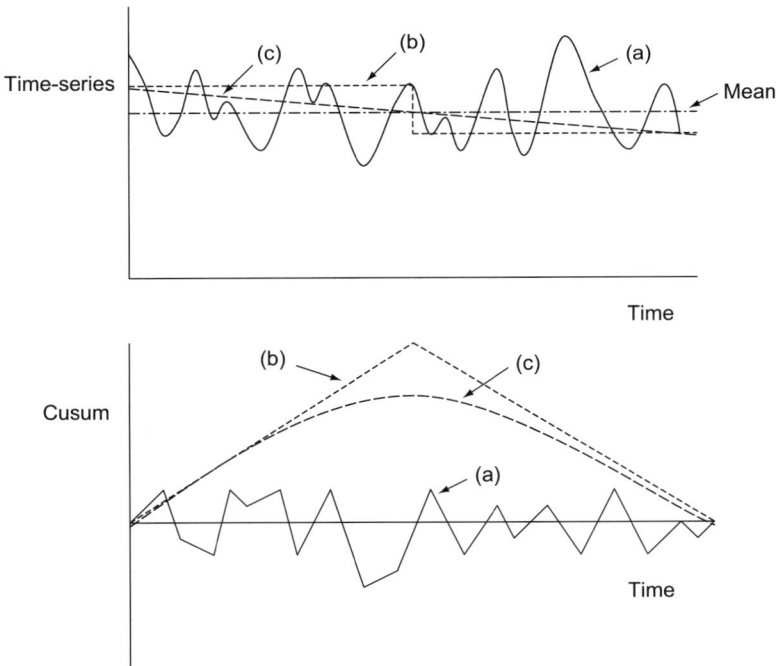

Figure 3.6   Examples of cusum curves: (a) random time-series; (b) step change; and (c) linear trend

3.3.6   Cusum curves

The cusum curve is a useful technique for detecting if there has been a change in a process or a record. The cusum is the cumulative sum of the deviations of the series about a given value, which is normally taken as the mean value, i.e.:

$$S_t = \sum_{i=1}^{t}(x_t - \mu) \qquad (3.40)$$

For a time-series with small, random fluctuations about the mean, the cusum will be close to zero throughout the time-series (Figure 3.6(a)). At the other extreme, if the time-series consists of two constant elements with a step change between them (Figure 3.6(b)), the cusum will consist of two linear portions increasing to a peak at the change. If there is a linearly decreasing trend, the cusum will be a curve with its apex where the trend crosses the mean line (Figure 3.6(c)).

It can be seen that the maximum amplitude of the cusum curve is an indication of the seasonality of an annual time-series, showing whether, for example, rainfall is evenly distributed throughout the year or whether it is concentrated at certain times.

3.3.7   The Hurst Coefficient

The Hurst Coefficient is another measure of persistence in a natural time-series. It is used in the calculation of the *range* of a series, which is defined as the difference between the greatest cumulative deficiency below the mean and the greatest cumulative excess above the mean (in other words the amplitude of the cusum curve). From an analysis of flows in the River Nile, Hurst (1951) found that the range (R) can be expressed as:

$$R = \sigma\left(\frac{n}{2}\right)^{h} \qquad (3.41)$$

where σ is the standard deviation of the series and n is the length of the series. The exponent h is known as the *Hurst Coefficient*, which, for a first order process (i.e. where each value depends only on the preceding value), can be shown to be 0·5. Hurst analysed over 800 different time-series and found the average value of h to be 0·73 and in some cases it approached 1, which suggests a higher than expected degree of persistence in the series.

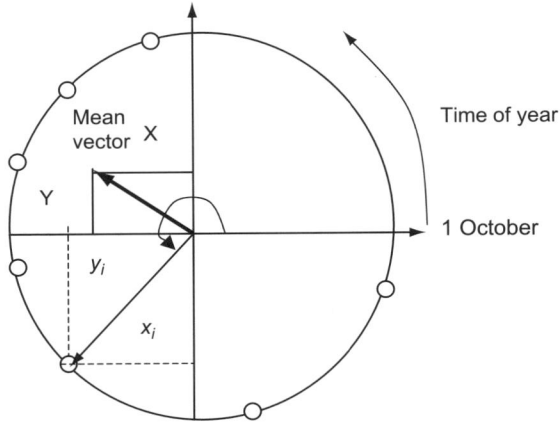

Figure 3.7 Analysis of seasonality

3.3.8 Measurement of seasonality

Seasonality is a common source of periodic variation in hydrological time-series. A graphical indication of the amount of seasonality in, say, records of flood peaks can be found by plotting the time of occurrence of the peaks on the circumference of a circle (Black and Werritty, 1997). To avoid discontinuities at the start and end of years, the year is conventionally started at a period of low flows, such as 1 October. Since the total circumference of the circle represents 365 days, the angular location of each peak can be determined (Figure 3.7). The circle has a unit radius and a given peak occurring at T days after the start of the year and will have coordinates x_i, y_i defined by:

$$x_i = \cos(\phi)$$

and:

$$y_i = \sin(\phi) \tag{3.42}$$

where $\phi = T \times 360 / 365$.

A mean vector can then be defined by coordinates X, Y, which are the means of the x and y coordinates of the individual observations. The direction of the mean vector is an indication the mean time of occurrence of the observations, and the length of the vector indicates the degree of variability of the data. A zero value indicates that the events were evenly distributed throughout the year, while a value close to 1 indicates that all the events occurred within a short period.

3.4   Models of time-series

Observed flow records are often shorter than is desirable to provide reliable data for design. Most flow records are less than 40 years, whereas estimates of flows with return periods of 100 years or more are often required. One solution is to generate synthetic flow records on the basis of the statistics of an observed record.

As has been described before, most hydrological time-series are a combination of deterministic and random components. Deterministic implies that the output of the model is uniquely defined by the input and the model parameters. Deterministic components, such as linear or periodic trends, can be identified as described above and can be programmed quite easily, if required. The random components can be modelled in a number of ways.

3.4.1   Purely random processes

A purely random process consists of a sequence of mutually independent variables which belong to a similar distribution. Such a series is commonly written as:

$$x_t = x_{t-1} + z_t \tag{3.43}$$

where z_t is a randomly distributed variable with a zero mean and a specified variance.

Most spreadsheets and other statistical software have random number generators that produce numbers based on a rectangular or uniform (0,1) distribution (zero mean, unit standard deviation). Random variables with a normal distribution can be generated from random numbers from a uniform distribution using the relationship:

$$z_N = \sqrt{-2 \log_{10}(X_{0,i})} \times \cos(2\pi X_{0,i+1}) \tag{3.44}$$

where $X_{0,i}$ and $X_{0,i+1}$ are random numbers with a uniform (0,1) distribution.

3.4.2 Moving average processes

In a moving average (MA) process, a value is calculated as a weighted average of a sequence of randomly distributed variables. Thus, a moving average process of order m is defined as:

$$x_t = (\beta_0 z_t) + (\beta_1 z_{t-1}) + \ldots (\beta_m z_{t-m}) \tag{3.45}$$

where z_t, z_{t-1}, etc., is a randomly distributed variable with a zero mean and β_0, β_1, etc., are constants.

3.4.3 Autoregressive processes

Where there is a degree of autocorrelation between successive values of the time-series, an autoregressive (AR) model may be appropriate. In general terms, an autoregressive model of order m is defined as:

$$x_t = (\alpha_1 x_{t-1}) + \ldots (\alpha_m x_{t-m}) + z_t \tag{3.46}$$

The simplest form of autoregressive model, a first order process which is often known as a *Markov Process*, after the Russian A. A. Markov, is written as:

$$x_t = \alpha x_{t-1} + z_t \tag{3.47}$$

The coefficient α can be taken as being approximately the lag 1 autocorrelation coefficient for the series.

3.4.4 Mixed autoregressive-moving average processes

Many hydrological time-series contain both autoregressive and moving average elements and a mixed autoregressive-moving average (ARMA) model may be appropriate. Such a model is written as:

$$x_t = (\alpha_1 x_{t-1}) + \ldots (\alpha_p x_{t-p}) + z_t + (\beta_1 z_{t-1}) + \ldots (\beta_q z_{t-q}) \tag{3.48}$$

where p and q are the orders of the autoregressive and moving average terms respectively.

3.4.5   Models of streamflow

In the case of streamflow, it is often sufficient to preserve only the mean, standard deviation and lag 1 serial correlation and a first-order Markov model is used. This implies that any event is dependent only on the preceding event. For example, the annual flow in a river could be represented by:

$$Q_t = \overline{Q} + \alpha(Q_{t-1} - \overline{Q}) + z_t \sigma \sqrt{1-\alpha^2} \qquad (3.49)$$

where z is a random variate from an appropriate distribution with a zero mean and unit variance, Q and σ are the mean and standard deviation of Q and α is the lag 1 autocorrelation coefficient. The synthetic time-series can be generated by assuming an initial value for Q and then calculating successive values using a randomly generated value for z, according to the chosen distribution.

If seasonal or monthly values are required, the model can be modified to incorporate the statistical properties corresponding to each month or season, i.e.:

$$Q_{i,j} = \overline{Q}_j + \alpha_j \frac{\sigma_j}{\sigma_{j-1}} (Q_{i-1,j-1} - \overline{Q}_{j-1}) + z_i \sigma_j \sqrt{1-\alpha_j^2} \qquad (3.50)$$

where the subscript j refers to the season or month and i denotes the sequence of the items in the time-series. To use the above model, values of Q, σ and α are required for each month (or season). It is advisable to start at the beginning of a water year when flows are low and, to avoid start-up bias, the model should be run for between 12 and 20 time-increments so that the influence of the assumed initial flow is removed. The choice of the probability distribution for z should accurately reflect the distribution of the original time-series, although it should be borne in mind that a short historical record will not clearly define the properties of the distribution. Also, the difficulty of generating random variables from the more complex distributions may preclude their use. A normal distribution is the simplest type unless the distribution is markedly skewed, in which case a log-normal distribution may be appropriate.

3.4.6   Fitting theoretical processes to observed time-series

If the mean and the deterministic components of a time-series are removed and an appropriate autocorrelation or moving average model fitted to the series, the closeness of fit can be measured by the sum of the squares of the residual differences, i.e.:

$$S = \sum_{t=1}^{N} (X_t - x_t)^2 \tag{3.51}$$

where X_t and x_t are the observed and generated values of the time-series. Thus, the various types of model (e.g. MA, AR, ARMA) can be compared in terms of the sum of the squares of the deviations. The residual sum of the squares decreases as the order of the model increases, but there is usually a point where the addition of extra terms does not lead to any significant further reduction in the sum of the squares.

Although the parameters of the autoregressive model can be estimated directly from the autocorrelation coefficients, the parameters β_0, β_1, ... , etc., for the moving average model cannot be so easily determined and in many cases a trial and error process has to be employed.

3.5 Estimation from spatially distributed point data

Hydrological parameters are generally observed at a limited number of recording stations and it is often necessary to estimate values at other discrete locations or at points on a regular grid in order to produce a contour map.

If it is required to estimate the value of a parameter at a given location based on records from a series of recording stations, it is logical to give more weight to those stations that are closest to the given location. Most methods of estimating from a series of point values involve applying a weight to each value, based on the closeness of the station, although this may not be in terms of a geographical distance.

One of the most straightforward approaches is the *inverse distance method*, where the weighting is proportional to the inverse of the distance between the given location and the recording station, i.e.:

$$\hat{z} = \frac{\displaystyle\sum_{i=1}^{N} \frac{1}{d_i^p} z_i}{\displaystyle\sum_{i=1}^{N} \frac{1}{d_i^p}} \tag{3.52}$$

where $d_1, \ldots, d_N$ are the distances between the recording stations and the point where the estimate is required, and $z_1, \ldots, z_N$ are the observed values. The exponent p is commonly set to 1, although increasing the value of p gives relatively more weight to the nearest samples and vice-versa. For $p = 0$, the method gives the simple mean, with equal weights for all observations.

A more sophisticated method known as *kringing* estimates the weights based on the covariance between the observed records. Each observation can be considered as a sample of a population with a given variance. The covariance between two samples, defined as:

$$C_{i,j} = \frac{\sum_{i=1}^{N}(z_i - \mu_i)(z_j - \mu_j)}{N} \tag{3.53}$$

will generally decrease as the distance between the samples increases. A typical model for the covariance might be:

$$C = \begin{cases} C_0 + C_1 \Rightarrow h = 0 \\ C_0 + C_1\left[1 - \exp\left(\frac{-kh}{a}\right)\right] \Rightarrow h > 0 \end{cases} \tag{3.54}$$

where h is the distance between the samples. The parameters C_0 and C_1 define the variance of the samples while α represents the maximum range over which samples are included and k defines the shape of the covariance distance model.

For a group of n recording stations a $(n+1) \times (n+1)$ covariance matrix can be established. The matrix of weights is then determined from the matrix relationship

$$\mathbf{C} \times \mathbf{w} = \mathbf{D}$$

$$\begin{bmatrix} C_{11} & \Rightarrow & C_{1n} & 1 \\ \Downarrow & \ldots & \Downarrow & \Downarrow \\ C_{n1} & \Rightarrow & C_{nn} & 1 \\ 1 & \ldots & 1 & 0 \end{bmatrix} \times \begin{bmatrix} w_1 \\ \Downarrow \\ w_n \\ \mu \end{bmatrix} = \begin{bmatrix} C_{10} \\ \Downarrow \\ C_{n0} \\ 1 \end{bmatrix} \tag{3.55}$$

This can be written in inverse form as:

$$\mathbf{w} = \mathbf{C}^{-1} \times \mathbf{D} \tag{3.56}$$

As can be seen, the process of kringing is quite complex and readers are recommended to consult textbooks in geostatistics, such as by Isaaks (1989). Standard computer software packages, such as *Surfer*, use kringing to generate surfaces and contour plots based on discrete spatial data.

3.6   Summary

This chapter describes some of the statistical methods that are available to assist any hydrological investigation. A hydrological time-series can be analysed in terms of a statistical distribution, which allows the probability or return period of a particular event to be estimated. The various types of distribution are described. However, time-series of flow do not generally consist of independent values, and methods of time-series analysis are often needed which take into account the sequence of the data. A time-series may include a trend and a periodic component, as well as impulse and random noise components. These components may be isolated and removed individually if necessary and may also be generated synthetically. Finally, the chapter considered the problem of dealing with spacially distributed data.

3.7   References

Black, A. R. and Werritty, A. (1997). Seasonality of Flooding: a case study of North Britain. *Journal of Hydrology*, **195**, 1–25.

Hurst, H. E. (1951). Long-Term Storage Capacity of Reservoirs. *Trans. ASCE*, **116**, 770–808.

Isaaks, E. H. (1989). *Applied Geostatistics*. Oxford University Press, Oxford.

Kendall, M. (1976). *Time-Series*. Charles Griffin, London.

Robson, A. and Reed, D. (1999). *Flood Estimation Handbook Vol. 3 — Statistical Procedures for Flood Estimation. Institute of Hydrology, Wallingford.*

Sneyers, R. (1990). *On the Statistical Analysis of a Series of Observations*. World Meteorological Organization, Geneva.

Appendix 3.1
Cumulative normal distribution

	0·00	0·01	0·02	0·03	0·04	0·05	0·06	0·07	0·08	0·09
−3·5	0·0002	0·0002	0·0002	0·0002	0·0002	0·0002	0·0002	0·0002	0·0002	0·0002
−3·4	0·0003	0·0003	0·0003	0·0003	0·0003	0·0003	0·0003	0·0003	0·0003	0·0002
−3·3	0·0005	0·0005	0·0005	0·0004	0·0004	0·0004	0·0004	0·0004	0·0004	0·0003
−3·2	0·0007	0·0007	0·0006	0·0006	0·0006	0·0006	0·0006	0·0005	0·0005	0·0005
−3·1	0·0010	0·0009	0·0009	0·0009	0·0008	0·0008	0·0008	0·0008	0·0007	0·0007
−3·0	0·0013	0·0013	0·0013	0·0012	0·0012	0·0011	0·0011	0·0011	0·0010	0·0010
−2·9	0·0019	0·0018	0·0018	0·0017	0·0016	0·0016	0·0015	0·0015	0·0014	0·0014
−2·8	0·0026	0·0025	0·0024	0·0023	0·0023	0·0022	0·0021	0·0021	0·0020	0·0019
−2·7	0·0035	0·0034	0·0033	0·0032	0·0031	0·0030	0·0029	0·0028	0·0027	0·0026
−2·6	0·0047	0·0045	0·0044	0·0043	0·0041	0·0040	0·0039	0·0038	0·0037	0·0036
−2·5	0·0062	0·0060	0·0059	0·0057	0·0055	0·0054	0·0052	0·0051	0·0049	0·0048
−2·4	0·0082	0·0080	0·0078	0·0075	0·0073	0·0071	0·0069	0·0068	0·0066	0·0064
−2·3	0·0107	0·0104	0·0102	0·0099	0·0096	0·0094	0·0091	0·0089	0·0087	0·0084
−2·2	0·0139	0·0136	0·0132	0·0129	0·0125	0·0122	0·0119	0·0116	0·0113	0·0110
−2·1	0·0179	0·0174	0·0170	0·0166	0·0162	0·0158	0·0154	0·0150	0·0146	0·0143
−2·0	0·0228	0·0222	0·0217	0·0212	0·0207	0·0202	0·0197	0·0192	0·0188	0·0183
−1·9	0·0287	0·0281	0·0274	0·0268	0·0262	0·0256	0·0250	0·0244	0·0239	0·0233
-1·8	0·0359	0·0351	0·0344	0·0336	0·0329	0·0322	0·0314	0·0307	0·0301	0·0294
−1·7	0·0446	0·0436	0·0427	0·0418	0·0409	0·0401	0·0392	0·0384	0·0375	0·0367
−1·6	0·0548	0·0537	0·0526	0·0516	0·0505	0·0495	0·0485	0·0475	0·0465	0·0455
−1·5	0·0668	0·0655	0·0643	0·0630	0·0618	0·0606	0·0594	0·0582	0·0571	0·0559
−1·4	0·0808	0·0793	0·0778	0·0764	0·0749	0·0735	0·0721	0·0708	0·0694	0·0681
−1·3	0·0968	0·0951	0·0934	0·0918	0·0901	0·0885	0·0869	0·0853	0·0838	0·0823
−1·2	0·1151	0·1131	0·1112	0·1093	0·1075	0·1056	0·1038	0·1020	0·1003	0·0985
−1·1	0·1357	0·1335	0·1314	0·1292	0·1271	0·1251	0·1230	0·1210	0·1190	0·1170

	0·00	0·01	0·02	0·03	0·04	0·05	0·06	0·07	0·08	0·09
−1·0	0·1587	0·1562	0·1539	0·1515	0·1492	0·1469	0·1446	0·1423	0·1401	0·1379
−0·9	0·1841	0·1814	0·1788	0·1762	0·1736	0·1711	0·1685	0·1660	0·1635	0·1611
−0·8	0·2119	0·2090	0·2061	0·2033	0·2005	0·1977	0·1949	0·1922	0·1894	0·1867
−0·7	0·2420	0·2389	0·2358	0·2327	0·2296	0·2266	0·2236	0·2206	0·2177	0·2148
−0·6	0·2743	0·2709	0·2676	0·2643	0·2611	0·2578	0·2546	0·2514	0·2483	0·2451
−0·5	0·3085	0·3050	0·3015	0·2981	0·2946	0·2912	0·2877	0·2843	0·2810	0·2776
−0·4	0·3446	0·3409	0·3372	0·3336	0·3300	0·3264	0·3228	0·3192	0·3156	0·3121
−0·3	0·3821	0·3783	0·3745	0·3707	0·3669	0·3632	0·3594	0·3557	0·3520	0·3483
−0·2	0·4207	0·4168	0·4129	0·4090	0·4052	0·4013	0·3974	0·3936	0·3897	0·3859
−0·1	0·4602	0·4562	0·4522	0·4483	0·4443	0·4404	0·4364	0·4325	0·4286	0·4247
0·0	0·5000	0·5040	0·5080	0·5120	0·5160	0·5199	0·5239	0·5279	0·5319	0·5359
0·1	0·5398	0·5438	0·5478	0·5517	0·5557	0·5596	0·5636	0·5675	0·5714	0·5753
0·2	0·5793	0·5832	0·5871	0·5910	0·5948	0·5987	0·6026	0·6064	0·6103	0·6141
0·3	0·6179	0·6217	0·6255	0·6293	0·6331	0·6368	0·6406	0·6443	0·6480	0·6517
0·4	0·6554	0·6591	0·6628	0·6664	0·6700	0·6736	0·6772	0·6808	0·6844	0·6879
0·5	0·6915	0·6950	0·6985	0·7019	0·7054	0·7088	0·7123	0·7157	0·7190	0·7224
0·6	0·7257	0·7291	0·7324	0·7357	0·7389	0·7422	0·7454	0·7486	0·7517	0·7549
0·7	0·7580	0·7611	0·7642	0·7673	0·7704	0·7734	0·7764	0·7794	0·7823	0·7852
0·8	0·7881	0·7910	0·7939	0·7967	0·7995	0·8023	0·8051	0·8078	0·8106	0·8133
0·9	0·8159	0·8186	0·8212	0·8238	0·8264	0·8289	0·8315	0·8340	0·8365	0·8389
1·0	0·8413	0·8438	0·8461	0·8485	0·8508	0·8531	0·8554	0·8577	0·8599	0·8621
1·1	0·8643	0·8665	0·8686	0·8708	0·8729	0·8749	0·8770	0·8790	0·8810	0·8830
1·2	0·8849	0·8869	0·8888	0·8907	0·8925	0·8944	0·8962	0·8980	0·8997	0·9015
1·3	0·9032	0·9049	0·9066	0·9082	0·9099	0·9115	0·9131	0·9147	0·9162	0·9177
1·4	0·9192	0·9207	0·9222	0·9236	0·9251	0·9265	0·9279	0·9292	0·9306	0·9319
1·5	0·9332	0·9345	0·9357	0·9370	0·9382	0·9394	0·9406	0·9418	0·9429	0·9441
1·6	0·9452	0·9463	0·9474	0·9484	0·9495	0·9505	0·9515	0·9525	0·9535	0·9545
1·7	0·9554	0·9564	0·9573	0·9582	0·9591	0·9599	0·9608	0·9616	0·9625	0·9633
1·8	0·9641	0·9649	0·9656	0·9664	0·9671	0·9678	0·9686	0·9693	0·9699	0·9706

	0·00	0·01	0·02	0·03	0·04	0·05	0·06	0·07	0·08	0·09
1·9	0·9713	0·9719	0·9726	0·9732	0·9738	0·9744	0·9750	0·9756	0·9761	0·9767
2·0	0·9772	0·9778	0·9783	0·9788	0·9793	0·9798	0·9803	0·9808	0·9812	0·9817
2·1	0·9821	0·9826	0·9830	0·9834	0·9838	0·9842	0·9846	0·9850	0·9854	0·9857
2·2	0·9861	0·9864	0·9868	0·9871	0·9875	0·9878	0·9881	0·9884	0·9887	0·9890
2·3	0·9893	0·9896	0·9898	0·9901	0·9904	0·9906	0·9909	0·9911	0·9913	0·9916
2·4	0·9918	0·9920	0·9922	0·9925	0·9927	0·9929	0·9931	0·9932	0·9934	0·9936
2·5	0·9938	0·9940	0·9941	0·9943	0·9945	0·9946	0·9948	0·9949	0·9951	0·9952
2·6	0·9953	0·9955	0·9956	0·9957	0·9959	0·9960	0·9961	0·9962	0·9963	0·9964
2·7	0·9965	0·9966	0·9967	0·9968	0·9969	0·9970	0·9971	0·9972	0·9973	0·9974
2·8	0·9974	0·9975	0·9976	0·9977	0·9977	0·9978	0·9979	0·9979	0·9980	0·9981
2·9	0·9981	0·9982	0·9982	0·9983	0·9984	0·9984	0·9985	0·9985	0·9986	0·9986
3·0	0·9987	0·9987	0·9987	0·9988	0·9988	0·9989	0·9989	0·9989	0·9990	0·9990
3·1	0·9990	0·9991	0·9991	0·9991	0·9992	0·9992	0·9992	0·9992	0·9993	0·9993
3·2	0·9993	0·9993	0·9994	0·9994	0·9994	0·9994	0·9994	0·9995	0·9995	0·9995
3·3	0·9995	0·9995	0·9995	0·9996	0·9996	0·9996	0·9996	0·9996	0·9996	0·9997
3·4	0·9997	0·9997	0·9997	0·9997	0·9997	0·9997	0·9997	0·9997	0·9997	0·9998
3·5	0·9998	0·9998	0·9998	0·9998	0·9998	0·9998	0·9998	0·9998	0·9998	0·9998

Appendix 3.2
$\aleph^2$ distribution

	0·995	0·99	0·975	0·95	0·5	0·2	0·1	0·05	0·025	0·01	0·005
1	0·00	0·00	0·00	0·00	0·45	1·64	2·71	3·84	5·02	6·63	7·88
2	0·01	0·02	0·05	0·10	1·39	3·22	4·61	5·99	7·38	9·21	10·60
3	0·07	0·11	0·22	0·35	2·37	4·64	6·25	7·81	9·35	11·34	12·84
4	0·21	0·30	0·48	0·71	3·36	5·99	7·78	9·49	11·14	13·28	14·86
5	0·41	0·55	0·83	1·15	4·35	7·29	9·24	11·07	12·83	15·09	16·75
6	0·68	0·87	1·24	1·64	5·35	8·56	10·64	12·59	14·45	16·81	18·55
7	0·99	1·24	1·69	2·17	6·35	9·80	12·02	14·07	16·01	18·48	20·28
8	1·34	1·65	2·18	2·73	7·34	11·03	13·36	15·51	17·53	20·09	21·95
9	1·73	2·09	2·70	3·33	8·34	12·24	14·68	16·92	19·02	21·67	23·59
10	2·16	2·56	3·25	3·94	9·34	13·44	15·99	18·31	20·48	23·21	25·19
11	2·60	3·05	3·82	4·57	10·34	14·63	17·28	19·68	21·92	24·73	26·76
12	3·07	3·57	4·40	5·23	11·34	15·81	18·55	21·03	23·34	26·22	28·30
13	3·57	4·11	5·01	5·89	12·34	16·98	19·81	22·36	24·74	27·69	29·82
14	4·07	4·66	5·63	6·57	13·34	18·15	21·06	23·68	26·12	29·14	31·32
15	4·60	5·23	6·26	7·26	14·34	19·31	22·31	25·00	27·49	30·58	32·80
16	5·14	5·81	6·91	7·96	15·34	20·47	23·54	26·30	28·85	32·00	34·27
17	5·70	6·41	7·56	8·67	16·34	21·61	24·77	27·59	30·19	33·41	35·72
18	6·26	7·01	8·23	9·39	17·34	22·76	25·99	28·87	31·53	34·81	37·16
19	6·84	7·63	8·91	10·12	18·34	23·90	27·20	30·14	32·85	36·19	38·58
20	7·43	8·26	9·59	10·85	19·34	25·04	28·41	31·41	34·17	37·57	40·00
21	8·03	8·90	10·28	11·59	20·34	26·17	29·62	32·67	35·48	38·93	41·40
22	8·64	9·54	10·98	12·34	21·34	27·30	30·81	33·92	36·78	40·29	42·80
23	9·26	10·20	11·69	13·09	22·34	28·43	32·01	35·17	38·08	41·64	44·18
24	9·89	10·86	12·40	13·85	23·34	29·55	33·20	36·42	39·36	42·98	45·56
25	10·52	11·52	13·12	14·61	24·34	30·68	34·38	37·65	40·65	44·31	46·93
26	11·16	12·20	13·84	15·38	25·34	31·79	35·56	38·89	41·92	45·64	48·29
27	11·81	12·88	14·57	16·15	26·34	32·91	36·74	40·11	43·19	46·96	49·65
28	12·46	13·56	15·31	16·93	27·34	34·03	37·92	41·34	44·46	48·28	50·99

	0·995	0·99	0·975	0·95	0·5	0·2	0·1	0·05	0·025	0·01	0·005
29	13·12	14·26	16·05	17·71	28·34	35·14	39·09	42·56	45·72	49·59	52·34
30	13·79	14·95	16·79	18·49	29·34	36·25	40·26	43·77	46·98	50·89	53·67
40	20·71	22·16	24·43	26·51	39·34	47·27	51·81	55·76	59·34	63·69	66·77
50	27·99	29·71	32·36	34·76	49·33	58·16	63·17	67·50	71·42	76·15	79·49
60	35·53	37·48	40·48	43·19	59·33	68·97	74·40	79·08	83·30	88·38	91·95
70	43·28	45·44	48·76	51·74	69·33	79·71	85·53	90·53	95·02	100·43	104·21
80	51·17	53·54	57·15	60·39	79·33	90·41	96·58	101·88	106·63	112·33	116·32
90	59·20	61·75	65·65	69·13	89·33	101·05	107·57	113·15	118·14	124·12	128·30
100	67.33	70.06	74.22	77.93	99.33	111.67	118.50	124.34	129.56	135.81	140.17

Appendix 3.3
t distribution

	0·1	0·05	0·025	0·01	0·005	0·001
1	3·078	6·314	12·706	31·821	63·656	318·289
2	1·886	2·920	4·303	6·965	9·925	22·328
3	1·638	2·353	3·182	4·541	5·841	10·214
4	1·533	2·132	2·776	3·747	4·604	7·173
5	1·476	2·015	2·571	3·365	4·032	5·894
6	1·440	1·943	2·447	3·143	3·707	5·208
7	1·415	1·895	2·365	2·998	3·499	4·785
8	1·397	1·860	2·306	2·896	3·355	4·501
9	1·383	1·833	2·262	2·821	3·250	4·297
10	1·372	1·812	2·228	2·764	3·169	4·144
11	1·363	1·796	2·201	2·718	3·106	4·025
12	1·356	1·782	2·179	2·681	3·055	3·930
13	1·350	1·771	2·160	2·650	3·012	3·852
14	1·345	1·761	2·145	2·624	2·977	3·787
15	1·341	1·753	2·131	2·602	2·947	3·733
16	1·337	1·746	2·120	2·583	2·921	3·686
17	1·333	1·740	2·110	2·567	2·898	3·646
18	1·330	1·734	2·101	2·552	2·878	3·610
19	1·328	1·729	2·093	2·539	2·861	3·579
20	1·325	1·725	2·086	2·528	2·845	3·552
21	1·323	1·721	2·080	2·518	2·831	3·527
22	1·321	1·717	2·074	2·508	2·819	3·505
23	1·319	1·714	2·069	2·500	2·807	3·485
24	1·318	1·711	2·064	2·492	2·797	3·467
25	1·316	1·708	2·060	2·485	2·787	3·450

	0·1	0·05	0·025	0·01	0·005	0·001
26	1·315	1·706	2·056	2·479	2·779	3·435
27	1·314	1·703	2·052	2·473	2·771	3·421
28	1·313	1·701	2·048	2·467	2·763	3·408
29	1·311	1·699	2·045	2·462	2·756	3·396
30	1·310	1·697	2·042	2·457	2·750	3·385
40	1·303	1·684	2·021	2·423	2·704	3·307
60	1·296	1·671	2·000	2·390	2·660	3·232
100	1·290	1·660	1·984	2·364	2·626	3·174

Chapter 4

Precipitation

Precipitation data are often needed by the hydrologist in the forecasting of river flows or in estimating the runoff in urban drainage schemes. This chapter describes the formation of precipitation, and the various types of rainfall and the methods of measurement. It considers the techniques available for dealing with incomplete data, either spatially or temporally, and how rainfall data can be analysed to provide the basis for time-series and storm profiles used in the design of urban and rural drainage schemes.

4.1   Formation of precipitation

Any given air-mass can contain a certain amount of moisture in the form of water vapour, depending on the temperature of the air. When an air mass is forced to rise, the expansion due to the lower pressure results in a reduction in temperature. When the temperature of the air mass falls to a level where it is no longer able to contain all of its moisture as vapour, i.e. it becomes saturated, condensation usually occurs and clouds are formed. However, saturation does not always result in precipitation: small particles (generally $0\cdot1$ μm to 10 μm in diameter) are necessary to provide nuclei or seeds for the water vapour to condense on to. Naturally occurring dust particles generally provide adequate nuclei, although rainfall can be induced by providing artificial seed particles. The droplets of liquid water or ice which are formed on the nuclei are initially so small that they are prevented from falling by upward air-currents. The droplets grow by collision with other particles or by coalescence, and when they reach a size of about 400 μm ($0\cdot4$ mm) they are large enough to overcome most upward air-currents and then fall as precipitation. This growth in droplet size does not always occur, however, and some clouds dissipate without any precipitation.

Droplets less than 100 μm (0.1 mm) are generally known as *fog* and may be a significant source of precipitation in arid coastal areas. Droplets larger than this, but less than about 0·5 mm are generally referred to as *drizzle*, which typically falls from low stratus cloud with intensities of less than 1 mm/h. Larger droplets are classified as *rain* and may fall with intensities of up to 20 or 30 mm/h or more. The maximum rainfall intensity recorded is a depth of 38 mm, recorded over 1 minute in 1970 in Guadeloupe. However, the maximum long-term rainfall recorded is a value of 40 768 mm, recorded over a period of two years in 1860–61 in India. Droplets may also fall as solid ice (*hail*), with diameters ranging from 5 mm up to 100 mm or more, these being mostly associated with cumulo-nimbus clouds. Such clouds contain violent air-currents which transport the ice particles up and down repeatedly, allowing them to accumulate layers of ice. Precipitation may also occur as *snow*, which is formed from the collision and coalescence of ice crystals. Although snow has a specific gravity of only between 0·05 and 0·2, its effect can be serious in terms of flood risk because the stored water may be released without any further precipitation. *Frost*, which consists of ice crystals separated by trapped air, is also common, but is not a major form of precipitation.

4.2   Types of precipitation

As has been noted, precipitation is generally caused by the lifting of an air mass and it is the mechanism of lifting which provides the most convenient form of classification.

4.2.1   Frontal precipitation

A front is a boundary between two air-masses, usually with different temperatures, and frontal precipitation occurs where a warm air mass is forced to rise over a colder air-mass. In latitudes between 35° and 50°, frontal precipitation is normally associated with westerly moving depressions. The depressions start as small waves on the boundary between the polar air and the temperate air (Figure 4.1(a)). The waves often progress along the front with little disturbance but, sometimes, a

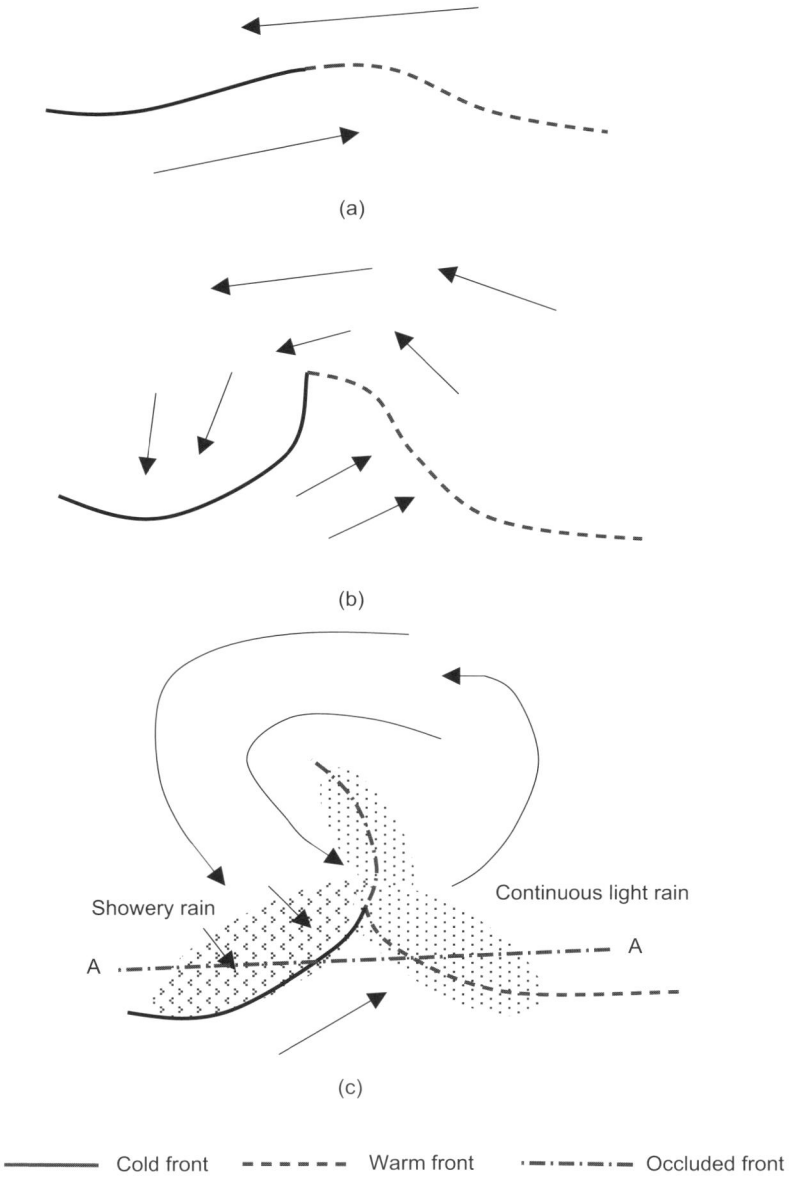

(a)

(b)

Showery rain

Continuous light rain

A

A

(c)

——————— Cold front   – – – – – Warm front   ·—·—·—·— Occluded front

Figure 4.1   Formation of a depression and frontal rainfall

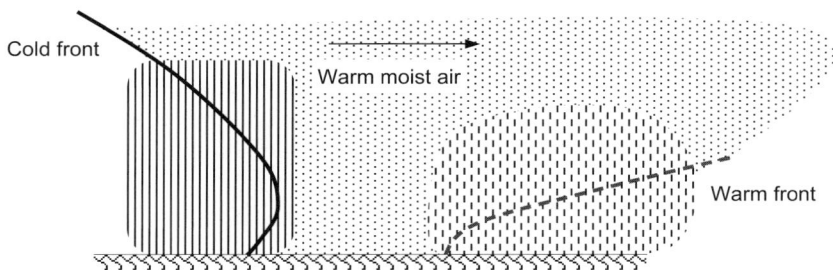

Figure 4.2   Section (A–A) through a frontal system

wave develops into an area of low pressure with winds generally circulating in an anticlockwise direction (Figure 4.1(b)). The depression, therefore, consists of a wedge of warm, moist temperate air intruding into a mass of colder, drier polar air. The leading edge of the warm sector is known as the warm front, and the warm air is forced to rise over the preceding cold air (Figure 4.2). The rate of ascent at the *warm front* is relatively slow since the slope of the interface between the air masses is quite low (usually between 1/100 and 1/300). Precipitation may extend 300 km to 500 km ahead of the front, and is typically continuous, but light to moderate in intensity.

Precipitation also occurs at the trailing edge of the warm sector known as the *cold front*. This front generally moves faster than the warm front and is also steeper, with a slope of 1/50 to 1/150. The rate of the rise of the air is thus much greater than at the warm front and consequently the rainfall intensity is higher, typically with heavy showers and blustery winds. Eventually the cold front catches up and overrides the warm front, in which case the warm sector is no longer in contact with the ground and the front is said to be *occluded*.

Near the equator, extremely low depressions (*cyclones* or *hurricanes*) can occur, where the water vapour from the warm ocean is condensed by the rising air at the centre of the storm. Thus, the heat of condensation is added to the energy of the storm, resulting in very strong winds and heavy rainfall. These storms tend to die away when they pass over land or over colder water.

4.2.2   *Orographic precipitation*

Orographic rainfall results when moist air is forced to rise by a land mass and it therefore tends to occur on the windward side of high ground, especially near the coast. Conversely, on the leeward side of mountains, there is typically an area of lower rainfall. In mountainous areas, therefore, the distribution of rainfall is largely controlled by the relief, and the rainfall will persist as long as the moist airflow continues.

In many areas of the UK, orographic factors tend to enhance precipitation resulting from weather fronts. Orographic enhancement of rainfall occurs particularly where the incoming air-stream has very high humidity in the lowest 1·5 km and there is pre-existing rainfall at higher levels, which can wash out the small droplets of the low-level cloud over the hills (Roy and Fox, 1995). Such conditions are most likely to be found ahead of warm fronts and in the warm sectors of depressions.

4.2.3   Convectional precipitation

Convectional precipitation occurs where there is local differential heating of an air mass, resulting in local convection currents. It typically occurs in tropical areas or in the summer in higher latitudes. It is characterised by short, intense thunderstorms over areas of less than 20 km^2, usually occurring in the afternoon or early evening, after the Earth's surface has been heated by the sun. In Africa, the convectional rainfall is often associated with the Inter-Tropical Convergence Zone (ITCZ), which is an area of converging air-flow moving between a few degrees north and south of the equator.

4.2.4   Artificially induced precipitation

Where sufficient moisture vapour exists in clouds, it may be possible to induce rainfall by providing artificial seeds to act as nuclei for the formation of droplets. Seeding can also be used to reduce the size of hailstones by providing more nuclei for the ice to freeze on. The common materials used for seeding are solid CO_2 and silver iodide. The latter, although cheaper, has the disadvantage of decaying in sunlight. The effectiveness of cloud seeding depends on many factors, such as the height of the cloud base and top, the cloud temperature, the updraft velocities and the amount of liquid water in the cloud. Increases of up to 10% have been claimed, although it is difficult to obtain reliable statistics. Cloud seeding is unlikely to be effective over large continental areas, since 90% of the moisture in the atmosphere comes from evaporation over the oceans. It can also sometimes have unpredictable consequences, with heavy rain occurring over 'unseeded' areas.

4.3   Geographical distribution of precipitation

Precipitation is not distributed evenly over the Earth's surface as it reflects the underlying patterns of air circulation. It tends to be concentrated in the equatorial regions and the temperate mid-latitudes, where there are general convergences of air masses. Between 20° and 30° north and south, where there is relatively dry descending air, there is generally a rainfall deficit. Amounts of precipitation also tend to be higher near coastlines and high ground exposed to the prevailing wind direction, since the main source of precipitation is the evaporation from the oceans.

Table 4.1   Rainfall regimes (Chorley, 1969)

Regime	Annual total: mm	Characteristics	Typical distribution
Equatorial	2500–3000	Rain throughout year Two maxima Low variability	
Tropical	250–1000	Summer maximum Dry winter Monsoons Convective storms	
Temperate oceanic	750–1000	All year precipitation Low variability Frontal depressions	
Mediterranean	600–750	Winter maximum Dry summer Westerly depressions	
Temperate continental	350–500	Spring/summer maximum Convective showers Considerable variability	
Arctic	120–400	Low total Summer rain Snow spring/autumn	

The distribution of precipitation between summer and winter also varies considerably. In general, six types of rainfall regime can be identified, which are described in Table 4.1.

4.4 Measurement of precipitation

4.4.1 Non-recording gauges

The simplest form of rain gauge is simply a container which collects the rainfall over a period of time. However, to ensure accurate measurement of rainfall, it is essential that rain falling on the gauge cannot splash out and that rain falling outside the gauge cannot splash into the container. It is also important that collected rain is not lost through evaporation and that no additional water is collected either by condensation or flooding. A

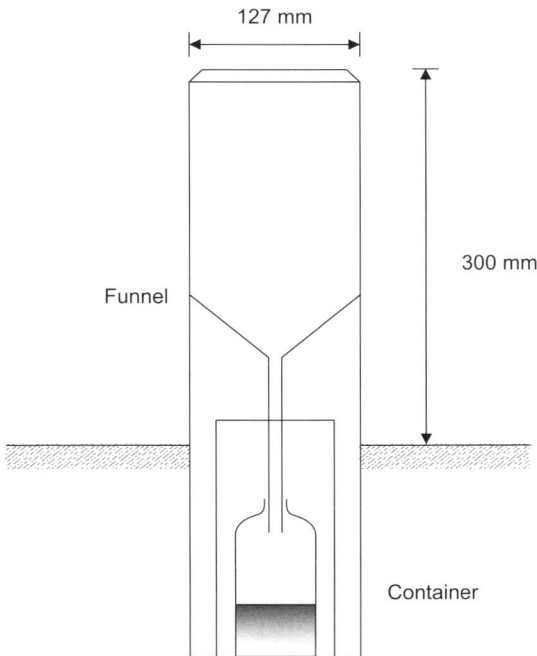

Figure 4.3   Typical manual rain gauge

typical gauge consists of a cylindrical container with a knife-edge rim set at 300 mm above ground level (Meteorological Office). The rain is collected through a funnel to reduce losses due to evaporation, and the measuring container is placed inside a larger container to collect any overflow. A space is provided above the funnel to collect any snow and to prevent the loss of rain after it has been collected. In the United Kingdom, the diameter of the opening is normally 127 mm, giving an area of 127 cm^2, although elsewhere the collection area varies between 200 cm^2 and 500 cm^2 (Figure 4.3).

One of the major errors in the measurement of rainfall is the effect of the wind. Local variations in wind speed and direction caused by nearby obstructions or the instrument itself can give errors of 10% or more. Gauges should therefore be placed, if possible, on level, open, unpaved ground at least twice the height of any obstruction away from it. In urban areas, it may not be possible to comply with these requirements, but an increased density of rain gauges can partly offset this. In exposed sites, it is recommended that the gauge is protected by a 300 mm high circular turf wall. Alternatively, the rim of the collector can be set at ground level with a special protection on the ground surface around the gauge to reduce the effect of splashing.

4.4.2 *Recording gauges*

The main disadvantage of non-recording gauges is that they give no information about the distribution of rainfall at time-intervals less than the reading interval, which is typically 24 h. For urban catchments, which often respond to rainfall in periods of minutes, this is clearly inadequate. Recording gauges can provide a continuous record of rainfall, which can be discretised into intervals of a few minutes if necessary.

The most common type of recording rain gauge uses a tipping bucket (Figure 4.4). This is a small bucket with two symmetrical compartments. It is constructed so that one of the compartments is always under the funnel outlet. When the compartment is full, the bucket rotates and spills so that the other compartment is under the outlet, and the time of tipping is recorded by a data logger. The two buckets therefore alternate in their position under the inlet. The buckets are designed so that they tip after a fixed depth of rainfall, typically 0·1 mm or 0·2 mm. The average rainfall intensity between each tip can easily be calculated by dividing the depth by the time interval between tips.

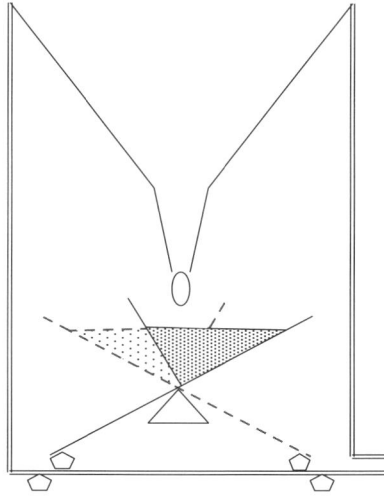

Figure 4.4   Tilting bucket rain gauge

4.4.3   Snowfall

Small quantities of snow can be measured by allowing the snow accumulated on a rain gauge to melt, if necessary by adding a measured quantity of warm water, which is then deducted from the observed total. In some cases, the depth of snowfall can be converted into an equivalent water depth. As a rough guide, 300 mm of fresh snow is approximately equivalent to 25 mm of rain, although the density of snow can vary by a factor of four or more (Shaw, 1994). For significant depths of compacted snow, the approximate water content given above can be modified by estimates of density obtained from samples at various depths.

4.4.4 Remote sensing

Remote sensing is the use of measurements of the electromagnetic radiation to infer certain environmental conditions. There are many different techniques involving different parts of the electromagnetic spectrum, measured either from satellite or airborne platforms or ground-based stations. Precipitation is difficult to observe from aircraft or satellites because clouds are generally opaque to most wavelengths of the spectrum. However, at certain microwave frequencies, clouds may be transparent and microwaves may be used to detect the thermal energy of the raindrops provided that there is a cooler background, such as that provided by the ocean. Although the spectral response of clouds is dependent on the rainfall intensity, the necessary algorithms relating the signal to rainfall intensity are still not fully developed.

Another technique uses the visible and infrared part of the spectrum to infer the temperature of the clouds and cloud tops; temperatures below a certain threshold being assumed to be rain clouds. Alternatively, rainfall can be inferred simply from the number, size and duration of the clouds. In all of these methods, empirical coefficients are required, which can only be estimated from calibration with conventional rainfall measurements. The calibration procedure is complicated by the fact that surface rainfall measurements are aggregated over periods of up to 24 h, while the remote-sensing images are instantaneous snapshots.

A more common approach uses ground-based radar stations, which measure the radar signal reflected from the raindrops. The return signal is proportional to the sixth power of the raindrop size and is also inversely proportional to the square of the distance. Therefore, nearby storms with drizzle may give similar signals to more distant and more intense storms, while the attenuation of the signal by the rain adds a further complication. Other problems include the difficulty of distinguishing between large raindrops and snow, and the evaporation of some rainfall detected by radar before it reaches the ground. Overall, typical errors associated with radar measurement of rainfall may be of the order of 50%, which are similar to the errors in estimating rainfall for a 2 km square based on a single rain gauge (Wood *et al.*, 2000). However, despite these limitations, if it is combined with point rainfall measurements, radar is able to give a useful two-dimensional visual representation of storms and the distribution of rainfall intensity. Radar measurement of precipitation is now also incorporated in satellite instrumentation.

4.5 Analysis of precipitation data

Raw rainfall data are measured as a depth, normally in millimetres over a period of time. The mean *intensity* is calculated as a depth per unit time, usually measured in millimetres per hour. In large rural catchments, daily totals may be quite adequate because of the large response time of such catchments: For smaller catchments, especially in urban areas, the time interval needs to be reduced to an hour or even a few minutes.

The main use of rainfall data for hydrologists is in the estimation of runoff or river flows using a rainfall–runoff model. In some cases, an historical rainfall sequence may be used, alternatively, an artificial rainfall time-series can be generated with statistical properties similar to an observed sequence. For most hydrological design work, it is more appropriate to investigate the response of a catchment to a single, idealised storm-event with a particular intensity, duration and frequency. Standard procedures exist in the United Kingdom to estimate the intensity, duration and frequency of such storm events based on historical records from a large number of stations over the whole country.

When rainfall over a catchment is to be estimated using observed data from rainfall stations, the data set is often limited by short record lengths, missing sections of a record or by a sparse geographical distribution of measuring stations. However, these limitations can be overcome by various statistical techniques.

4.5.1 Missing data

Precipitation records often have breaks due to equipment failure or other reasons. These gaps can be filled fairly reliably by reference to adjacent stations. For example, if there are three nearby stations with similar characteristics, a missing rainfall value can be estimated using the ratio of the mean values, e.g.:

$$I_m = \frac{1}{3}\left(\frac{\bar{I}_m}{\bar{I}_1}I_1 + \frac{\bar{I}_m}{\bar{I}_2}I_2 + \frac{\bar{I}_m}{\bar{I}_3}I_3\right) \tag{4.1}$$

where I_1, I_2 and I_3 are the corresponding values from nearby stations and $\bar{I}_1$, $\bar{I}_2$, etc., are the long-term averages from those stations and m refers to the missing station. If the long-term average precipitation from the three stations differs by less than 10% from that of the specified station, a simple arithmetic mean can be used. Alternatively, a regression analysis can be carried out between the station with the missing data and one or more adjacent stations. This will enable an empirical relationship to be established, which can be used to generate the missing data.

Occasionally, in a long record, conditions at a rainfall station may change because of the construction of a new building or the growth of vegetation, and this may influence the recorded precipitation values. It is possible to determine whether there has been a significant change in the site conditions by the use of a *double mass curve*. This is a plot of the cumulative total rainfall recorded at the station against the cumulative total from a nearby station, where it is known there has been no change. If there is a distinct change in the gradient of the graph, this will indicate the point at which there was a change in the site conditions (Figure 4.5) and the subsequent precipitation values can be adjusted by the ratio of the two slopes. However, it should be remembered that there will always be

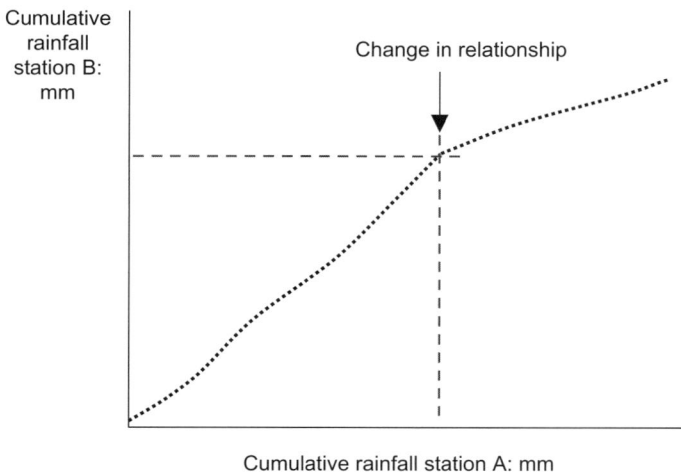

Figure 4.5   Double mass curve

minor variations in the slope of the double mass curve, and no change should be assumed unless the change in gradient is significant or there is other evidence (Example 4.1).

Example 4.1

The following annual rainfall totals were recorded at two stations. Determine, using the double mass curve analysis, whether any change has occurred in the conditions at either station and, if so, when it occurred.

					Year					
1	2	3	4	5	6	7	8	9	10
Station A									
1250	1061	990	1310	1296	1320	965	1085	1254	1173
Station B									
1430	1221	1139	1511	1491	1310	995	1075	1224	1182

Year	Station A	Cusum A	Station B	Cusum B
1	1250	1250	1430	1430
2	1061	2311	1221	2651
3	990	3301	1139	3790
4	1310	4611	1511	5301
5	1296	5907	1491	6792
6	1320	7227	1310	8102
7	965	8192	995	9097
8	1085	9277	1075	10 172
9	1254	10 531	1224	11 396
10	1173	11 704	1182	12 578

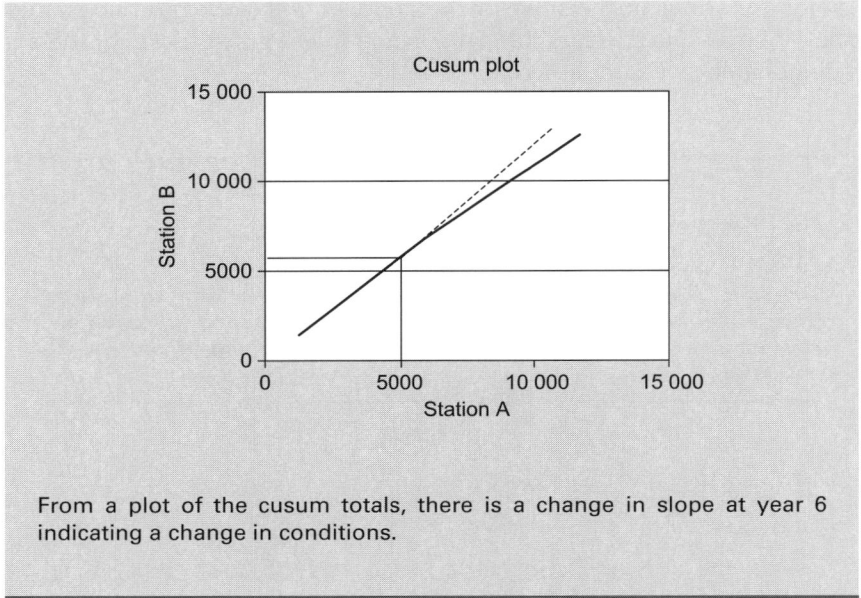

Cusum plot

From a plot of the cusum totals, there is a change in slope at year 6 indicating a change in conditions.

4.5.2 *Spatial variation*

Many hydrological problems require rainfall information over areas of several thousand square kilometres or more. There are various techniques for spatially averaging point-values of rainfall. For small areas with a reasonably high and uniform density of rainfall stations, a simple arithmetic mean of the observed rainfall $i_1, i_2, \ldots, i_N,$ can be used, i.e.:

$$I = \frac{i_1 + i_2 + i_3 + \ldots i_N}{N} \tag{4.2}$$

For larger areas, particularly where the density of rain gauges is low or irregular, a weighting should be applied to take account of the relative location of each gauge. The gauges near the centre of the catchment are likely to be more representative of the catchment rainfall than those near the boundary and should therefore have a higher weight.

The most common method of weighting is by the use of *Thiessen's Polygons* (Figure 4.6). Each station is connected to adjacent stations by straight lines and perpendicular bisectors to these lines are drawn. These bisectors form polygons around each station, and the area of each

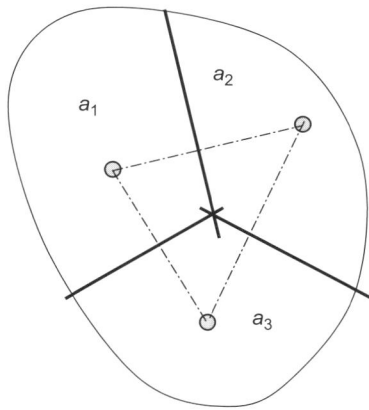

Figure 4.6   Thiessen's Polygons

polygon (a_1, a_2, etc.) in relation to the total area (A) gives the weighting factor for each station, i.e.:

$$I = \frac{i_1 a_1 + i_2 a_2 + i_3 a_3 + \ldots i_N a_N}{A} \qquad (4.3)$$

Where the distribution of rainfall is significantly modified by the topography, even the Thiessen method may give misleading results. A more accurate method is to construct an *isohyetal map* of the area with contour lines of equal rainfall (isohyets). The contour lines can be produced with standard software packages, which generally use more sophisticated methods of interpolating between point data, such as kringing, which are described in Chapter 3. Although the isohyets are based on the point rainfall values, they can be modified to allow for topographic influences in relation to known directions of storm movements. The average precipitation over a catchment area is then estimated by weighting the mean value of two adjacent isohyets by the area between the isohyets in relation to the total area (Figure 4.7). Although this method can give more realistic results, it requires more time and skill, and, if not used correctly, can give inaccurate results.

Spatial estimation of rainfall is particularly difficult in arid areas where the annual rainfall may comprise a few short, intense storms over very limited areas.

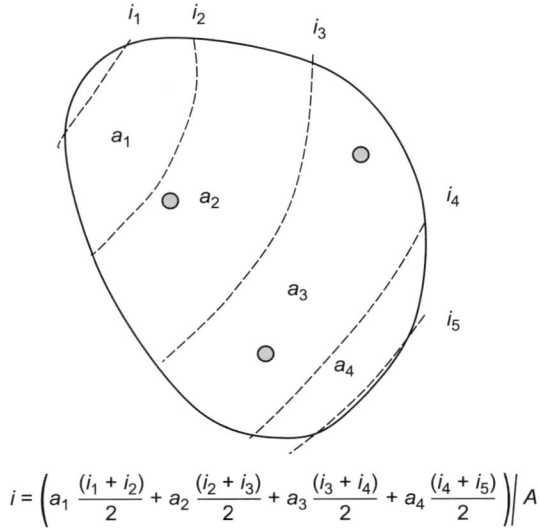

$$i = \left(a_1 \frac{(i_1 + i_2)}{2} + a_2 \frac{(i_2 + i_3)}{2} + a_3 \frac{(i_3 + i_4)}{2} + a_4 \frac{(i_4 + i_5)}{2} \right) \Big/ A$$

Figure 4.7   Isohyetal method

4.5.3   *Disaggregation of a rainfall time-series*

One of the problems with the use of observed rainfall data is that the time interval is often too long, particularly for urban catchments. A process known as *disaggregation* can be carried out to divide, say, an hourly rainfall value into, say, four values, each representing a 15-minute time-interval (Ormsbee, 1989). It is assumed that rainfall occurs in discrete pulses of, say, 0·25 mm. If three consecutive hourly rainfall values x, y, z are considered, the probability density function (PDF) of a single rainfall-pulse occurring in the central hour can be assumed to consist of two lines at equal angles to the horizontal which connect with the preceding and subsequent rainfall values (Figure 4.8). The value, $f(t)$, of the PDF at any time can therefore be found from simple geometry and hence the cumulative distribution function $F(t)$ can be found by integration.

The central time-interval is then divided into the desired disaggregation time-interval of t minutes, giving M time-intervals where $M = 60/t$. The probability of a pulse occurring within the ith interval of t minutes can be calculated as:

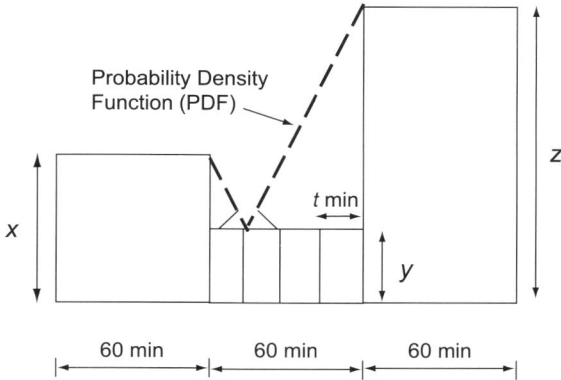

Figure 4.8   Disaggregation of central block of rainfall

$$P_i = F_{it} - F_{(i-1)t} \qquad (4.4)$$

where F_{it} and $F_{(i-1)t}$ are the values of the cumulative frequency distribution at times it and $(i–1)t$ minutes respectively. The depth of rainfall (y mm) in the central hour is then divided into $N –1$ equal pulses of say 0·25 mm and one pulse of $y – (N – 1) \times 0·25$ mm (thus ensuring that the total depth of rainfall in the central interval is y mm). Each of the N pulses of rainfall is then assigned to one of the m time-intervals using the probabilities calculated above together with a sequence of uniformly distributed (0,1) random numbers (Example 4.2).

Example 4.2

Successive hourly values of rainfall were recorded as 6 mm, 4 mm and 9 mm. Estimate values of rainfall at 15-minute intervals over the central hour of the record.

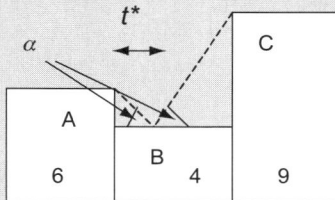

The equations of the lines A–B and B–C are:

$$F(t)_{A-B} = \frac{xt}{k} - \frac{(x-y)t^2}{2kt^*} \qquad\qquad 0 < t < t^*$$

$$F(t)_{B-C} = \frac{(y+x)t^*}{2k} + \frac{y(t-t^*)}{k} - \frac{(y-z)(t-t^*)^2}{2k(60-t^*)} \qquad\qquad t^* < t < 60$$

where x, y, z are 6, 4, 9 respectively.

$$k = 30(y+z) - \frac{t^*(z-x)}{2}$$

Constructing lines at equal angles (α) to the horizontal the distance t^* can be calculated as:

$$\mathrm{Tan}\,\alpha = \frac{2}{t^*} = \frac{5}{60-t^*}$$

Hence:

$$5t^* = 2(60 - t^*) = 120 - 2t$$

$$t^* = \frac{120}{7} = 17 \cdot 1\,\mathrm{min}$$

$$k = 30(4+9) - \frac{17 \cdot 1(9-6)}{2} = 364 \cdot 35$$

For $t = 15$

$$F(15) = \frac{6 \times 15}{364 \cdot 35} - \frac{(6-4)15^2}{2 \times 364 \cdot 35 \times 17 \cdot 1} = 0 \cdot 210$$

For $t = 30$

$$F(30) = \frac{(4+6) \times 17 \cdot 1}{2 \times 364 \cdot 35} + \frac{4 \times (30 - 17 \cdot 1)}{364 \cdot 35} - \frac{(4-9)(30 - 17 \cdot 1)^2}{(2 \times 364 \cdot 35(60 - 17 \cdot 1))} = 0 \cdot 403$$

For $t = 45$

$$F(45) = \frac{(4+6) \times 17 \cdot 1}{2 \times 364 \cdot 35} + \frac{4 \times (45 - 17 \cdot 1)}{364 \cdot 35} - \frac{(4-9)(45 - 17 \cdot 1)^2}{(2 \times 364 \cdot 35(60 - 17 \cdot 1))} = 0 \cdot 640$$

pr1 = 0·210 − 0 = 0·210
pr2 = 0·403 − 0·210 = 0·193
pr3 = 0·640 − 0·403 = 0·237
pr4 = 1·0 − 0·640 = 0·360

The central value (4 mm) is divided into 16 pulses of 0·25 mm. A sequence of 16 uniformly distributed random numbers is generated with zero mean and unit variance, and the pulses are allocated to the time interval according value of the random number. The number of pulses in each time interval multiplied by 0·25 mm gives the depth allocated to that time interval.

Random number	1st	2nd	3rd	4th
0·202	1			
0·161	1			
0·174	1			
0·127	1			
0·210	1			
0·814				1
0·864				1
0·600			1	
0·205	1			
0·538			1	
0·348		1		
0·909				1
0·575			1	
0·553			1	
0·022	1			
0·214		1		
Total	6	3	4	3
Depth: mm	1·5	0·75	1·0	0·75

4.6   The use of design rainfall

The design of new drainage or flood protection schemes in urban or rural areas requires the input of a rainfall time-series that represents the most critical event which could be expected to occur over a reasonable time-period. The selection of such a time-series is clearly crucial to the engineering and economic success of the scheme, but is often the most uncertain factor. The design rainfall input can be derived from:

- a long historical record
- a synthetic record based on parameters from historical data
- a specific storm-profile with given intensity, duration and frequency.

4.6.1   Historical records

Although rainfall records have been collected for up to 100 years or more, very long records are not widely available and are becoming increasingly expensive. In most cases, such a record will need to be modified to reflect local conditions as well as to have any missing periods filled in, as described above. Moreover, running hydrological models with several long time-series is rather inefficient computationally, if the main interest is on a few extreme events. However, using a continuous time-series has the advantage that the antecedent conditions of the catchment at the beginning of a storm are likely to be realistic.

4.6.2   Synthesis of rainfall time-series

A synthetic rainfall time-series can, in principle, be generated either using a deterministic or a stochastic model. However, since rainfall is the result of many complex processes, it is not possible to develop a fully deterministic model. A stochastic model uses combinations of random variables with certain parameters, whose values are determined by matching certain statistical properties of the generated time-series with a chosen historical record. The advantages of such models are that their length is not limited by the length of historical records and their time intervals can be smaller than that of observed records.

One method of rainfall synthesis is that proposed by (Cowpertwait *et al.*, 1991) based on the Neyman-Scott Rectangular Pulses Model. The

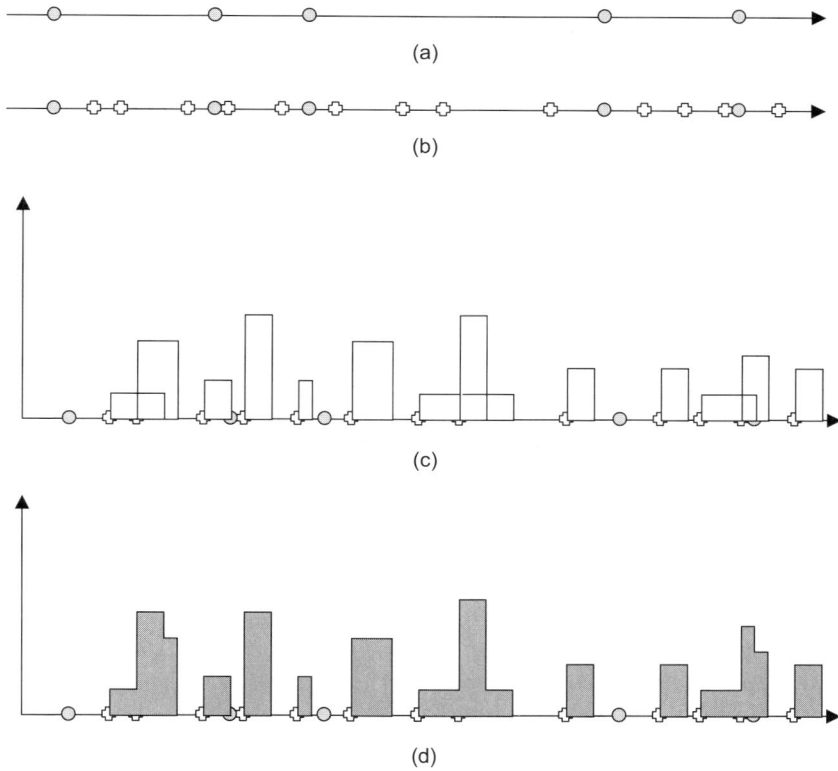

Figure 4.9   Neyman-Scott Rectangular Pulses Model (Cowpertwait et al., 1991)

model has a conceptual basis in that it assumes that rainfall is generated at one of a series of storm origins, which arrive at random intervals (Figure 4.9(a)). Each origin generates a random number of rain cells and each rain cell is assumed to have a uniform intensity which can be considered as a rectangular pulse of rainfall (Figure 4.9(b) and (c)). The rainfall time-series is then formed by aggregating all the pulses of rainfall at each time-interval (Figure 4.9(d)). The intervals between the storm origins and the number of cells per storm origin are assumed to be randomly distributed following the Poisson distribution. The interval between rain cells, the duration of each rain cell and the rainfall intensity of each cell are all assumed to be exponentially distributed random variables. The five parameters of the model are therefore:

- the mean time between storm origins
- the mean delay for cells after the storm origin
- the mean number of rain cells per storm
- the mean cell duration
- the mean cell intensity.

The values of these parameters are estimated by fitting certain statistics of the generated time-series to corresponding statistics of the required observed time-series. The recommended statistics are:

- the mean
- the variance
- the proportion of dry intervals
- the proportion of wet intervals preceded by a wet interval
- the proportion of dry intervals preceded by a dry interval.

Ideally, these statistics should be preserved in the generated time-series for a range of time intervals. In practice, it is not possible to preserve all the statistics for all time-steps and it is suggested that the variance and the proportion of wet intervals preceded by a wet interval are matched at 1, 3, 6, 12 and 24 h intervals and the remainder at 24 h intervals.

The optimum values of the model parameters are those which minimise the function:

$$S = \sum_{i=1}^{N} \left(1 - \frac{f_i}{k_i} \right)^2 \tag{4.5}$$

where f_i is a statistic of the synthetic time-series and k_i is the corresponding statistic for the historical time-series.

The above method is based on an hourly time-interval. For shorter time-intervals, the time-series can be disaggregated using the method described in Section 4.5.3 above.

4.6.3 Design storms

An alternative to using a long historical or synthetic time-series is to use a standard design storm with a depth or intensity associated with a given duration and frequency. Such storms can be derived from long rainfall records by isolating the storm events within the record. However, it is sometimes difficult to define the precise limits of a given rainfall event since there may be short periods of no rainfall within a storm. The end of a storm is therefore normally defined by an arbitrary period during which the rainfall did not exceed a certain depth, although more sophisticated procedures using autocorrelation analysis may be used.

4.6.4 Depth– and intensity–duration–frequency curves

Having defined a storm, it can then be described by a *depth, mean intensity, peak intensity* or *duration*. It can be seen that there are only three independent variables, since the mean intensity is depth divided by duration. Any of these parameters can be associated with a particular frequency of occurrence, which is normally measured in terms of a *return period*. This is defined as the average interval between occurrences of an intensity (or other characteristic) greater than or equal to the specified value. Thus, a 10-year return period storm will occur on average once in 10 years, with a probability of occurring in any one year of 0·1. (It naturally does not mean that such a storm will always occur every 10 years.) A frequency or return period can also be associated with a combination of independent parameters, and the most common parameters are either depth and duration or mean intensity and duration. Therefore, any storm event can have several frequencies of occurrence depending on which parameters or combination of parameters are used.

If storms of a given duration are abstracted from the record and arranged in order of magnitude of, say, depth, the return period (T) can be estimated using the Weibull, or similar, plotting position:

$$T = \frac{(n+1)}{m} \tag{4.6}$$

where n is the number of data points and m is the rank order. The data can be plotted on probability paper on which the relationship is linearised and can be extrapolated to estimate return periods of more extreme storms, although it is not recommended to extrapolate beyond about twice the record length.

If a long rainfall record is examined it will be seen that there is an inverse relationship between the storm duration and the mean intensity. This follows from the fact that the average intensity for a given storm depends on the period over which the rainfall is recorded. The maximum intensity for a storm recorded at 15-minute intervals will be higher than the maximum intensity for the same storm if it were recorded at 1 h intervals. Thus, a graph relating intensity and duration would be

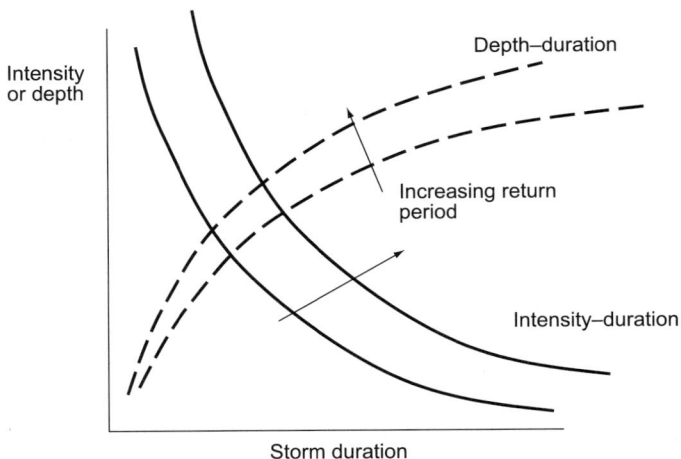

Figure 4.10   Intensity–duration and depth–duration curves

typically of the form shown in Figure 4.10. A typical depth–duration curve is also shown. Normally, the most severe storm for each year is identified and, therefore, if a reasonable length of record is available, the return period or frequency of a given combination of intensity and duration can be determined. Thus, sets of Intensity–Duration–Frequency (IDF) or Depth–Duration–Frequency (DDF) curves for a given location can be produced (Examples 4.3 and 4.4).

It might be thought that the choice between IDF and DDF curves is immaterial since, for any combination of intensity and duration, a depth can be calculated and vice versa. However, it can be shown (Adams and Papa, 2000) that the frequency of a depth calculated from an IDF curve is not the same as the same depth derived from a DDF curve. In other words, a given storm will have a different frequency for its intensity and its depth. The question of whether to use IDF or DDF curves depends on the object of the procedure. In the case of an urban catchment, the design of the pipes is largely a function of the flow rate, which depends on the rainfall intensity, while for attenuation or storage structures, the volume of flow (i.e. the depth of rainfall) is important.

A further complication arises when considering the frequency of runoff peaks, since a given return period rainfall-intensity can produce floods with a wide range of return periods, depending on the antecedent conditions and the spatial and temporal variability of the rainfall. In the

Example 4.3

A rainfall station recorded the following hourly rainfall totals. Deduce the maximum average intensity for a duration of (a) 1 h, (b) 2 h and (c) 4 h.

Hour	1	2	3	4	5	6	7	8	9	10
Rainfall: mm	2	4	0	5	1	4	0	8	4	2

Cumulative rainfall for different periods:

1 h	2 h	4 h
2		
4	6	
0	4	
5	5	11
1	6	10
4	5	10
0	4	10
8	8	13
4	**12**	**16**
2	6	14

Peak intensity of 1 h duration = 8 mm/h.

Peak intensity of 2 h duration = 12/2 = 6 mm/h.

Peak intensity of 4 h duration = 16/4 = 4 mm/h.

Table 4.2 Flood and rainfall return periods for rural catchments (Houghton-Carr, 1999) — reproduced with the permission of the Centre for Ecology and Hydrology, Wallingford)

Flood return period: years	2·33	10	30	50	100	1000
Rainfall return period: years	2	17	50	81	140	1000

case of small urban catchments, the normal practice is to assume that the flood return period is the same as the rainfall return period, but in the case of rural catchments, it may be necessary to use a range of storm profiles and antecedent conditions to find the appropriate design storm. The *Flood Estimation Handbook* (Faulkner, 1999) recommends the storm return period for given flood return period shown in Table 4.2.

4.6.5 Drought–duration–frequency curves

The frequency and duration of varying degrees of drought can be described by drought–duration–frequency curves in a similar way to depth–duration–frequency curves. A definition of drought is obviously required and this can be, for example, the cumulative difference between expected and actual rainfall (Rugumayo, 2002). A given drought is therefore characterised by its *duration*, which is the period between daily rainfall totals exceeding the mean daily total, and its *volume*, which is the sum of the differences between the mean actual daily totals. In tropical countries which regularly experience several consecutive months without rainfall, a more specific definition of drought has been proposed (Herbst *et al.*, 1966) which relates the observed rainfall deficits to the average deficits.

The average drought *intensity* is defined as:

$$Y = \frac{\sum_{t=1}^{D} \left[(E_t - M_t) - (MMD)_t \right]}{\sum_{t=1}^{D} (MMD)_t} \tag{4.7}$$

where E_t is the *effective* rainfall, M_t is the mean monthly rainfall, and MMD is the *mean monthly drought*; t is the month number ($t = 1, 2, 3, ..., 12$) and D is the drought duration in months. The effective rainfall is defined as:

$$E_t = R_t + (R_{t-1} - M_{t-1})W_t \tag{4.8}$$

where W_t is a carry-over factor, and the mean monthly drought is calculated by:

$$MMD_t = \sum_{t=1}^{N} \frac{(M_t - R_t)}{N} \tag{4.9}$$

A *severity index* is defined as $Y \times D$.

4.6.6 Rainfall estimation for urban catchments

In general terms, a design rainfall intensity or depth for urban catchments depends on

- the location of the site
- the return period to be used
- the duration of the design storm.

One of the earliest attempts to produce a systematic method to predict rainfall frequency for urban areas in the United Kingdom was made by Bilham (1936), who produced a formula based on 10 years of rain gauge data:

$$N = 1{\cdot}25D\left(\frac{I}{25{\cdot}4} + 0{\cdot}1\right)^{-3{\cdot}55} \tag{4.10}$$

where N is the number of occurrences of a storm event of depth I (mm) and duration D (hours). Although the Bilham formula is valid only for durations up to 2 h, it has been extended, but tends to overestimate the probability of high-intensity storms. It also does not reflect the geographical variation of rainfall.

A more general method is that described in the Wallingford Procedure (Hydraulics Research, 1983). The procedure uses a map which has been produced from meteorological records (Figure 4.11) to estimate rainfall depths for a reference return period (five years) and a reference duration (60 minutes). (The Wallingford Procedure uses a standard notation for rainfall data whereby MT_{-D} represents the depth of rainfall in millimetres for a storm occurring with a return period of T years and with a duration of D minutes.) The value of $M5_{-60}$ obtained from the map is modified for the required return period and duration by two factors, Z_1 and Z_2,

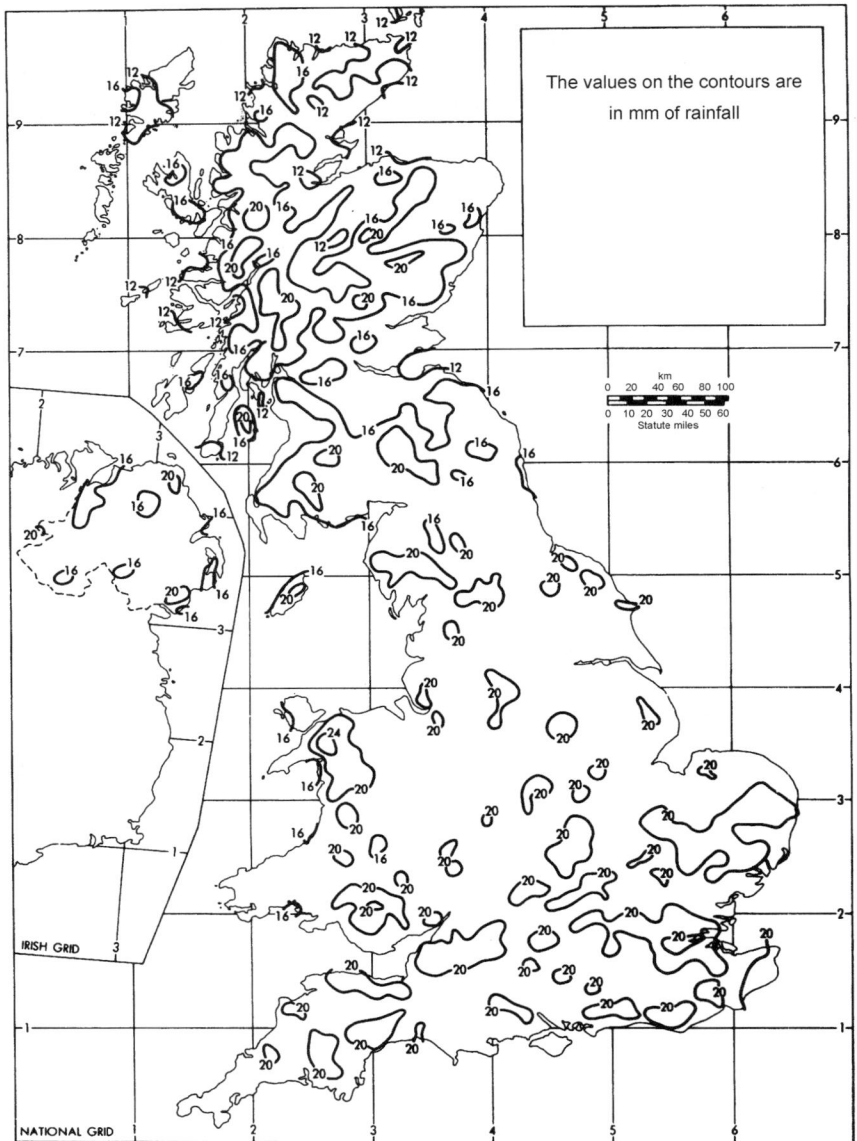

The values on the contours are in mm of rainfall

Figure 4.11   Rainfall depths for five-year return period and 60-minute duration (reproduced with the permission of HR Wallingford)

reflecting the required return period and duration respectively. Z_1 is estimated from a chart (Figure 4.12) using a value of the ratio of 60 minute to two-day rainfall (*r*), which is read from a second map (Figure 4.13) and the required duration. The factor Z_2 is found from Table 4.3 (according to the region of the UK) with the required return period. The required rainfall depth (MT_{-D}) for a return period of *T* years and a duration of *D* minutes is then given by:

$$M_{T-D} = M_{5-D} \times Z_2 \text{ (mm)} \tag{4.11}$$

where:

$$M_{5-D} = M_{5-60} \times Z_1 \text{ (mm)} \tag{4.12}$$

The average intensity (mm/h) is then:

$$i = M_{T-D} \times \frac{60}{D} \tag{4.13}$$

The intensity may be reduced by an *areal reduction factor* (ARF), which reflects the fact that it is less likely that extreme rainfall depths will occur simultaneously over an extensive area. The *ARF* can be calculated for a given duration (*D*) and catchment area (*A*) from Figure 4.14, or from the following equation:

$$ARF = 1 - f_1 D^{-f_2} \tag{4.14}$$

where, for small catchments:

$$f_1 = 0 \cdot 0394 A^{0 \cdot 354}$$

and:

$$f_2 = 0 \cdot 40 - 0 \cdot 0208 \ln(4 \cdot 6 - \ln[A])$$

The *ARF* is not significant for catchment areas less than about 5 km^2.

It is therefore possible, using this method, to produce an intensity–duration–frequency curve for any given location in the United Kingdom.

For small urban areas, the critical storm duration is normally taken as equal to the *time of concentration*, which is the time for flow to travel from the farthest part of the catchment to the outlet (see Section 7.2). For larger urban catchments, a range of durations of, say, 15 minutes, 30 minutes, 60 minutes and 120 minutes are used to obtain the critical peak runoff.

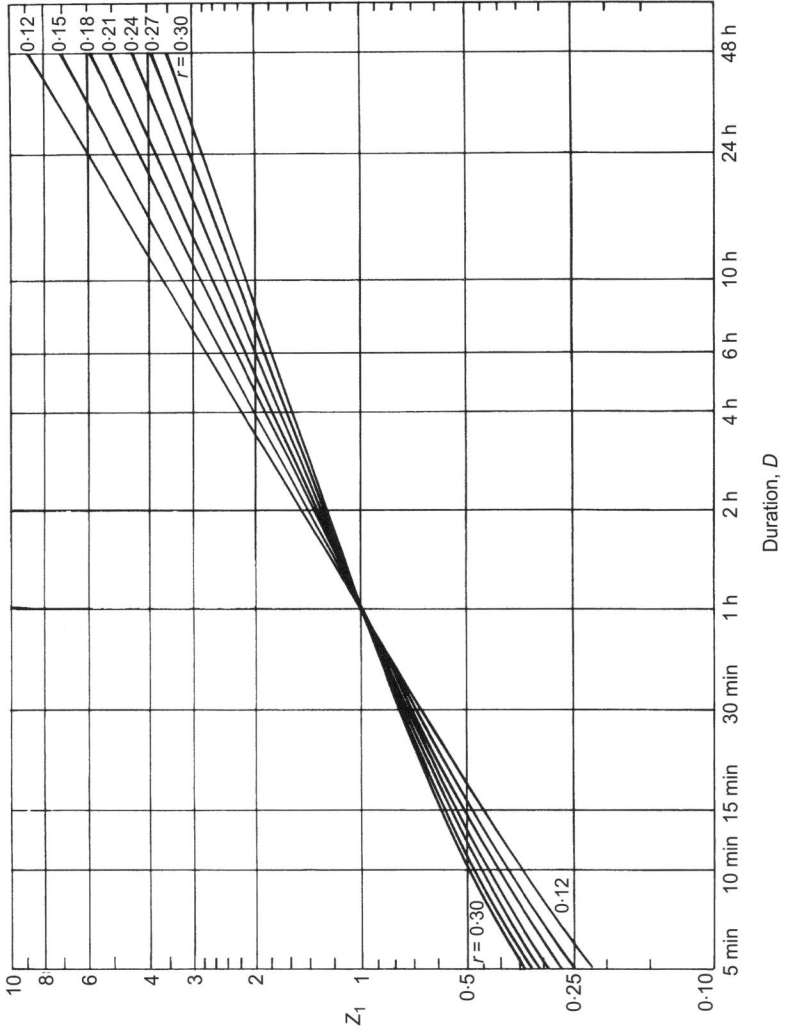

Figure 4.12   Correction factor for duration (Z_1) (reproduced with the permission of HR Wallingford)

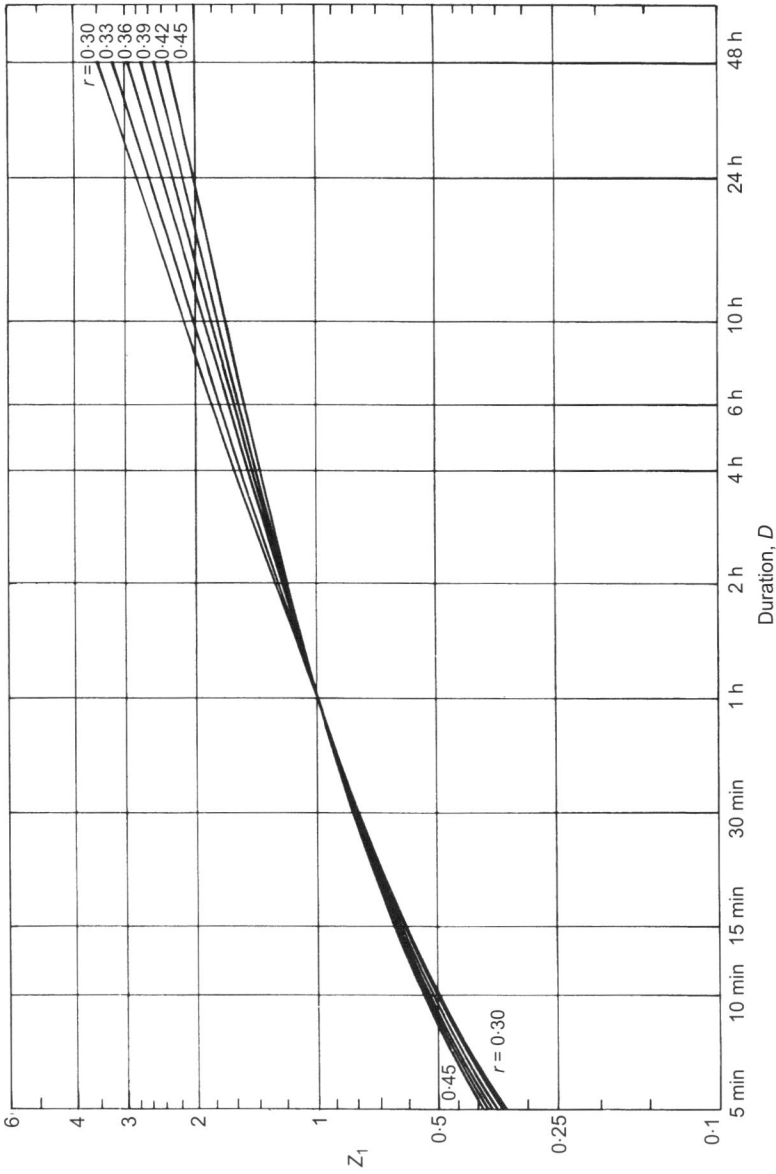

Figure 4.12   continued

145

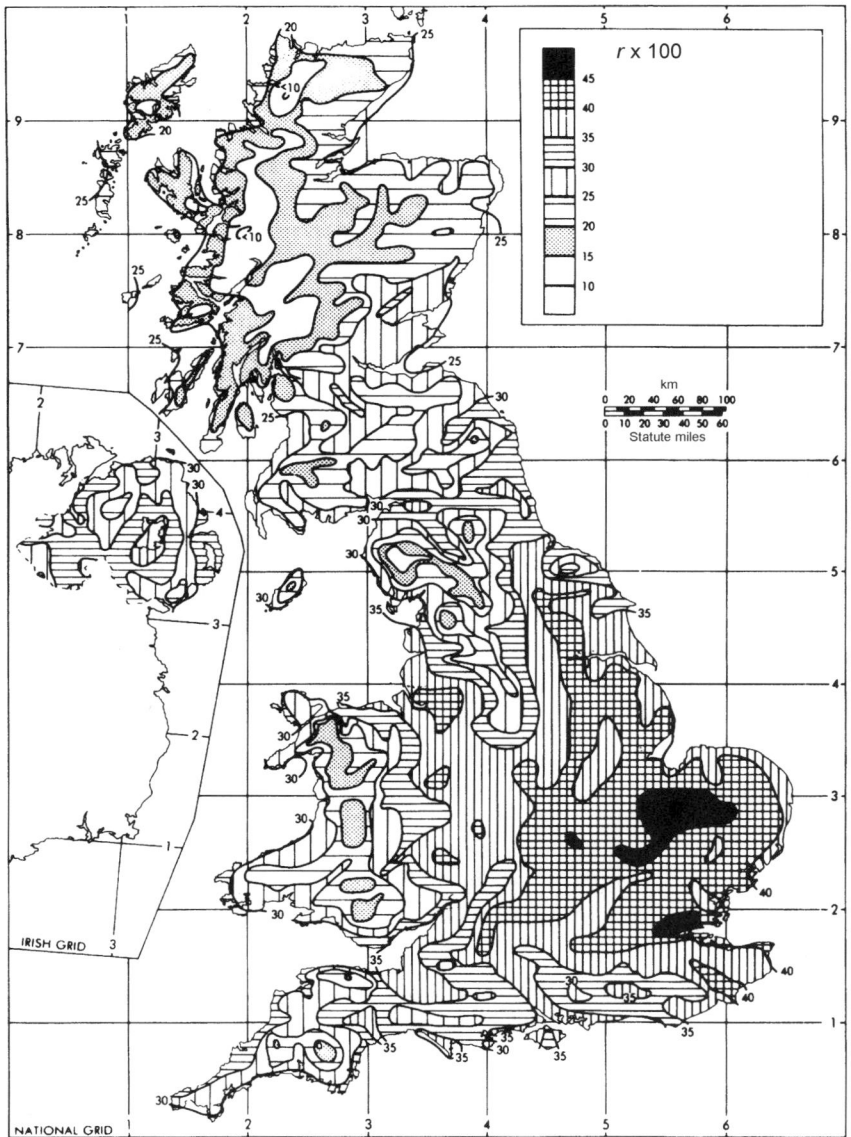

Figure 4.13   Ratio of 60-minute to two-day rainfall depths for five-year return period (reproduced with the permission of HR Wallingford)

Table 4.3 *Relationship between rainfall and M5 (Hydraulics Research, 1983) — reproduced with the permission of HR Wallingford (continued overleaf)*

M5 rainfall mm	M1	M2	M3	M4	M5	M10	M20	M50	M100
England and Wales									
5	0·62	0·79	0·89	0·97	1·02	1·19	1·36	1·56	1·79
10	0·61	0·79	0·90	0·97	1·03	1·22	1·41	1·65	1·91
15	0·62	0·80	0·90	0·97	1·03	1·24	1·44	1·70	1·99
20	0·64	0·81	0·90	0·97	1·03	1·24	1·45	1·73	2·03
25	0·66	0·82	0·91	0·97	1·03	1·24	1·44	1·72	2·01
30	0·68	0·83	0·91	0·97	1·03	1·22	1·42	1·70	1·97
40	0·70	0·84	0·92	0·97	1·02	1·19	1·38	1·64	1·89
50	0·72	0·85	0·93	0·98	1·02	1·17	1·34	1·58	1·81
75	0·76	0·87	0·93	0·98	1·02	1·14	1·28	1·47	1·64
100	0·78	0·88	0·94	0·98	1·02	1·13	1·25	1·40	1·54
150	0·78	0·88	0·94	0·98	1·01	1·12	1·21	1·33	1·45
200	0·78	0·88	0·94	0·98	1·01	1·11	1·19	1·30	1·40

Table 4.3   *continued*

M5 rainfall: M1 mm	M2	M3	M4	M5	M10	M20	M50	M100
Scotland								
5	0.67	0.82	0.91	0.98	1.02	1.17	1.35	1.62	1.86
10	0.68	0.82	0.91	0.98	1.03	1.19	1.39	1.69	1.97
15	0.69	0.83	0.91	0.97	1.03	1.20	1.39	1.70	1.98
20	0.70	0.84	0.92	0.97	1.02	1.19	1.38	1.66	1.93
25	0.71	0.84	0.92	0.98	1.02	1.18	1.37	1.64	1.89
30	0.72	0.85	0.92	0.98	1.02	1.18	1.36	1.61	1.85
40	0.74	0.86	0.93	0.98	1.02	1.17	1.34	1.56	1.77
50	0.75	0.87	0.93	0.98	1.02	1.16	1.30	1.52	1.72
75	0.77	0.88	0.94	0.98	1.02	1.14	1.27	1.45	1.62
100	0.78	0.88	0.94	0.98	1.02	1.13	1.24	1.40	1.54
150	0.79	0.89	0.94	0.98	1.02	1.11	1.20	1.33	1.45
200	0.80	0.89	0.95	0.99	1.01	1.10	1.18	1.30	1.40

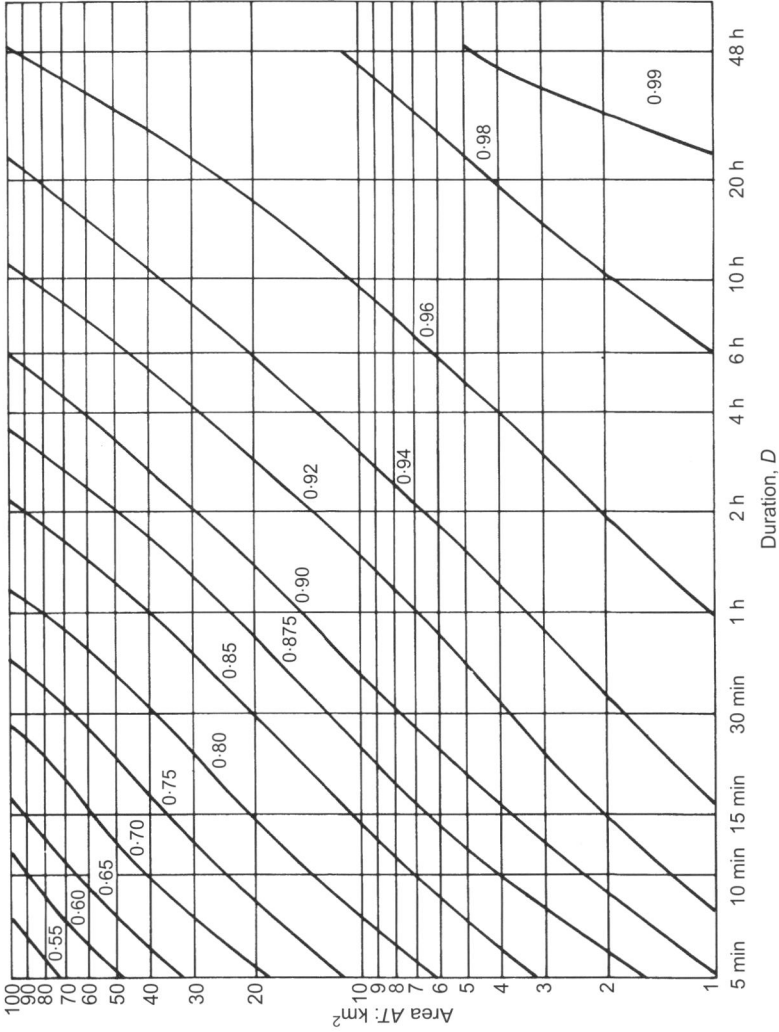

Figure 4.14 Areal reduction factor (Hydraulics Research, 1983) — reproduced with the permission of HR Wallingford

Example 4.4

Construct an intensity–duration curve for Glasgow for a 10-year return period using durations of 15, 30, 60 and 120 minutes.

From map (Figure 4.11), M5 – 60 for Glasgow = 16 mm
From map (Figure 4.13), *r* for Glasgow = 0·25
Z_1 is read from chart (Figure 4.12(b)):

$$M5 - D = M5 - 60 \times Z_1$$

Z_2 is read from Table 4.3(b):

$$M10 - D = M5 - D \times Z_2$$

$$\text{Intensity} = \frac{M10 - D}{D}$$

Duration	Z_1	M5 – D	Z_2	M10 – D	*i*
15	0·58	9·28	1·18	10·95	43·80
30	0·75	12·00	1·19	14·28	28·56
60	1·00	16·00	1·20	19·20	19·20
120	1·25	20·00	1·19	23·80	11·90

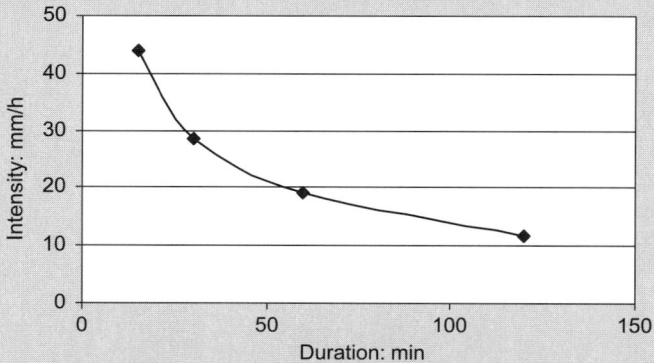

4.6.7 *Rainfall estimation for rural catchments*

For rural catchments, the *Flood Estimation Handbook* (Faulkner, 1999) proposes the following depth–duration–frequency model, based on the analysis of data from over 6000 daily total stations and over 100 hourly records in the UK.

The method determines the depth–duration–curve as the product of a non-dimensional *growth curve* and an *index variable* (Figure 4.15(a) and (b)). The growth curve relates the ratio of rainfall depth/index depth to the duration for different return periods and, since it is non-dimensional, it is independent of the annual rainfall total; in other words, similar sites will have similar growth curves. It is therefore possible to combine the records from several sites to extend the effective length of record and thereby increase the confidence of the estimates of longer period rainfall. The index variable scales the growth curve according to the annual rainfall total and in the *Flood Estimation Handbook* it is taken as the median of the annual maximum rainfalls (RMED).

By plotting the growth curve on a log–log scale, the curve is idealised as three contiguous linear segments (Figure 4.15(c)) over the ranges $D<12$ h, $12<D<48$ h and $48<D$ h. The line can therefore be defined by four parameters, namely the slope of each segment and the intercept of the first segment on the rainfall axis. These four parameters are assumed to be related to the return period in the following way:

$$a_1 = cy + d_1$$

$$a_2 = cy + d_2$$

$$a_3 = cy + d_3 \hspace{3cm} (4.15)$$

and:

$$b = ey + f$$

where a_1, a_2 and a_3 are the slopes of the three segments of each curve, b is the intercept of the first segment, and y is the Gumbel reduced variate defined such that the Gumbel Distribution plots as a straight line, i.e.:

$$y = -\ln\left[-\ln\left(1 - \frac{1}{T}\right)\right] \hspace{3cm} (4.16)$$

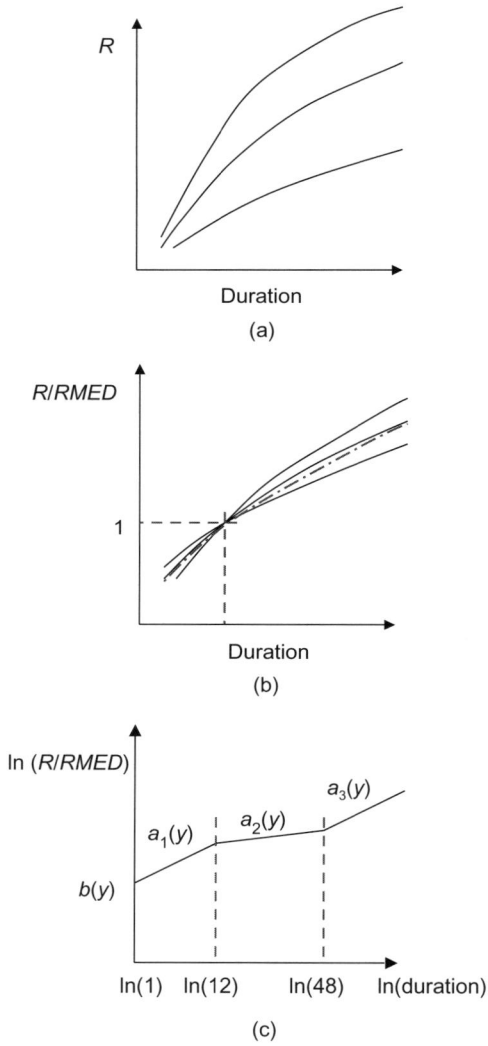

Figure 4.15   Derivation of rainfall depth–duration curve in the Flood Estimation Handbook *(Faulkner, 1999) — reproduced with the permission of the Centre for Ecology and Hydrology, Wallingford*

The six parameters c, d_1, d_2, d_3, e and f therefore define the set of growth curves for any duration and return period. These six parameters, together with values of RMED, have been interpolated from the rainfall stations onto a 1 km grid. To obtain a design rainfall over the area of a catchment, the method averages the six parameters estimated for each 1 km square of the catchment area (Example 4.5).

Example 4.5

Estimate the 7 h and 20 h rainfall for a return period of 100 years for a location where the DDF parameters are:

c	d_1	d_2	d_3	e	f
−0·016	0·430	0·394	0·383	0·248	2·368

$$y = -\ln\left[-\ln\left(1 - \frac{1}{T}\right)\right] = -\ln\left[-\ln\left(1 - \frac{1}{100}\right)\right] = 4.60$$

$$a_1 = cy + d_1 = -0.016 \times 4.60 + 0.430 = 0.356$$

$$a_2 = cy + d_2 = -0.016 \times 4.60 + 0.394 = 0.320$$

$$a_3 = cy + d_3 = -0.016 \times 4.60 + 0.383 = 0.309$$

$$b = ey + f = 0.248 \times 4.60 + 2.368 = 3.509$$

For $D = 7$

$$\ln(r) = b + a_1 \times \ln(7) = 3.509 + 0.356 \times 1.946 = 4.20$$

$$r = 66.8 \text{ mm}$$

For $D = 20$

$$\ln(r) = b + a_1 \times \ln(12) + a_2 \times \{\ln(20) - \ln(12)\}$$
$$= 3.509 + 0.356 \times 2.485 + 0.320 \times 0.511 = 4.56$$

$$r = 95.3 \text{ mm}$$

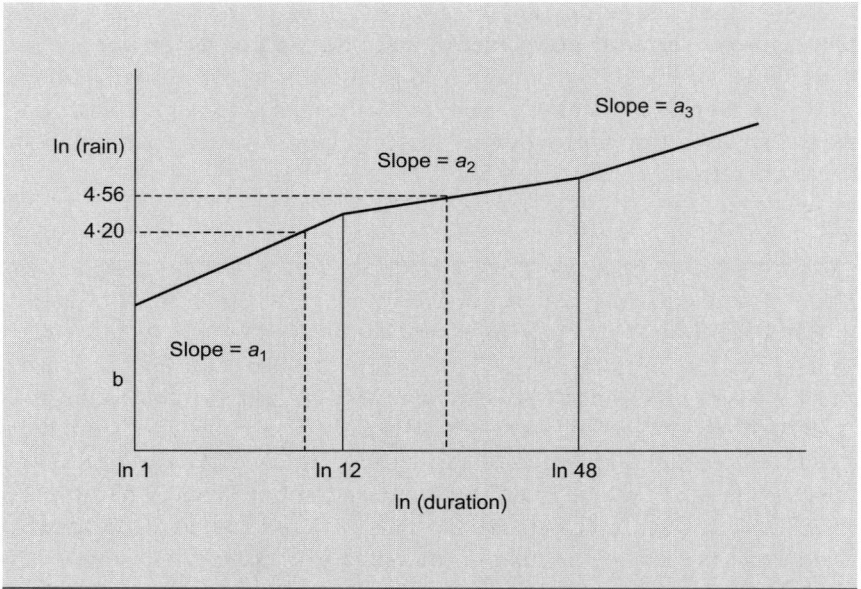

The model is intended to produce design rainfall for durations from 1 h to eight days and return periods from one year up to 1000 years. The duration of the most critical storm will obviously depend on the nature of the catchment, in particular on the response time. The *Flood Estimation Handbook* recommends that the design storm duration is calculated from:

$$D = Tp\left(1 + \frac{SAAR}{1000}\right)$$
(4.17)

where Tp is the time to peak of the unit hydrograph for the catchment, which is a measure of the catchment response time and $SAAR$ is the standard average annual rainfall available from a published map. The rainfall is significant since, in catchments with high rainfall, the floods tend to be longer than in the drier areas where short convective storms are more common.

For moderate or large catchments an aerial reduction factor (*ARF*) may need to be applied as described above in relation to urban areas. Values of *ARF* for rural areas in the United Kingdom are contained in the *Flood Studies Report* (NERC, 1975), and depend on the catchment size and the rainfall duration. However, there is some debate about the validity of the values of the areal reduction factor, especially since they are based on a relative small amount of data mainly from the south of England.

4.6.8 Design storm profiles

In addition to specifying a design rainfall depth and duration for a given return period, it is sometimes required to specify a design storm profile, i.e. a description of the variation of intensity with time during the storm. The main feature of a storm profile is its peakedness, which is the ratio of the peak intensity to the mean intensity. It has been found that the average peakedness of summer storms is generally higher than that of winter storms. This follows from the preponderance of short, heavy convective storms in summer compared with the lighter frontal-type rainfall in winter. The profiles recommended by the *Flood Estimation Handbook* are the 75% winter profile for rural catchments and the 50% summer profile for urban catchments. These are the profiles with peakedness exceeded by 25% of winter storms and 50% of summer storms respectively.

In both cases, the profile is symmetrical about the peak and can be derived from an approximate formula:

$$y = \frac{1-a^z}{1-a} \tag{4.18}$$

where y is the proportion of depth, $z = x^b$, x is the proportion of duration centred on the peak, and a and b have the values shown in Table 4.4, depending on the profile. The depth–duration relationship is shown in Figure 4.16.

The procedure is first to estimate the rainfall in the central hour of the storm from the appropriate storm profile and then for the three hours

Table 4.4 Constants a and b in Equation (4.18)

Profile	a	b
75% winter	0·060	1·026
50% summer	0·100	0·815

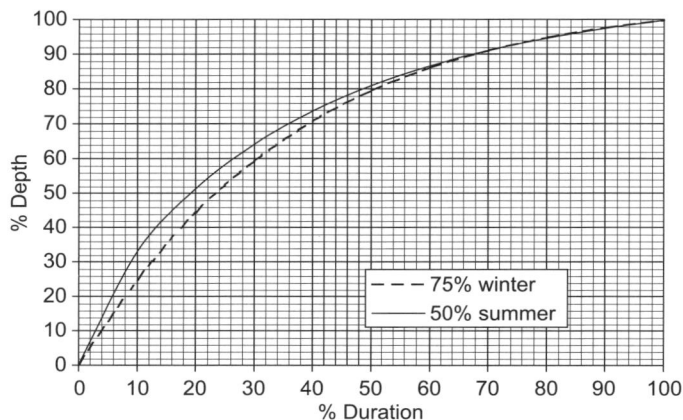

Figure 4.16   Design profiles for summer and winter storms

centred on the mid-point of the storm. The rainfall for each hour before and after the central hour is then given by:

$$i_{m-1} = i_{m+1} = \frac{(i_3 - i_1)}{2}$$
(4.19)

where i_1 and i_3 are the rainfall for the central hour and the central three hours, respectively, obtained from the storm profile, and m is the index of the central hour ($= D/2 + \frac{1}{2}$, since the duration should be an odd integer). The rainfall for the five central hours, etc., is then estimated in a similar way (Example 4.6).

The use of a standard symmetrical profile can lead to unreliable estimates of design floods for very long durations, for example six or eight days, which might be the critical duration for estimating floods in large reservoired catchments. In these cases, it is more appropriate to use the most severe sequence of storms which has been observed locally as the design rainfall.

4.6.9   Probable maximum precipitation

Probable maximum precipitation (PMP) is 'theoretically the greatest depth of precipitation that is physically possible over a given sized storm area at a particular location at a certain time of year (with no allowance for

Example 4.6

Determine the design storm profile (75% winter) for the location in Example 4.5 for a return period of 100 years and a duration of 7 hours.

Using:

$$y = \frac{1 - 0.06^{z}}{0.94}$$

where $z = x^{1.026}$, y is the proportion of depth, and x is the proportion of duration.

Total rainfall depth = 66·8 mm

For the central hour:

$$x = \frac{1}{7} = 0.143$$

$$z = 0.143^{1.026} = 0.136$$

$$y = \frac{1 - 0.06^{0.136}}{0.94} = 0.338$$

Rainfall depth $= 0.338 \times 66.8 = 22.11\,\text{mm}$

For hours 3 and 5:

$$x = \frac{3}{7} = 0.429$$

$$z = 0.429^{1.026} = 0.419$$

$$y = \frac{1 - 0.06^{0.419}}{0.94} = 0.737$$

Rainfall depth $= 0.737 \times 66.8 = 49.23$ mm

Depth per hour:

$$= \frac{49.23 - 22.11}{2} = 13.6 \text{ mm}$$

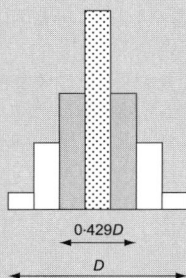

0·429D

D

The next two hours are calculated in the same way, summarised in the table below.

Hour	x	z	y	Cumulative depth: mm	Depth per h: mm
4	0·143	0·136	0·338	22·1	22·1
3 and 5	0·429	0·419	0·737	49·2	13·6
2 and 6	0·714	0·708	0·919	61·4	6·1
1 and 7	1·00		1·00	66·8	2·7

long-term climatic trend)' (WMO, 1986). The concept is used mainly in the design of dam spillways that must be adequate to pass the largest flood which might ever occur. This so-called *probable maximum flood* (PMF) is a combination of extreme rainfall together with the worst values of wetness and other conditions. The return periods associated with the PMP and PMF are not normally specified, although nominal values of between 10^6 and 10^8 years have been proposed. Methods have also been suggested to link flood frequency curves estimated for return periods of up to 1000 years to these limiting values using spline curves (Lowing and Law, 1995).

In the *Flood Estimation Handbook*, the PMP is estimated using maps of *estimated maximum precipitation* (EMP), which are based on an analysis of observed extreme events and an estimate of the theoretical maximum precipitable water in a vertical column above the catchment. Maps of

EMP have been produced for durations of 2 h, 24 h and 25 days (Houghton-Carr, 1999).

Values of EMP are estimated for durations ranging from the unit hydrograph time-interval (ΔT) to $5D$, where D is the design storm duration. The duration may be calculated using Equation (4.17), but if the reservoir has a significant attenuation, this should be included and the duration estimated from:

$$D = (Tp + RLAG)\left(1 + \frac{SAAR}{1000}\right) \tag{4.20}$$

$RLAG$ is the lag time of the reservoir, which is defined as the time between the peak of the rainfall and the peak of the hydrograph. As this is initially unknown, it should be estimated at first and the process repeated until the value becomes stable. Tp and $SAAR$ are the *time to peak* of the hydrograph and the *standard average annual rainfall*, as described earlier.

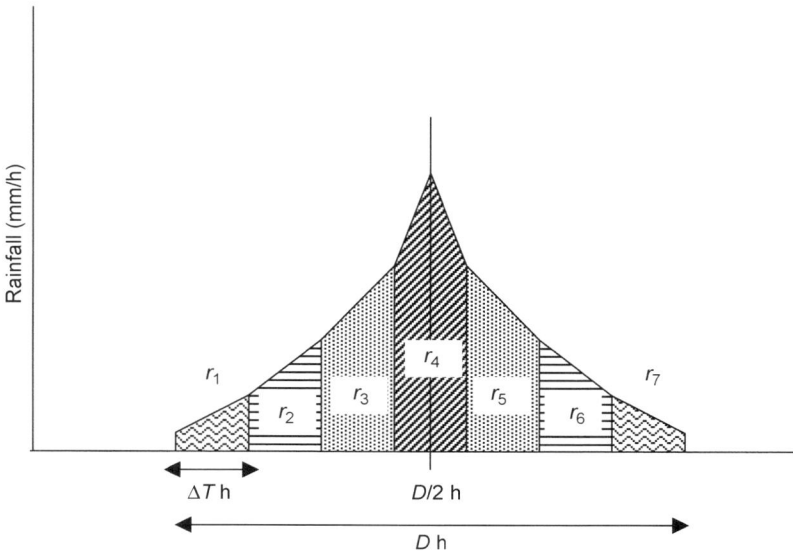

Figure 4.17   Construction of rainfall hyetograph for a PMP (reproduced with the permission of the Centre for Ecology and Hydrology, Wallingford)

Using a logarithmic relationship between EMP and duration, additional values of EMP are then interpolated for durations of $3\Delta T$, $5\Delta T$, $7\Delta T$, etc. The series of EMP values is then converted to catchment rainfall by applying an ARF appropriate to each duration. The rainfall profile is constructed by nesting the EMP values for a storm length of ΔT within that for a storm length of $3\Delta T$, which is within the $5\Delta T$ storm, etc. Thus the rainfall depth over the central interval from $0.5D - 0.5\Delta T$ to $0.5D + 0.5\Delta T$ is the EMP depth corresponding to a duration of ΔT hours (Figure 4.17). The depth over the central three intervals (from $0.5D - 1.5\Delta T$ to $0.5D + 1.5\Delta T$) is the EMP depth corresponding to a duration of $3\Delta T$ hours and the depth in the interval before and after the central interval will be $0.5(\text{EMP}_{\Delta T} - \text{EMP}_{3\Delta T})$ and likewise with the remaining intervals. Therefore, it can be seen that the total rainfall of the storm increases with duration, but with no compensating decrease in maximum intensity (Table 4.5).

It is normal practice to consider summer and winter conditions separately and Table 4.6 can be used to convert the all-year value of EMP to the winter or summer value depending on the average rainfall and the storm duration (Example 4.7).

4.7 Summary

This chapter has described the design methods for urban and rural rainfall estimation, and how they are derived from historical rainfall records allowing for incomplete data. It has also provided a background to the formation and measurement of precipitation, and how rainfall data can be analysed to provide the basis for time-series and storm profiles used in the design methods.

Table 4.5 Ratio of estimated maximum (EM) rainfall to EM 2 h, EM 24 h and EM 48 h (reproduced with the permission of the Centre for Ecology and Hydrology, Wallingford)

Duration	SAAR: mm							
	500–600	600–800	800–1000	1000–1400	1400–2000	2000–2800	2800–4000	>4000
1 min	0·06	0·06	0·06	0·06	0·06	0·06	0·06	0·06
2 min	0·11	0·11	0·11	0·11	0·11	0·11	0·10	0·10
5 min	0·23	0·23	0·23	0·23	0·22	0·22	0·21	0·21
10 min	0·36	0·36	0·36	0·36	0·34	0·34	0·32	0·32
15 min	0·47	0·47	0·47	0·47	0·45	0·45	0·43	0·43
30 min	0·65	0·65	0·65	0·65	0·62	0·62	0·59	0·59
1 h	0·83	0·83	0·83	0·83	0·79	0·79	0·75	0·75
48 h	*1·10*	*1·10*	*1·10*	*1·11*	*1·12*	*1·14*	*1·20*	*1·23*
72 h	*1·13*	*1·13*	*1·14*	*1·16*	*1·18*	*1·23*	*1·31*	*1·35*
96 h	*1·17*	*1·17*	*1·18*	*1·20*	*1·24*	*1·32*	*1·42*	*1·48*
192 h	**0·84**	**0·80**	**0·76**	**0·71**	**0·68**	**0·65**	**0·62**	**0·60**

Note: For durations up to 1 h the ratios are with respect to EM 2 h (normal type); for durations of 48, 72 and 96 h, the ratios are with respect to EM 24 h (italics) and for the duration of 192 h the ratio is with respect to EM 48 h (bold).

Table 4.6 Seasonal PMP as a percentage of all-year value (reproduced with the permission of the Centre for Ecology and Hydrology, Wallingford)

Duration	Season	SAAR: mm					
		500–600	600–800	800–1000	1000–1400	1400–2000	>2000
1 min	Winter*	13	15	19	26	30	33
2 min	Winter*	17	19	24	32	38	42
5 min	Winter*	21	24	30	40	47	53
10 min	Winter*	24	27	35	47	55	61
15 min	Winter*	26	30	38	50	59	66
30 min	Winter*	30	33	42	57	67	74
1 h	Summer	100	100	100	100	100	100
	Winter	33	37	47	63	74	82
2 h	Summer	100	100	100	100	100	100
	Winter	38	42	50	69	86	90

6 h	Summer	100	100	100	100	100	100
	Winter	45	51	61	79	93	96
1 day	Summer	100	100	100	100	100	100
	Winter	55	62	70	79	99	100
2 days	Summer	100	100	100	100	100	100
	Winter	63	69	78	85	90	84
4 days	Summer	100	100	100	100	100	100
	Winter	64	73	84	92	92	88
8 days	Summer	100	100	100	100	100	100
	Winter	67	80	91	96	89	83

* For duration less than 1 h, summer PMPs are the same as the all-year values.

Example 4.7

Determine the duration, depth and profile for a winter PMP storm for a 10 km² catchment where the SAAR is 1650 mm and the time to peak (*Tp*) of the unit hydrograph is 3.3 h. The estimated maximum 2 h rainfall (*EM* – 2 h) is 135 mm and *EM* – 24 h is 250 mm.

For duration = 1 h
From Table 4.5:

$$\frac{EM - 60 \text{ min}}{EM - 2\text{ h}} = 0.79$$

$$EM - 60 \text{ min} = 0.79 \times 135 = 106.7 \text{ mm}$$

From Table 4.6:

$$\frac{\text{Winter } EM - 1\text{h}}{\text{All year } EM - 1\text{h}} = 0.74$$

$$\text{Winter } EM - 1\text{h} = 0.74 \times 106.7 = 78.9 \text{ mm}$$

For duration = 2 h
From Table 4.6:

$$\frac{\text{Winter } EM - 2\text{h}}{\text{All year } EM - 2\text{h}} = 0.86$$

$$\text{Winter } EM - 2\text{ h} = 0.86 \times 135 = 116.1 \text{mm}$$

For duration = 24 h
From Table 4.6:

$$\frac{\text{Winter } EM - 24\text{ h}}{\text{All year } EM - 24\text{ h}} = 0.99$$

$$\text{Winter } EM - 24 \text{ h} = 0.99 \times 250 = 247.5 \text{ mm}$$

Design duration

$$= Tp1 + \frac{SAAR}{1000} = 3.3(1 + 1.65)$$
$$= 8.75 \text{ h (take } D = 9 \text{ h as nearest odd number)}$$

The values of estimated maximum precipitation for durations of 1 h, 2 h and 24 h are plotted against duration on a linear-log graph which is then used to estimate the precipitation for durations from 1 h (Δt) to 9 h (D).

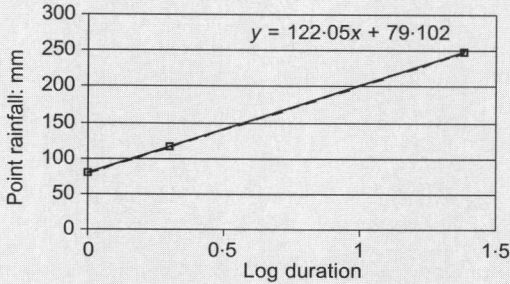

The point rainfall is modified by the respective *ARF* for the catchment area and duration. The *EM* for 1 h (71·8 mm) is applied to the central hour (5) of the storm and the value for 3 h (129·2 mm) is applied to the central 3 h. The central value is deducted, leaving 57·4 mm to be divided between hours 4 and 6. The procedure is continued to give the final profile.

Duration: h	1	3	5	7	9
Point rain: mm	78·9	137·3	164·3	182·2	195·5
ARF	0·910	0·941	0·952	0·956	0·960
Catchment rain: mm	71·8	129·2	156·4	174·1	187·7
Difference: mm	71·8	57·4	27·3	17·7	13·5

Profile

Hours	1	2	3	4	5	6	7	8	9
Depth: mm	6·8	8·9	13·6	28·7	71·8	28·7	13·6	8·9	6·8

4.8   References

Adams, B. J. and Papa, F. (2000). *Urban Stormwater Management Planning with Analytical Probabilistic Models*. John Wiley.

Bilham, E. G. (1936). *Classification of Heavy Falls of Rain in Short Periods*. HMSO, London.

Chorley, R. J. (1969). *Water, Earth and Man*. Methuen, London.

Cowpertwait, P. S. P., Metcalfe, A. V. *et al*. (1991). *Stochastic Generation of Rainfall Time-series*. WRc.

Faulkner, D. (1999). *Flood Estimation Handbook Vol. 2 — Rainfall Frequency Estimation*. Institute of Hydrology, Wallingford.

Herbst, P. H., Bredenkamp, D. B., *et al*. (1966). A technique for the evaluation of drought from rainfall data. *Jounal of Hydrology*, **4**, 264–272.

Houghton-Carr, H. (1999). *Flood Estimation Handbook Vol. 4 — Restatement and Application of the FSR rainfall–runoff Method*. Institute of Hydrology, Wallingford.

Hydraulics Research (1983). *Design and Analysis of Urban Storm Drainage*. Hydraulics Research, Wallingford.

Lowing, M. J. and Law, F. M. (1995). Reconciling Flood Frequency Curves with Probable Maximum Flood. *Proceedings of the BHS Fifth National Hydrology Symposium*, Edinburgh.

Meteorological Office. *Rules for Rainfall Observers*. http://www.met.office.gov.uk

NERC (1975). *Flood Studies Report*. Natural Environment Research Council, London, UK.

Ormsbee, L. E. (1989). Rainfall Disaggregation Model for Continuous Hydrologic Modelling. *Journal of Hydraulic Engineering*, **115**, 507–525.

Roy, G. R. and Fox, I. (1995). Flooding in the Tweed Catchment — weather and orographic controls. *Proceedings of the BHS Fifth National Hydrology Symposium*, Edinburgh.

Rugumayo, A. I. (2002). Drought Intensity Duration and Frequency Analysis: A Case Study of Western Uganda. *Journal of CIWEM*, **16**, 111–115.

Shaw, E. M. (1994). *Hydrology in Practice*. Chapman and Hall, London.

World Meteorological Organization (1986). *Manual for Estimation of Probable Maximum Precipitation*. WMO, Geneva.

Wood, S. J., Jones, D. A. *et al*. (2000). Accuracy of Rainfall Measurement for Scales of Hydrological Interest. *Hydrology and Earth System Sciences*, **4**, 531–541.

Chapter 5

Evaporation and other losses

Estimating the *effective* rainfall, i.e. the proportion of rainfall which appears as direct runoff, is critical to any hydrological forecasting and it is also the most uncertain part of flood estimation. The effective rainfall can be considered as the total rainfall, less the losses due to interception, infiltration, evaporation and transpiration from vegetation. These losses are, therefore, partly a function of the catchment land use, soil type and topography, and partly a function of the catchment wetness. The former can be taken as essentially constant while the wetness obviously varies with time. The term 'losses' may be misleading in that it represents water which is not permanently lost, but which passes back into the atmosphere before completing the land phase. It may also include water which is delayed within the groundwater. This chapter describes the physical basis of these processes and considers the various mathematical models which are used to represent them.

5.1   Interception

Interception is the storage of precipitation on the vegetation canopy and stems. Intercepted water is mostly evaporated and can represent up to 20% of total rainfall, depending on the type of vegetation. Interception storage is satisfied early in a storm and, when the stores are full, there is overflow to the ground. When there is a strong wind, interception storage capacity is reduced, although higher wind speeds encourage evaporation and therefore wind may actually increase overall losses for longer storms. Figure 5.1 shows a simplified model of interception.

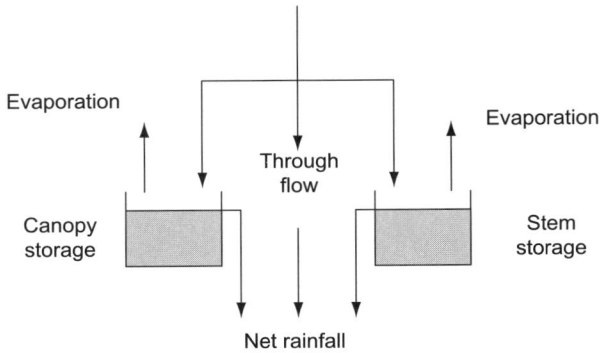

Figure 5.1   Simplified model of interception

5.2   Depression storage

Depression storage is the initial storage within depressions on the ground surface and also within the surface layers of nominally impervious materials (surface wetting). The depressions vary considerably in size and, as each depression is filled, flow passes to other depressions and eventually directly into channels. Water is lost from the depressions by subsequent evaporation, although some water may infiltrate into the ground. The amount of depression storage varies widely, depending on the surface micro-topography, slope and permeability, as well as the antecedent rainfall conditions. The Wallingford Procedure (Hydraulics Research, 1983) incorporates the following regression equation for estimating depression storage:

$$DEPSTOG = 0.71 \times SLOPE^{-0.48} \tag{5.1}$$

where *DEPSTOG* is the average depth of depression storage (mm) and *SLOPE* is the average ground slope (%). Typical values of depression storage vary from less than 2 mm for paved areas, 3–7 mm for flat roofs and up to 10 mm for gardens.

In urban drainage analysis, the depression storage is normally deducted progressively from the initial values of the rainfall hyetograph, as shown in Figure 5.2

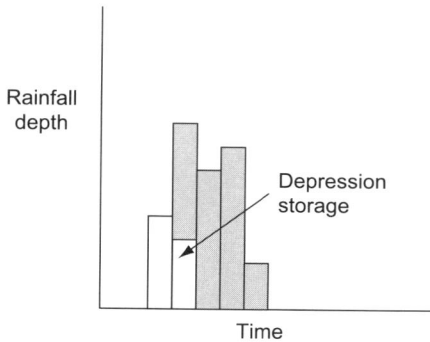

Figure 5.2   Depression storage

5.3   Infiltration

Infiltration is the passage of water into the soil. The infiltration capacity of a particular land surface depends on many factors including:

- *ground slope* — larger slopes give less opportunity for infiltration
- *stream pattern* — a dense stream pattern means that water does not travel far over the surface and hence reduces the chances of infiltration
- *soil type and geology* — sandy soils have a texture which encourages infiltration
- *vegetation* — dense vegetation traps surface water and also the root systems provide pathways for the water to infiltrate the soil surface. In addition, the leaf canopy protects the soil from compaction by the impact of raindrops
- *area of paved surface* — paved surfaces inhibit infiltration, although most paved surfaces permit some infiltration.

The amount of infiltration also depends on the wetness of the ground and typically decreases during the period of a storm. Exceptionally, there can be an increase in filtration when, for example, a crust of hard soil is broken up by the rainfall. The initial high infiltration is due to the capillary attraction as well as gravity, and actual rates of infiltration at the start of a storm are generally limited by the rainfall intensity. Once the soil becomes saturated and surface ponding starts, infiltration is limited by the soil properties and excess rainfall becomes runoff. The most common mathematical representation of infiltration is the exponential decay model proposed by Horton (1940):

$$f_t = f_c + (f_0 - f_c)e^{-kt} \tag{5.2}$$

where f_t is the infiltration capacity at time t, f_c is a residual constant infiltration rate and f_0 is the infiltration capacity at the start of a storm. The actual infiltration rate will, of course, only equal the infiltration capacity when the supply rate (i.e. the precipitation) is equal to, or greater than, the infiltration capacity. This model is illustrated in Figure 5.3(a).

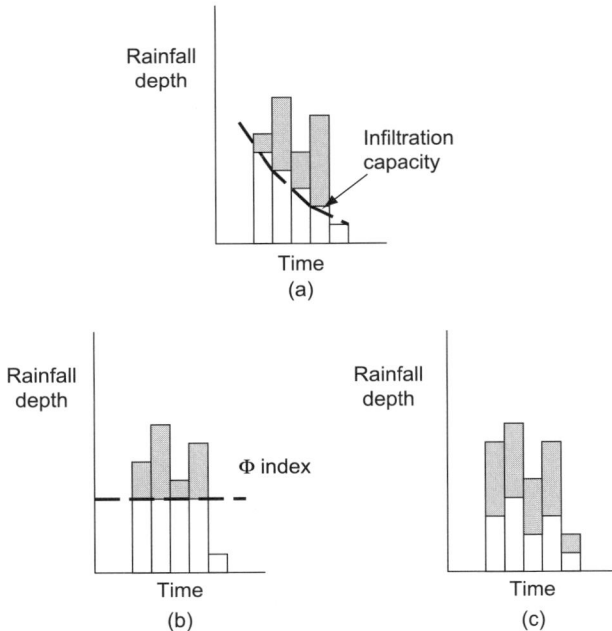

Figure 5.3   Representation of infiltration losses: (a) Horton model; (b) constant infiltration; and (c) proportional model

5.4 Evaporation and transpiration

5.4.1 General principles

Evaporation is the transfer of water from a liquid to a gaseous state as a result of the difference between the vapour pressure at the land surface and in the overlying air. It can range from less than 300 mm per year in temperate areas to over 1000 mm per year in arid areas, and over 2000 mm per year over open water. Evaporation is driven by the incoming solar energy which overcomes the forces between the water molecules. As water evaporates into the atmosphere, the vapour pressure in the atmosphere increases and it become more difficult for water to pass into the air (unless the air is replaced with drier air). Eventually the air becomes saturated and is unable to take any more moisture. The vapour pressure at which this occurs is known as the *saturated vapour pressure* (e_a) and the saturated vapour pressure decreases with decreasing temperature (Figure 5.4). The *dew point* (T_d) of an air mass is the temperature at which it becomes saturated. Therefore, if an air mass with a vapour pressure e_a is reduced in temperature to T_d, the air will reach saturation. If the air mass were at a temperature T_a, the difference between the saturation vapour pressure and the actual vapour pressure of the air $(e_a - e_d)$ would be the *saturation deficit*. The ratio e_d/e_a is known as the *relative humidity* and is normally expressed as a percentage. The capacity of the air to absorb moisture therefore depends on its temperature and humidity.

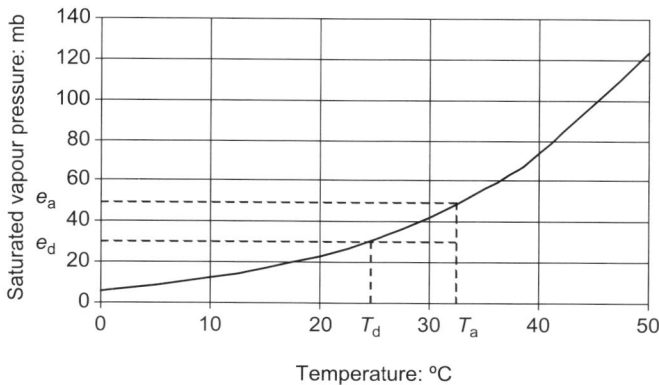

Figure 5.4   Saturated vapour pressure and temperature

Three elements are required for evaporation to take place:

- a source of energy to overcome the latent heat of vaporisation (approximately 2 kJ per kg)
- a mechanism for removing the overlying air before it has become saturated as a result of evaporation (i.e. wind)
- a supply of moisture (i.e. evaporation will reduce as the ground surface becomes drier).

Evaporation may also be affected by the salinity of the water or by the presence of pollutants such as oil films, but these are of relatively minor importance.

Transpiration is a similar process to evaporation, except that water is transferred from the soil to plant tissue and then passes into the atmosphere as water vapour through the small openings (stomata) in the leaves of plants. A mature broadleaf tree consumes about 70 l/day, but nearly all water taken up by plants is lost through transpiration; only a very small fraction is used within the plant. Transpiration, like evaporation, depends on the incoming energy, air temperature and humidity and wind speed, as well as the availability of water within the soil. It also depends on the physiological properties of the plant, which include the hydraulic capacity of the stomata and the aerodynamic resistance of the crop canopy. Both of these parameters vary with the type of crop and its stage of growth. In addition, the stomatal resistance increases with the degree of water stress which the plant is experiencing and is also sometimes related to the concentration of CO_2 in the atmosphere. In other words, increases in greenhouse gases such as CO_2, which are related to global warming, may also reduce rainfall losses due to transpiration.

Evaporation and transpiration occur simultaneously and it is not easy to separate the two processes. The combination of evaporation and transpiration is called *evapotranspiration*. For crops at an early stage of growth, most of the loss is through evaporation from the ground surface. As the crop begins to cover more of the ground, there is less evaporation from the soil surface and more transpiration from the leaves.

The concept of *potential evapotranspiration* is sometimes used to eliminate the uncertainties of soil moisture. Potential evapotranspiration is that which occurs where there is no limitation on the supply of moisture, in effect evaporation from open water. (The evaporation from saturated ground surfaces is generally about 90% of potential evaporation.) *Actual evapotranspiration* is that which is limited by the availability of moisture supply. Moisture which evaporates from the

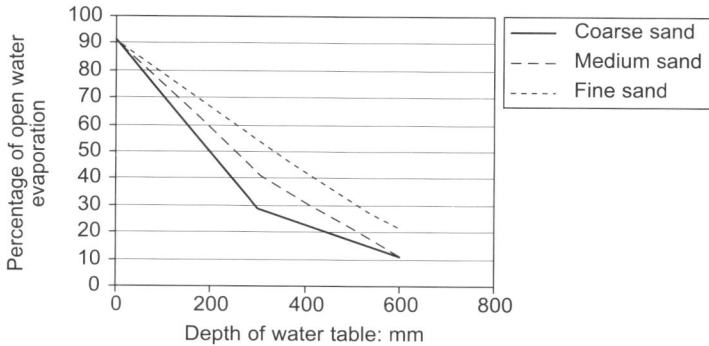

Figure 5.5   Evaporation from sand (Hellwig, 1973)

surface of a soil is replaced by moisture which rises by capillary action, and the availability of moisture depends on the soil-particle size as well as the depth to the water table. For coarse sand there is a fairly rapid decrease in evaporation as the depth of the water table increases, but for fine sands the decrease with the depth of the water table is less (Figure 5.5). This is because, in fine sands, there is much more upward flow due to capillary action.

Potential evapotranspiration is a function of both climatic and biological factors, and the concept of *reference crop evapotranspiration* (ET_0) is used to distinguish between these two factors. ET_0 refers to the evapotranspiration from a specific reference surface, which is defined as 'an extensive surface of green grass between 8 cm and 15 cm tall, actively growing, completely shading the ground and not short of water' (Doorenbos and Pruit, 1977). A particular *crop evapotranspiration* is found from the product of ET_0 and a crop coefficient (K), which depends on the nature of the crop and its stage of growth.

5.4.2   Measurement of evaporation

An estimation of potential evaporation or reference evapotranspiration can be obtained using an evaporation pan. A class A pan is 1207 mm diameter and 250 mm deep, and is mounted on a timber frame 150 mm above ground level. The water level should be within the range 50 mm to 75 mm below the level of the rim. Pans provide an integrated measurement of the effects of radiation, wind, temperature and humidity. However, there are several factors which produce differences between pan evaporation and that from soil surfaces or large water-bodies. The reflection of solar energy is different in a pan than a larger water-body, causing differences in temperature, and there are also differences in the turbulence patterns and humidity of the air immediately above the water in the pan. If pan evaporation is compared with grass or other evapotranspiration, the storage of heat in the water and the consequent higher evaporation at night also needs to be taken into account. In general, these factors are allowed for by the use of a pan coefficient (K_p), i.e.:

$$E = K_p E_{pan} \tag{5.3}$$

The value of the pan coefficient depends on the wind speed and relative humidity at the pan site as well as on the type of pan. The coefficient typically varies between 0·4 and 0·8.

A simpler instrument to measure evaporation is the *piche*, which consists of a tube 14 mm in diameter and 225 mm long with one end closed. A circular disc of absorbent blotting paper is fitted against the open end and provides the evaporating surface which is fed by water in the tube. The tube is held vertically and the amount of evaporation is found from the change in water level in the tube. It has been found that the results of annual evaporation from a piche are approximately equivalent to those from a class A pan.

Another instrument for measuring evapotranspiration is a *lysimeter* which contains a volume of soil hydrologically isolated from the surrounding soil and which is planted with representative vegetation. The amount of evaporation is measured by the difference between the precipitation and the change in soil moisture. It is necessary to ensure that the thermal, hydrological and mechanical properties of the sample soil are similar to those of the parent material, and the level of maintenance they require usually restricts their use to research applications.

5.4.3   Estimation of evapotranspiration using energy balance

Evaporation requires relatively large amounts of energy, either in the form of sensible heat or radiant energy. It is therefore possible, in principle, to apply the principle of energy conservation and to assume that the energy arriving at a surface is equal to that leaving over the same period. The main energy fluxes are related by:

$$R - G - E - H = 0 \tag{5.4}$$

where R is the net incoming radiation, G the soil heat flux, E the latent heat flux of vaporisation, and H the sensible heat. R and G can be estimated from climatic parameters but H requires accurate measurements of temperature gradients above the surface. If these parameters can be determined, then E can be estimated by rearranging the equation. It should be noted that only vertical fluxes are considered in the equation and any advection horizontally is ignored. In other words, it only applies to large extensive surfaces with homogeneous characteristics.

A similar method of estimating evapotranspiration is the mass transfer method. This considers small parcels of air above a large homogenous surface. These parcels, or eddies, transport water vapour and energy from and to the evaporating surface. By assuming steady state conditions and using various eddy transfer coefficients, the evapotranspiration rate can be calculated from the vertical gradients of air temperature and water vapour. This method requires accurate measurements of vapour pressure and temperature at various heights and is generally only used in research.

5.4.4   FAO Penman-Monteith Equation

In 1948, Penman combined the energy balance and the mass transfer method and derived the equation below to compute the evaporation from an open water surface using standard meteorological records of radiation, temperature, humidity and wind speed:

$$E_0 = \frac{\Delta H + \gamma E_a}{\Delta + \gamma} \tag{5.5}$$

where Δ is the slope of the saturated vapour pressure–temperature curve, H is the net incoming energy, E_a is the hypothetical evaporation for equal air and water temperature, and γ is the psychometric constant. This was subsequently developed by Monteith and later combined with equations for aerodynamic and surface resistances to give the FAO Penman-Monteith Equation which is one of the generally accepted methods for estimating reference evapotranspiration:

$$ET_0 = \frac{0{\cdot}408\Delta(R_n - G) + \gamma \dfrac{900}{T+273} u_2(e_s - e_a)}{\Delta + \gamma(1 + 0{\cdot}34u_2)} \tag{5.6}$$

where ET_0 is the reference evapotranspiration (mm/day), R_n is the net radiation at the crop surface (MJ/m^{2}day), G is the soil heat flux density

(MJ/m^{2}day), T is the mean daily temperature at 2 m height (°C), u_2 is the wind speed at 2 m height (m/s), e_s is the saturation vapour pressure (kPa), given by:

$$e_s = \frac{e(T_{max}) + e(T_{min})}{2}$$

where:

$$e(T) = 0.6108 \exp\left[\frac{17.27T}{T + 237.3}\right]$$

$e_s - e_a$ is the saturation vapour pressure deficit (kPa), where:

$$e_a = e(T_{dew})$$

T_{dew} is the dewpoint temperature (for well-watered sites, T_{dew} can be approximated to the minimum air temperature), Δ is the slope of vapour pressure curve (kPa/°C), given by:

$$\Delta = \frac{4098\left[0.6108 \exp\left(\dfrac{17.27T}{T + 237.3}\right)\right]}{(T + 237.3)^2}$$

γ is the psychometric constant (kPa/°C), given by:

$$\gamma = \frac{c_p P}{\epsilon \lambda} = 0.665 \times 10^{-3} P$$

where P is the atmospheric pressure (kPa/°C), c_p is the specific heat at constant pressure = 1.013 MJ kg^{-1} °C^{-1}, ϵ is the ratio molecular weight of water vapour/dry air = 0.622, and λ is the latent heat of vaporisation = 2.45 MJ kg^{-1}.

The equation can be used with daily, weekly or monthly data. To ensure consistency, the meteorological observations should be made at a level of 2 m above an extensive surface of green grass, shading the ground and not short of water. It has been pointed out (Jensen *et al.*, 1997) that the measurements of weather conditions are often not made under the reference conditions assumed in the procedure. In particular, the vapour-pressure deficit, air temperature and wind speed may be modified when measured in a non-reference environment. Simpler procedures (e.g. Hargreaves and Samani, 1985) are available for estimating ET_0 from only temperature and/or precipitation.

The radiation term (R_n) can be estimated from the global extraterrestrial radiation which is the radiation striking the top of the Earth's atmosphere and is generally taken to be about 82 kJ/m^2 min. This extraterrestrial radiation term is then corrected for various factors, including:

- the scattering and reflection of radiation within the atmosphere
- the angle of the Earth's surface to the sun's rays, i.e. the latitude
- the cloudiness or duration of sunshine
- the emission of long-wave radiation by the atmosphere due to the energy received from long-wave emissions from the Earth's surface.

If net radiation data are not available, it can be estimated from geographical and sunshine duration data using the method described in Appendix 5.1.

The stages in the calculation of ET_0 are illustrated in Example 5.1 and the results are summarised in Appendix 5.2 for a monthly time-step. The resulting grass reference evapotranspiration is plotted as potential evaporation in Figure 5.6, showing the relationship with temperature and wind speed.

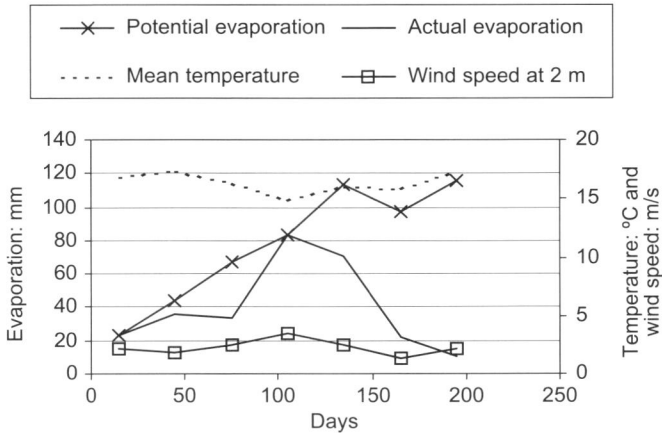

Figure 5.6   Potential and actual evaporation (see Appendix 5.2)

Example 5.1

Estimate the daily grass reference evapotranspiration for a location with the following details:

Net radiation (R_n) = 1·9
Soil heat flux (G) = 0
Maximum temperature = 21·5°C
Minimum temperature = 12·3°C
Dewpoint temperature = 12·1°C
Altitude (H) = 100 mASL
Wind speed at 2 m (u_2) = 2·08 m/s

Mean temperature $Tm = \dfrac{21 \cdot 5 + 12 \cdot 3}{2} = 16 \cdot 9°C$

Slope of vapour pressure – temperature curve Δ

$$= \dfrac{4098 \left[0 \cdot 6108 \exp\left(\dfrac{17 \cdot 27T}{T + 237 \cdot 3} \right) \right]}{\left(T + 237 \cdot 3 \right)^2} = \dfrac{4098 \left[0 \cdot 6108 \exp\left(\dfrac{17 \cdot 27 \times 16 \cdot 9}{16 \cdot 9 + 237 \cdot 3} \right) \right]}{\left(16 \cdot 9 + 237 \cdot 3 \right)^2} = 0 \cdot 12$$

Atmospheric pressure P

$$= 101 \cdot 3 \left[\dfrac{293 - 0 \cdot 0065H}{293} \right]^{5 \cdot 26} = 101 \cdot 3 \left[\dfrac{293 - 0 \cdot 0065 \times 100}{293} \right]^{5 \cdot 26} = 100 \cdot 12\,\text{kN/m}^2$$

Psychometric constant $\gamma = \dfrac{c_P P}{\epsilon \lambda} = 0 \cdot 665 \times 10^{-3} P = 0 \cdot 665 \times 10^{-3} \times 101 \cdot 12 = 0 \cdot 07$

SVP at mean temperature $e_o = e(16 \cdot 9) = 0 \cdot 6108 \exp\left[\dfrac{17 \cdot 27 \times 16 \cdot 9}{16 \cdot 9 + 237 \cdot 3} \right] = 1 \cdot 93$

Mean SVP $e_s = \dfrac{e(21 \cdot 5) + e(12 \cdot 3)}{2} = 2 \cdot 00$

Actual VP $e_a = e(12 \cdot 1) = 1 \cdot 41$

Grass reference evapotranspiration

$$ET_0 = \frac{0 \cdot 408 \Delta (R_n - G) + \gamma \dfrac{900}{T + 273} u_2 (e_s - e_a)}{\Delta + \gamma (1 + 0 \cdot 34 u_2)}$$

$$= \frac{0 \cdot 408 \times 0 \cdot 12(19 - 0) + 0 \cdot 07 \dfrac{900}{16 \cdot 9 + 273} 2 \cdot 08(2 \cdot 0 - 1 \cdot 41)}{0 \cdot 12 + 0 \cdot 07(1 + 0 \cdot 34 \times 2 \cdot 08)} = 1 \cdot 47 \text{ mm}$$

5.4.5   Crop evapotranspiration

To estimate the actual evapotranspiration from other crops, a crop coefficient (K_c) is introduced so that:

$$ET_c = K_c ET_0 \tag{5.7}$$

where ET_C is the crop evapotranspiration. The crop coefficient represents the characteristics which distinguish a crop from grass, including:

- the crop height, which influences the aerodynamic resistance term and the transfer of water vapour from the crop to the atmosphere
- the albedo or reflectance of the crop, which affects the level of radiation and is largely determined by the proportion of ground covered by vegetation and the soil surface wetness
- the resistance of the leaves to the transfer of vapour, which depends on the nature, age and condition of the leaves
- the evaporation from the soil, which depends on the proportion of bare soil.

Typical values of the crop coefficient for various crops at three stages of growth are given in Table 5.1.

Table 5.1   Typical crop coefficients (Doorenbos and Pruit, 1977)

Crop	Growth state		
	Initial	Mid	End
Vegetables	0·15	0·95	0·85
Roots and tubers	0·15	1·00	0·85
Legumes	0·15	1·10	0·50
Cereals	0·15	1·10	0·25

5.5   Water balance

The water balance technique is often used to estimate losses due to evapotranspiration, etc., especially in rural catchments. It basically equates the rainfall inputs with the sum of the runoff and losses, and any change in the storage of the catchment. The water balance approach treats the catchment as a closed system and accounts for all the different inflows and outflows from the ground on a daily, weekly or monthly basis.

5.5.1   Water balance in rural catchments

In a rural catchment the major flows include precipitation (P), runoff (Q), percolation to deep groundwater (D), evapotranspiration (E), groundwater inflow (G), and changes in soil moisture storage (ΔS). A water balance model would be of the form:

$$P - E - Q - D + G \pm \Delta S = 0 \qquad (5.8)$$

where all the parameters are in, say, millimetres. Although the approach is conceptually appealing, there are considerable problems in evaluating the various components. Rainfall can be estimated by integrating point measurements although there are uncertainties in the measurement and interpolation of the data. Over long periods, such as a year, the changes in soil moisture and groundwater storage can be neglected and the water balance then becomes:

$$P - E - Q = 0 \qquad (5.9)$$

Rather than measuring actual evapotranspiration directly, use can be made of the models described above which estimate potential evapotranspiration by representing the physical processes involved. However, the relationship between actual evapotranspiration and potential evapotranspiration is a complex function of soil moisture storage, vegetation and climatic factors. In general, it can be assumed that evaporation will take place at the potential rate in a 'surplus' situation, i.e. when precipitation (*P*) exceeds potential evapotranspiration (*PET*), during which time the difference between precipitation and *PET* results in a change in soil moisture (ΔS). This continues until the maximum soil storage capacity (S_{max}) is reached; thereafter, the excess precipitation above *PET* goes to runoff or deep groundwater. Where the rainfall is less than *PET*, actual evapotranspiration is less than *PET* and some water is lost from soil moisture storage. This continues until the remaining soil moisture cannot be accessed by plants, at which point the soil moisture is said to be at *wilting point*. The relationships can be expressed mathematically as (Thornthwaite and Mather, 1957):

$$S_i = \min[(P_i - PET_i + S_{i-1}), S_{max}] \quad \text{if } P_i \geq PET_i \text{ (surplus condition)}$$

$$S_i = S_{i-1} \exp[-(PET_i - P_i)/S_{max}] \quad \text{if } P_i \geq PET_i \text{ (deficit condition)} \quad (5.10)$$

where S_i is the soil moisture at the end of month *i*, S_{i-1} is the soil moisture at the end of the preceding month, P_i is the precipitation for month *i*, PET_i is the potential evapotranspiration for month *i*, and S_{max} is the maximum storage capacity of the soil in the root zone. The soil moisture deficit at the end of month *i* (SMD_i), is defined as:

$$SMD_i = S_{max} - S_i \qquad (5.11)$$

The actual evapotranspiration for month *i* (AET_i), is calculated as:

$$AET_i = PET_i \quad \text{if } P_i \geq PET_i$$

$$AET_i = P_i + S_{i-1} - S_i \quad \text{if } P_i < PET_i \qquad (5.12)$$

The runoff in month i, (Q_i), is given by the precipitation less the actual evapotranspiration and any loss or gain to or from the catchment storage (subject to $Q > 0$), i.e.:

$$Q_i = \max[(P_i - AET_i - \Delta S_i), 0] \tag{5.13}$$

where $\Delta S_i = S_i - S_{i-1}$

The maximum available soil moisture in the root zone depends on the soil type and the effective root zone depth for the plant. It can be estimated from:

$$S_{max} = 1000(\vartheta_{FC} - \vartheta_{WP})D_r \tag{5.14}$$

where ϑ_{FC} is the moisture content at field capacity (m^3/m^3) and ϑ_{WP} is the moisture content at permanent wilting point, and D_r is the root zone depth (m), which is typically 0·5 m to 1·5 m. Typical values for ϑ_{FC} and ϑ_{WP} are given in Table 5.2.

However, plants begin to suffer water-stress at moisture contents above wilting point and it has been suggested that a fraction of readily available water (*RAW*) should be used in the water balance instead of S_{max} where:

$$RAW = pS_{max} \tag{5.15}$$

and p is often taken as 0·5.

An example of the monthly water balance calculations is given in Example 5.2 and is summarised in Appendix 5.3. The results are shown in Figures 5.6 and 5.7, demonstrating the effect of water availability on the actual evapotranspiration.

Other similar models which deal with the relationship between actual and potential evaporation include the Daily Soil Moisture Accounting

Table 5.2   Soil moisture characteristics (Allen et al.*, 1998)*

Soil type	ϑ_{FC}	ϑ_{WP}
Sand	0·07–0·17	0·02–0·07
Clay	0·32–0·40	0·20–0·24

Example 5.2

Using a water balance approach, estimate the storage, actual evapotranspiration and runoff for a site with the following characteristics:

Root zone depth = 0·5 m

m/c at field capacity = 0·5

m/c at wilting point = 0·07

Readily available moisture = 0·5

Initial soil moisture = 100 mm

Precipitation month 1 = 100 mm

Precipitation month 2 = 0 mm

Potential evaporation month 1 = 23·0mm

Potential evaporation month 2 = 44·4 mm

$$S_{max} = 1000(\vartheta_{FC} - \vartheta_{WP})D_r = 1000(0·5 - 0·07)0·5 = 215·0 \text{ mm}$$

$$RAW = 0·5 \times 215 = 107·5 \text{ mm}$$

Month 1

$$SMD_i = S_{max} - S_{i-1} = 107·5 - 100 = 7·5 \text{ mm}$$

$$S_i = \min\left[(P_i - PET_i + S_{i-1}), S_{max}\right](P_i \geq PET_i)$$

$$= \min\left[(100 - 23·0_i + 100), 107·5\right] = 107·5 \text{ mm}$$

$$AET_i = PET_i(P_i \geq PETi) = 23·0 \text{ mm}$$

$$Q_i = \max\left[(P_i - AET_i - \Delta S_i), 0\right] = \max\left[100 - 23·0 - (107·5 - 100), 0\right]$$

$$= 69·5 \text{ mm}$$

Month 2

$$SMD_i = S_{max} - S_{i-1} = 107·5 - 107·5 = 0 \text{ mm}$$

$$S_i = S_{i-1} \exp\left[\frac{-(PET_i - P_i)}{S_{max}}\right] \qquad (P_i < PET_i)$$

$$= 107 \cdot 5 \exp\left[\frac{-(44 \cdot 4 - 0)}{107 \cdot 5}\right] = 71 \cdot 1 \, mm$$

$$AET_i = P_i + S_{i-1} - S_i \qquad (P_i < PET_i)$$
$$= 0 + 107 \cdot 5 - 71 \cdot 1 = 36 \cdot 4 \, mm$$

$$Q_i = \max\left[(P_i - AET_i - \Delta S_i), 0\right] = \max\left[0 - 36 \cdot 4 - (107 \cdot 5 - 71 \cdot 1), 0\right] = 0$$

(DSMA) model (Holmes *et al.*, 2002), which is particularly suited to dry catchments. In this method, the actual evaporation for time-step $i + 1$ depends on whether or not the soil moisture deficit exceeds a rooting constant (*RC*), i.e.:

$$AE_{i+1} = \left[1 - \frac{(SMD_i - RC)}{A \times RC}\right] \qquad SMD_i > RC$$

$$= PE_{i+1} \qquad SMD_i \le RC \qquad (5.16)$$

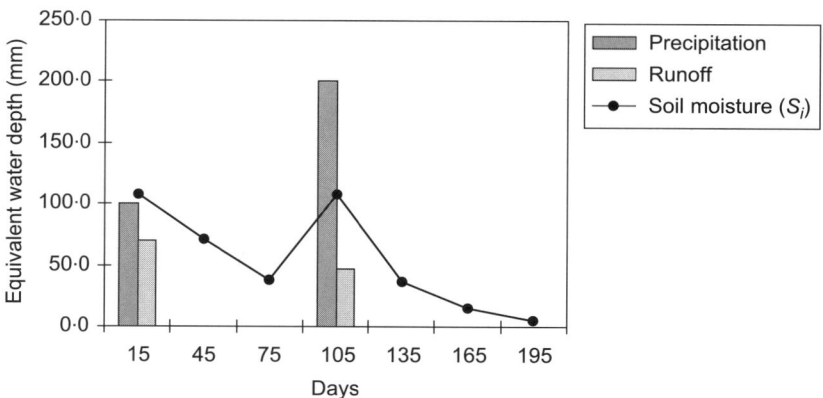

Figure 5.7   Soil moisture and runoff (see Appendix 5.3)

The shape parameter (A) effectively determines the maximum SMD, after which point evaporation ceases, rather than using a wilting point as in other models. The value of the rooting constant (RC) can be estimated by reference to the Hydrology of Soil Types (HOST) classification.

When the precipitation is sufficient to overcome the soil moisture deficit and satisfy the evaporative demand, runoff is generated and the SMD is reset to zero, otherwise no runoff is generated and the SMD is increased, i.e.:

$$\left.\begin{array}{l} Q_{i+1} = P_{i+1} - AE_{i+1} - SMD_i \\ SMD_{i+1} = 0 \end{array}\right\} P_{i+1} > AE_{i+1} + SMD_i$$

$$\left.\begin{array}{l} Q_{i+1} = 0 \\ SMD_{i+1} = P_{i+1} - AE_{i+1} - SMD_i \end{array}\right\} P_{i+1} \leq AE_{i+1} + SMD_i \qquad (5.17)$$

The DSMA model was developed to derive values of mean flow in catchments where no flow records exist.

In the United Kingdom, the Meteorological Office has developed a method for monitoring evaporation and soil moisture deficit over the whole country. The Meteorological Office Rainfall and Evaporation Calculation System (MORECS) collects and analyses weather data and estimates potential and actual evaporation, soil moisture deficit and excess rainfall based on a 40 km × 40 km grid on a seven-day period. Allowance can be made for different crop water requirements at different stages of growth.

5.5.2   *Water balance in urban catchments*

A water balance approach can also be useful in investigating the hydrological processes occurring in urban catchments. Urban areas are obviously characterised by a large proportion of paved surface, although it should be remembered that even in the centres of cities there can be extensive grassed and planted areas. The same processes of evaporation and infiltration etc. occur in urban catchments, but the relative significance of them is different. Transpiration by vegetation will obviously be less important and there will also be lower interception losses, although buildings can intercept a certain proportion of rainfall. Infiltration will also be less, but there can be significant losses of rainfall due to the initial wetting of paved surfaces, and also through cracks etc. in the surfaces (Davies, 1981). In contrast to unpaved areas, however, the

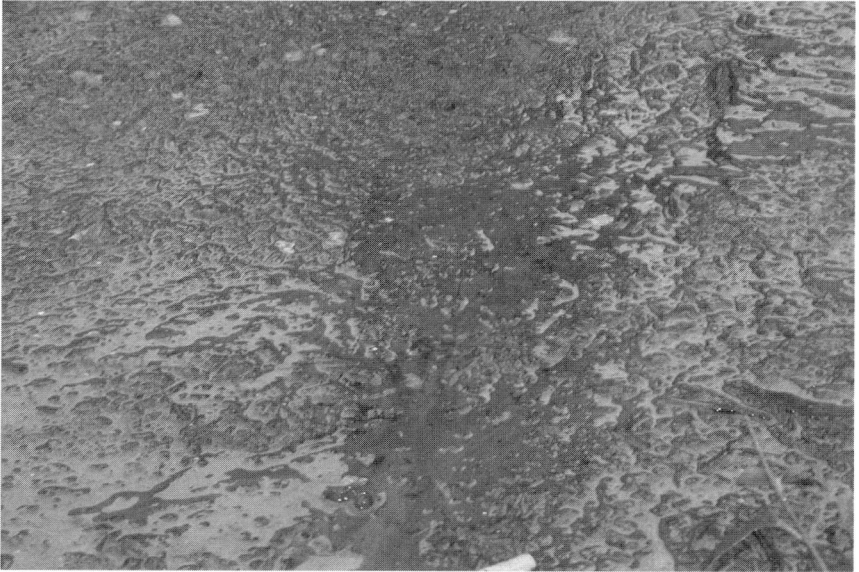

Figure 5.8   Depression storage and evaporation on a paved surface

infiltration rate of paved surfaces is much less than that of the underlying soil and therefore the infiltration rate is generally limited by the surface capacity and does not decrease significantly during a rainstorm. Likewise, losses due to evaporation from paved surfaces can be considerable. It is commonly assumed that the losses due to evaporation during a storm are negligible but there may be significant evaporation losses between storm events. The overall losses from paved surfaces depend largely on the amount of depression storage, since it is the water lying on the surface that is most likely to be lost through evaporation or infiltration. The amount of depression storage is largely a function of the surface type and the average slope of the catchment, decreasing markedly as the slope increases (Figure 5.8).

5.6   Approximate methods of estimating losses

In many cases it is not possible, or practical, to estimate the amount of evapotranspiration or infiltration losses on a continuing basis. If rainfall and runoff data are available concurrently, an estimate of the overall losses can be made by comparing observed volumes of rainfall and runoff. Although, as has been noted, losses tend to decrease with time, a simple approach to estimating losses is to assume a constant infiltration rate which is represented by the horizontal line on a typical rainfall hyetograph (Figure 5.3(b)). The Φ index line is such a line drawn so that the area of the hyetograph above the line (the effective rainfall) is equal to the observed runoff volume divided by the contributing area of the catchment. The Φ index does, in fact, vary with the distribution of rainfall and therefore it has limited value unless the average of many estimations is used (Example 5.3).

Another simple approach is to assume that the losses are proportional to the rainfall depth in each time interval, i.e. that the proportions of effective rainfall remain constant throughout a rainfall event (Figure 5.3(c)).

Example 5.3

The observed runoff volume from a storm event was estimated as 205 000 m³. The hourly rainfall totals were 6, 10, 12, 8 and 9 mm. Estimate the ϕ index for the catchment which has an area of 25 km².

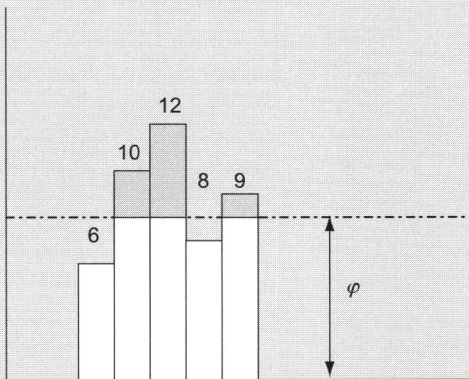

$$\text{Depth of effective runoff} = \frac{205\,000}{25 \times 10^6} = 0.0082\,\text{m} = 8.2\,\text{mm}$$

$$[6 - \phi] + [10 - \phi] + [12 - \phi] + [8 - \phi] + [9 - \phi] = 8.2$$

(terms are neglected if less than zero)

Try $\phi = 7$

  $0 + 3 + 5 + 1 + 2 = 11\,\text{mm}$

Try $\phi = 8$

  $0 + 2 + 4 + 0 + 1 = 7\,\text{mm}$

Try $\phi = 7.7$

  $0 + 2.3 + 4.3 + 0.3 + 1.3 = 8.2\,\text{mm}$

5.7 Estimation of losses using the *Flood Estimation Handbook*

The *Flood Estimation Handbook* (2000) proposes a model for estimating effective rainfall, which is based on a constant proportion of the total rainfall, although this proportion of runoff is estimated on the basis of both physical catchment characteristics and dynamic time-varying catchment wetness.

The method is based on a notional *standard percentage runoff* (*SPR*), which is the normal runoff from a catchment and is constant for all storms. The *SPR* is then modified by terms relating to the catchment wetness and rainfall intensity to produce a notional proportion of runoff for a rural catchment (PR_{RURAL}). Finally, the percentage runoff for a general catchment (*PR*) is estimated by introducing terms relating to the extent of urban development on the catchment.

5.7.1 Standard percentage of runoff

The *SPR* can be estimated either from observed rainfall and flow records using a method described later in the chapter, or from a characteristic of a catchment known as the *Baseflow Index (BFI)*. This index is the ratio of the baseflow component of a storm hydrograph to the total runoff and therefore reflects the geology, soil type and topography of the catchment. Areas which are underlain by permeable aquifers will have baseflow indices of 0·7 or more, while more 'flashy' catchments with clay soils, impermeable bedrock and a steeper topography, will have indices of 0·3 or less. The *BFI* for a catchment can be estimated from the analysis of a sufficiently long hydrograph (Chapter 6) or use can be made of published values of *BFI* for major watercourses in the United Kingdom (Gustard *et al.*, 1992). The value of *SPR* is then estimated from the *BFI* using the following regression equation:

$$SPR = 72 \cdot 0 - 66 \cdot 5 BFI \qquad (5.18)$$

Where it is not possible to determine the *BFI*, the *SPR* can be estimated from a soil-type classification, although this is not as reliable as estimation from hydrometric data. The soils of the United Kingdom have been classified into 29 classes under a system known as the Hydrology of Soil Types (HOST) classification (see Chapter 6). Values of the *SPR* for each of the HOST soil classes have been produced and the overall percentage runoff for the catchment is based on the proportion of the catchment covered by the various soil types, i.e.:

$$SPRHOST = \frac{1}{AREA} \sum_{N=1}^{29} a_n SPR_n \qquad (5.19)$$

where a_n is the area covered by HOST soil type *n*, and SPR_n is the standard percentage runoff for HOST soil type *n*. The values of SPR for different soil types are given in Table 5.3.

5.7.2 The effect of catchment wetness and rainfall intensity

Having determined a value for the *SPR* by one of the methods outlined above, the next stage is to include the effect of catchment wetness and rainfall. The parameter reflecting the catchment wetness (DPR_{CWI}) is based on the *catchment wetness index (CWI)*, which defines the state of the catchment at the start of the storm. The *CWI*, in turn, is dependent on the

Table 5.3   Standard percentage of runoff for HOST soil types (reproduced with the permission of the Centre for Ecology and Hydrology, Wallingford)

HOST class	SPR: %	HOST class	SPR: %	HOST class	SPR: %
1	2·0	11	2·0	21	47·2
2	2·0	12	60·0	22	60·0
3	14·5	13	2·0	23	60·0
4	2·0	14	25·3	24	39·7
5	14·5	15	48·4	25	49·6
6	33·8	16	29·2	26	58·7
7	44·3	17	29·2	27	60·0
8	44·3	18	47·2	28	60·0
9	25·3	19	60·0	29	60·0
10	25·3	20	60·0		

soil moisture deficit (*SMD*) and the *antecedent precipitation index* (*API*). The *SMD* represents the difference between the soil moisture content and its field capacity and, in wet conditions in winter, will be close to zero (see Chapter 6). *API5* describes the rainfall over the five days preceding the storm event and is defined as:

$$APHI5 = 0.5 \left[P_{d-1} + \sum_{n=2}^{5} 0.5^{n} P_{d-n} \right]$$ (5.20)

where P_{d-n} is the rainfall n days before the storm event.

The decay factor of 0·5 means that the effect of the more recent rainfall on wetness is greater than the effect of the rainfall occurring several days previously.

An additional parameter (DPR_{RAIN}) is included to reflect the dependence of runoff on rainfall intensity. It increases with the intensity of rainfall, but is only incorporated where the depth of rainfall is greater than 40 mm.

Adjusting the standard percentage of runoff for the effects of catchment wetness and rainfall intensity gives the notional percentage runoff for a rural area (PR_{RURAL}), which is estimated from:

$$PR_{RURAL} = SPR + DPR_{CWI} + DPR_{RAIN} \qquad (5.21)$$

where:

SPR = standard percentage runoff

$DPR_{CWI} = 0{\cdot}25(CWI - 125)$, where $CWI = 125 + API5 - SMD$

$DPR_{RAIN} = 0 \qquad\qquad\qquad (P <= 40\ \text{mm})$

$\qquad\quad = 0{\cdot}45(P - 40)^{0{\cdot}7} \qquad (P > 40\ \text{mm})$

5.7.3 Catchment wetness for extreme precipitation

In estimating the *probable maximum precipitation* (*PMP*) for reservoir safety analysis, the method of estimating losses is modified to represent the more extreme conditions of the catchment. For a winter storm, the catchment is assumed to be frozen and the percentage runoff associated with the soil type is given a value of 53%, unless the soil type indicates a value greater than this. The catchment is also allowed to 'wet up' by applying an additional period of 'antecedent rainfall' equal to twice the storm duration. The amount of this antecedent rainfall (*EMa*) is determined by extending the design storm profile of duration D hours to a duration of $5D$ hours (Figure 5.9). The rainfall occurring in the antecedent period (the first $2D$ hours) is found by subtracting the rainfall over the central period of D hours from the total and dividing by 2 as illustrated in Figure 5.9. Areal reduction factors (*ARF*) corresponding to durations of $5D$ and D are applied so that the antecedent rainfall is calculated from:

$$EMa = 0{\cdot}5[(ARF_{5D} \times (EM - 5D)) - (ARF_D \times (EM - D))] \qquad (5.22)$$

where $EM - 5D$ and $EM - D$ are the values of Estimated Maximum Precipitation (*EMP*) corresponding to durations of $5D$ hours and D hours respectively, interpolated from the logarithmic relationship, as described in Chapter 4.

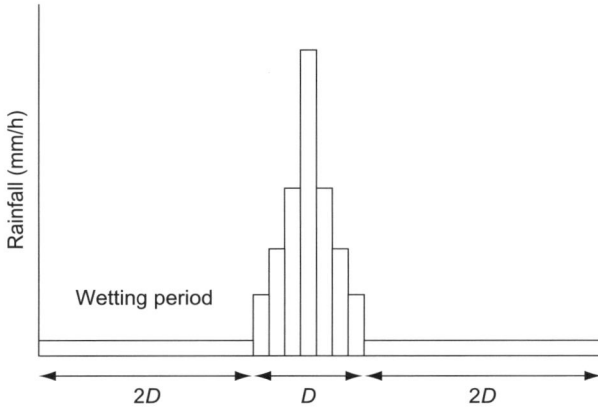

Figure 5.9   Antecedent conditions for estimating probable maximum precipitation (reproduced with the permission of the Centre for Ecology and Hydrology, Wallingford)

The *CWI* is then calculated by:

$$CWI = 125 + EMa(0.5^{D/24})\tag{5.23}$$

and the catchment wetness term of the dynamic runoff component is given by:

$$DPR_{CWI} = 0.25(CWI - 125)\tag{5.24}$$

The value of PR_{RURAL} is calculated as above using the rainfall term for normal storms, i.e.:

$$PR_{RURAL} = SPR + DPR_{CWI} + DPR_{RAIN}\tag{5.25}$$

where:

$$DPR_{RAIN} = 0.45(P - 40)^{0.7}$$

Where a winter storm-profile is used, snowmelt is often added to both the storm rainfall and the antecedent rainfall. A uniform melt rate of 1·75 mm/h is normally used but a check should be made on the 100-year total snow depth to calculate the duration of this melt rate.

5.7.4   The effect of urbanisation

The final step in estimating the percentage runoff for a general catchment is to include the effect of urbanisation. The extent of urbanisation is represented by the parameter $URBEXT$, which represents the proportion of developed land on a catchment. This has been estimated for the United Kingdom from digital remote-sensing data by distinguishing urban and suburban areas from agricultural and other land uses on a grid square basis. $URBEXT$ is an overall parameter representing the proportion of both urban and suburban development. Urban areas are areas where most of the pixels are covered by concrete or other impervious surface, whereas suburban areas have a mixture of paved areas and vegetation, such as parks as gardens. $URBEXT$ is defined as:

$$URBEXT = URB_{EXT} + 0.5\, SUBURB_{EXT}$$ (5.26)

where $URBEXT$ is the proportion of catchment covered by urban development and $SUBURB_{EXT}$ is the proportion of catchment covered by suburban development. Since most of the remote-sensing data used for this analysis was obtained around 1990, a method for adjusting the parameter for a notional growth in urban development is available.

The overall percentage runoff is then estimated from:

$$PR = PR_{RURAL} (1 - 0.615\, URBEXT) + 70(0.615\, URBEXT)$$ (5.27)

Example 5.4

Estimate the proportion of runoff for a catchment with the following proportion of soil types:

Host soil type	4	8	10
Proportion	0·30	0·42	0·28

SMS = 0, Proportion of Urban Area = 0·05, Proportion of Suburban Area = 0·08. The antecedent rainfall is 8 mm, 5 mm, 10 mm, 0 mm and 4 mm, respectively.

$$PR_{RURAL} = SPR + DPR_{CWI} + DPR_{RAIN}$$

From Table 5.5

$$SPR = 0.3 \times SPR_4 + 0.42 \times SPR_8 + 0.28 \times SPR_{10}$$

$$= (0.3 \times 2.0) + (0.42 \times 44.3) + (0.28 \times 25.3) = 26.29$$

$$CWI = 125 + API5 - SMD$$

$$API5 = 0.5 \left[P_{d-1} + \sum_{n=2}^{5} 0.5^n P_{d-n} \right]$$

$$API5 = 0.5 \left[8 + 0.5^2 \times 5 + 0.5^3 \times 10 + 0.5^4 \times 0 + 0.5^5 \times 4 \right] = 5.31 \, \text{mm}$$

$$CWI = 125 + 5.31 - 0 = 130.31$$

$$DPR_{CWI} = (0.25(CWI - 125)) = 0.25(130.3 - 125) = 1.33$$

$$DPR_{RAIN} = 0 \qquad (P <= 40 \, \text{mm})$$

$$PR_{RURAL} = 26.29 + 1.33 + 0 = 27.62\%$$

$$PR = PR_{RURAL}(1.0 - 0.615 \, URBEXT) + 70(0.615 \, URBEXT)$$

$$URBEXT = URB_{EXT} + 0.5 \, SUBURB_{EXT} = 0.05 + 0.5 \times 0.08 = 0.09$$

$$PR = 27.62 \times (1.0 - 0.615 \times 0.09) + 70(0.615 \times 0.09) = 29.97\%$$

5.7.5 Standard percentage of runoff from observed data

Where existing flow and rainfall data are available for a catchment, the *SPR* can be determined by using the above process in reverse. The procedure involves estimating the actual proportion of runoff (*PR*) for individual storms and then working back to estimate PR_{RURAL} and hence *SPR*. If possible, individual storm events are identified and the observed hydrographs separated into the baseflow and direct runoff components (see Chapter 6). The volume of effective rainfall is then equivalent to the total volume of direct runoff and the *PR* for that storm will be the ratio of this volume to the total rainfall volume. The value of PR_{RURAL} can then be

---

Example 5.5

The average proportion of direct runoff from the analysis of rainfall and flow records from a catchment is 40·0%. The values of DPR_{CWI} and DPR_{RAIN} are estimated to be 1·33 and 0 respectively, and the value of $URBEXT$ is estimated to be 0·13. Estimate the SPR.

$$PR_{RURAL} = SPR + DPR_{CWI} + DPR_{RAIN}$$

$$SPR = PR_{RURAL} - DPR_{CWI} - DPR_{RAIN}$$

$$PR = PR_{RURAL}(1\cdot0 - 0\cdot615\ URBEXT) + 70(0\cdot615\ URBEXT)$$

$$40\cdot0 = PR_{RURAL}(1\cdot0 - 0\cdot615 \times 0\cdot13) + 70(0\cdot615 \times 0\cdot13) = 0\cdot92 \times PR_{RURAL} + 5\cdot597$$

$$PR_{RURAL} = 37\cdot40\%$$

$$SPR = 37\cdot40 - 1\cdot33 - 0$$

$$SPR = 36\cdot07\%$$

---

estimated by rearranging Equation (5.27) using estimates of *URBEXT*. *SPR* is then calculated by rearranging Equation (5.25) with DPR_{CWI} and DPR_{RAIN} estimated from corresponding rainfall and soil-moisture data (Example 5.5).

5.7.6 Summary of the **Flood Estimation Handbook** *method of estimating effective rainfall*

The main stages in the estimation of effective rainfall using the *Flood Estimation Handbook* are summarised in Figure 5.10. The gross rainfall depth is estimated for a given duration and return period and distributed using a standard profile following the method described in Chapter 4 to give the gross rainfall hyetograph. The effective rainfall is determined from the gross rainfall by multiplying by a percentage of runoff. The percentage of runoff is a function of the antecedent catchment conditions, the rainfall depth, the extent of urban development and the soil type/geology. The latter is reflected in the BFI, which can be determined

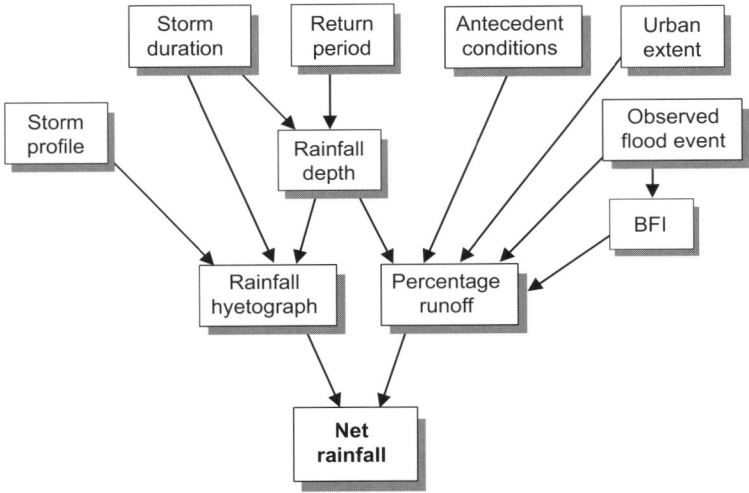

Figure 5.10   Summary of Flood Estimation Handbook *method for estimating effective rainfall*

from observed flow records or from published maps. The percentage of runoff may also be estimated directly from rainfall and flow data allowing for the antecedent conditions and the other factors mentioned above.

5.8   Summary

This chapter describes the processes through which rainfall is 'lost' in the sense that it does not become direct runoff. These processes include interception, depression storage, evaporation, transpiration and infiltration. These losses are complex functions of the physical characteristics of the catchment and the time varying moisture content of the soil. Mathematical models can be used to represent these processes but in most cases there is insufficient data available to justify their use. A more practical approach is that recommended by the *Flood Estimation Handbook.* This assumes a constant proportion of losses where this proportion is based on the physical nature of the catchment, the antecedent moisture conditions and the rainfall intensity.

5.9 References

Allen, R. G., Pereira, L. S., Raes, D. and Smith, M. (1998). Crop Evapotranspiration — Guidelines for Computing Crop Water Requirements. *FAO Irrigation and Drainage Paper 56*, Food and Agricultural Organisation, 300pp.

Davies, H. *Water Balance of Impermeable Surfaces*. PhD Thesis. Department of Geography.

Doorenbos, J. and Pruit, W. O. (1977). *Crop Water Requirements*. FAO.

Gustard, A., Bullock, A. *et al*. (1992). *Low Flow Estimation in the United Kingdom*. Institute of Hydrology, Wallingford.

Hargreaves, G. H. and Samani, Z. A. (1985). Reference Crop Evapotranspiration from Temperature. *Applied Engineering in Agriculture*, **1**, No. 2, 96–99.

Hellwig, D. H. R. (1973). Evaporation of Water from Sand. *Journal of Hydrology*, **18**, 317–327.

Holmes, M. G. R., Young, A. R., *et al*. (2002). A New Approach to estimating Mean Flow in the UK. *Hydrology and Earth System Sciences*, **64**, No. 4, 709–720.

Horton, R. E. (1940). An Approach towards a Physical Interpretation of Infiltration Capacity. *Proceedings of the Soil Science Association of America*, **5**, 399–417.

Hydraulics Research (1983). *Design and Analysis of Urban Storm Drainage*. Hydraulics Research, Wallingford.

Jensen, D. T., Hargreaves, G. H. *et al*. (1997). Computation of ETo under Non-ideal Conditions. *American Society of Civil Engineers Journal of Irrigation and Drainage Engineering*, **123**, No. 5, 394–400.

Thornthwaite, C. W. and Mather, J. R. (1957). Instructions and Tables for Computing Potential Evapotranspiration and Water Balance. *Publications in Climatology*, **10**, No. 3, 311pp.

Appendix 5.1
Calculation of net solar radiation from geographical data

The extraterrestrial radiation (Ra) can be calculated from:

$$Ra = \left(\frac{24 \times 60}{\pi}\right) G_{sc} d_r \left[\omega_s \sin(\phi)\sin(\delta) + \cos(\phi)\cos(\delta)\sin(\omega_s)\right]$$

where G_{sc} is the solar constant = 0.0820 MJ/m^2 min, d_r is the inverse relative distance from the Earth to the sun = $1 + 0.033\cos(2\pi J/365)$ (where J is the number of the day in the year ($1 = 1$ January)), ω_s is the sunset hour angle = $\cos^{-1}[-\tan(\phi)\tan(\delta)]$, ϕ is the latitude (rads), and δ is the solar declination (rads) = $0.049\sin(2\pi J/365 - 1.39)$.

The short-wave solar radiation (Rs) incident on the Earth's surface can be calculated from:

$$Rs = \left(a_s + \frac{b_s n}{N}\right) Ra$$

where a_s is the regression constant (= 0.25 unless empirical data are available), b_s is the regression constant (= 0.50 unless empirical data are available), N is the daylight hours = $24\,\omega_s/\pi$, and n is the actual sunshine hours.

The clear sky solar radiation (Rso) incident on the Earth's surface can be calculated from:

$$Rso = \left(a_s + b_s\right) Ra$$

The net short-wave radiation (Rns) is a balance between incoming and reflected solar radiation, and is given by:

$$Rns = (1 - \alpha)Rs$$

where α is an albedo coefficient which is 0.23 for a grass reference crop.

The net long-wave radiation (Rnl) is calculated from:

$$Rnl = \sigma\left[\frac{T_{max,K}^4 - T_{min,K}^4}{2}\right]\left(0.34 - 0.14\sqrt{e_a}\right)\left(1.35\frac{Rs}{Rso} - 0.35\right)$$

where σ is the Stefan-Boltzmann Constant ($4 \cdot 903 \ 10^{-9}$ MJ $K^{-4} m^{-2}$ day^{-1}), $T_{max,K}$ is the maximum absolute temperature during a 24 h period (°K), $T_{min,K}$ is the minimum absolute temperature during a 24 h period (°K), and e_a is the actual vapour pressure (kPa).

The net radiation (Rn) is the difference between the incoming net short-wave radiation and the outgoing net long-wave radiation, i.e.:

$$Rn = Rns - Rnl$$

Appendix 5.2
Example of evapotranspiration using the FAO Penman-Monteith Equation

Altitude = 100
Latitude = 50.8
Time interval = 30 days

Bold type indicates input data and Roman type indicates calculated values.

		15	45	75	105	135	165	195
Day of the year	Day	**15**	**45**	**75**	**105**	**135**	**165**	**195**
Maximum temperature	T_{max}	**21·5**	**22·7**	**20·1**	**19·4**	**23·1**	**21·6**	**22**
Minimum temperature	T_{min}	**12·3**	**11·9**	**12·5**	**10·5**	**9·1**	**10·3**	**12·8**
Mean temperature	T_{mean}	16·9	17·3	16·3	14·95	16·1	15·95	17·4
Dewpoint temperature	T_{dew}	**12·1**	**12·1**	**12·1**	**12·1**	**12·1**	**12·1**	**12·1**
Wind speed at 2 m	u_2	**2·08**	**1·80**	**2·50**	**3·40**	**2·40**	**1·30**	**2·10**
Slope of vapour pressure curve	δ	0·12	0·12	0·12	0·11	0·12	0·12	0·13
Atmospheric pressure	P	100·12	100·12	100·12	100·12	100·12	100·12	100·12
Psychometric constant	γ	0·07	0·07	0·07	0·07	0·07	0·07	0·07
Sat. vapour pressure at mean temperature	e_0	1·93	1·97	1·85	1·70	1·83	1·81	1·99
Mean saturated vapour pressure	e_s	2·00	2·08	1·90	1·76	1·99	1·92	2·06
Actual vapour pressure	e_a	1·41	1·41	1·41	1·41	1·41	1·41	1·41
Inverse Earth–sun distance	d_r	1·03	1·02	1·01	0·99	0·98	0·97	0·97
Solar declination	Decl	−0·37	−0·24	−0·04	0·17	0·33	0·41	0·38
Sunset hour angle	s_{ha}	1·07	1·27	1·52	1·78	2·00	2·13	2·08
Extraterrestrial radiation	Ra	8·41	13·72	22·12	31·23	38·37	41·65	40·26
Daylight hours	N	8·21	9·71	11·62	13·58	15·30	16·24	15·88
Actual sunshine duration	n	**9·25**	**9·1**	**8·6**	**9·5**	**9·8**	**6·4**	**8·5**
Solar radiation	Rs	6·84	9·86	13·71	18·73	21·88	18·62	20·84
Clear sky solar radiation	Rso	6·32	10·32	16·63	23·49	28·85	31·32	30·28

Net solar radiation	Rns	5·27	7·59	10·56	14·43	16·85	14·34	16·05
Net long-wave radiation	Rnl	6·68	5·69	4·56	4·26	4·02	2·69	3·51
Net radiation	Rn	−1·42	1·90	6·00	10·17	12·83	11·64	12·54
Soil heat flux	G	0·00	0·00	0·00	0·00	0·00	0·00	0·00
Grass reference evapotranspiration	Eto	0·77	1·48	2·25	2·77	3·78	3·24	3·85

The results are shown in Figure 5.6.

Appendix 5.3
Example of soil moisture calculation using water balance

Root zone depth	$= 0{\cdot}5$
m/c at field capacity	$= 0{\cdot}5$
m/c at wilting point	$= 0{\cdot}07$
Readily available moisture	$= 0{\cdot}5$
S_{max}	$= 107{\cdot}5$
Time interval	$= 30$ days

Day	15	45	75	105	135	165	195
Soil moisture (S_{i-1})	**100·0**	107·5	71·1	38·0	107·5	37·4	15·1
SMD	7·5	0·0	36·4	69·5	0·0	70·1	92·4
Potential evaporation	23·0	44·4	67·5	83·1	113·5	97·2	115·6
Precipitation	**100·0**	**0·0**	**0·0**	**200·0**	**0·0**	**0·0**	**0·0**
Soil moisture (S_i)	107·5	71·1	38·0	107·5	37·4	15·1	5·2
Actual evaporation	23·0	36·4	33·2	83·1	70·1	22·3	10·0
Runoff	69·5	0·0	0·0	47·3	0·0	0·0	0·0

The results are shown in Figure 5.7.

Chapter 6

Natural flow processes

This chapter is concerned with the processes that transform effective precipitation into the outflow from a catchment. It considers the flow through the ground and in open channels. Whereas the previous chapter was concerned mainly with the processes in which rainfall is 'lost' or diverted from the main path of the hydrological cycle, this chapter is principally concerned with the attenuation or storage effect, which occurs in or on the ground or in channels.

6.1   Subsurface flow

6.1.1   Properties of soils and rocks

Soil is essentially the result of the weathering and erosion of the surface layers of rock, together with some organic material derived from plants and other organisms. The surface layer of soil, usually termed topsoil, consists largely of soft organic or humus material. Below this is usually well-weathered parent material modified by root systems, etc., although in some cases the material may have been transported some distance by wind, river or glacial action.

Soil basically consists of three elements — the soil particles, the void space between the particles and the water, which may occupy all or part of the void space. Soils are generally classified according to the mean particle size and the usually accepted descriptions are given in Table 6.1.

The main physical parameters used to describe the soil properties are:

Table 6.1   Soil classification

Soil type	Particle size: mm
Clay	<0·002
Silt	0·002–0·05
Sand	0·05–2·0
Gravel	>2·0

- *bulk density* = mass of solids/total volume (typically 1·1–1·6 t/m^3)
- *solid density* = mass of solids/volume of solids (typically 2·6–2·7 t/m^3)
- *voids ratio* = volume of voids/volume of solids (typically 0·25–2·0).

The amount of water contained in a soil may be expressed in terms of volume or mass, i.e.:

- *water content by mass* = mass of water/mass of soil
- *water content by volume* = volume of water/volume of soil.

The water content by volume can also be regarded as the equivalent depth of free water per metre depth of soil, and can therefore be related to precipitation and evaporation depths.

The amount of water stored in a rock or soil is governed by its *porosity*, which is defined as the ratio of the volume of voids to the total volume, and is typically between 20 and 60%. However, the amount of water that can be extracted from a rock or soil (the *specific yield*) is generally less than the porosity, as some water is attached to the rock grains and cannot be removed by gravity. In fine-grained clay soils there is a lot of electrostatic attraction between the water and the soil particles and, although the volume of voids is relatively high, the specific yield tends to be very low (Figure 6.1). In free-draining material, very little water is retained and the specific yield approaches the porosity.

The flow of water in soils and rocks depends mainly on the *permeability* of the material, which is a measure of not only the volume of voids but also the degree of connection between them. In some highly porous rocks, the permeability may be quite low, if there is little connection between the voids. In the case of fissured rocks, the flow rate depends more on the distribution and size of the fissures in the rock and it may be many times

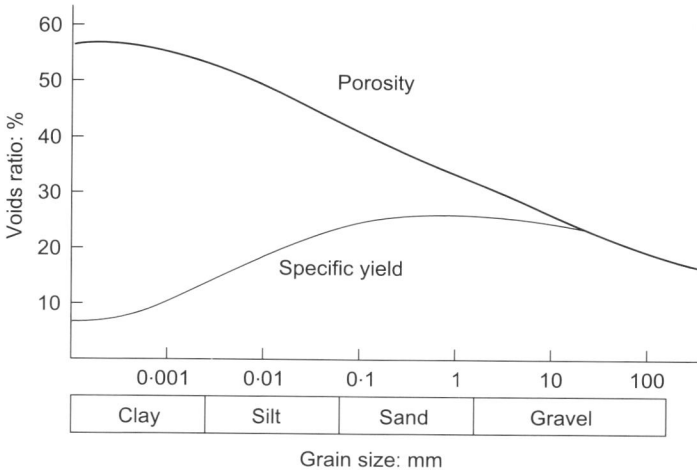

Figure 6.1   Variation of porosity and specific yield with grain size

the flow rate based on the rock permeability. Typical values of permeability for different soil and rock types are given in Table 6.2.

The hydrological characteristics of soils also depend on other factors apart from permeability. In many isotropic soils, water drains vertically until it reaches the water table, after which there is a slow horizontal movement towards springs and watercourses. The response of such soils to rainfall will generally be very slow. Where the soil structure is more layered, the water flow will be more shallow and horizontal, and the response to rainfall will be more rapid. Soils have been classified according to their hydrological properties under the Hydrology of Soil Types (HOST) classification (Boorman *et al.*, 1995), which uses the following key indicators of the hydrological nature of a soil/substrate:

- the permeability of the substrate,
- whether the water table is less than or greater than 2 m below the surface, or whether there is no significant groundwater
- whether there is a an impermeable layer within 1 m of the surface (or a 'gleyed' layer within 0·4 m of the surface)
- whether there is a peat layer.

The various combinations of the above indicators, which are commonly found in the United Kingdom, results in an overall classification consisting of 29 classes (Appendix 6.1).

Table 6.2   Hydraulic properties of typical soils and rocks

Soil/rock type	Porosity: %	Specific yield: %	Overall permeability: m/day
Clay	45	3	0·0004
Sand	35	25	40
Gravel	25	22	4000
Sandy gravel	20	16	400
Sandstone	15	8	4
Limestone, shale	5	2	0·04
Quartzite, granite	1	0·5	0·0004

6.1.2   Subsurface water

Water exists in soil in various forms. Below the water table is the *saturated zone*, where the voids are completely filled with water. However, the amount of water stored in the saturated zone progressively decreases with depth as the overburden pressures reduces the pore space and most groundwater is circulating in the top 100 to 200 m of the saturated zone. In some cases, the saturated zone reaches the surface, resulting in springs that supply watercourses: conversely, the saturated zone may be fed by the watercourses.

Above the water table is the *vadose zone*, where the voids generally contain a mixture of air and water. Within the vadose zone there are three distinct moisture regions (Figure 6.2). Immediately above the water table is a *capillary zone*, where water is permanently held by capillary action in small voids and cracks, and the pressure is less than atmospheric. In clay soils the suction pressures can be significant because of the small capillary size, and this zone can extend up to a metre or more above the saturated zone. In sandy soils the capillary forces are less but the amount of water stored may be greater because of the larger pore-size. Near the ground surface is the *root zone* where the soil water fluctuates according to the demands of vegetation and the effects of evaporation and precipitation.

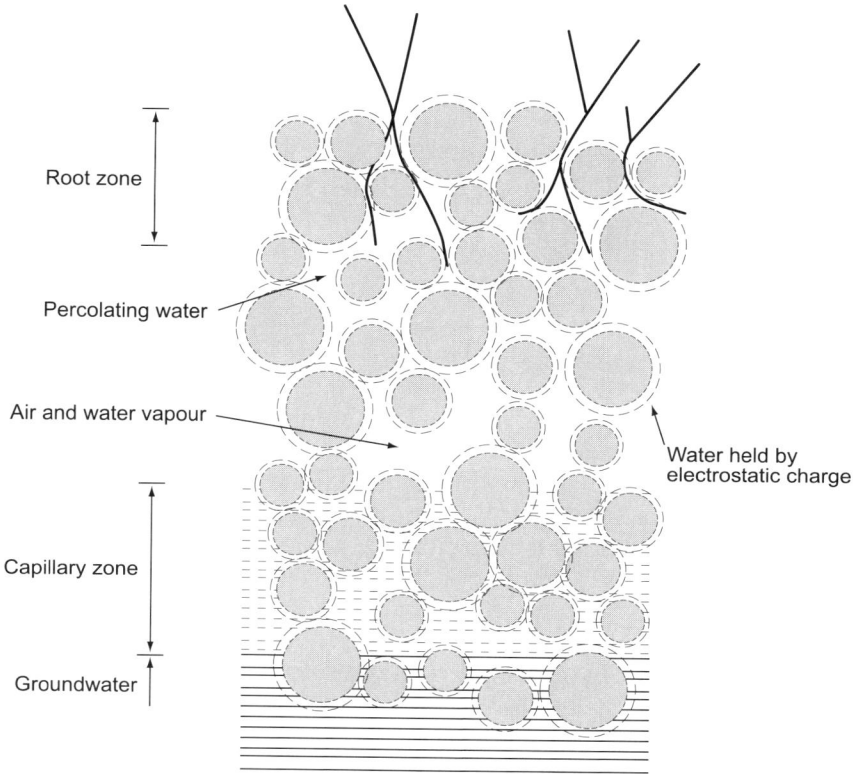

Figure 6.2   Zones of subsurface water

Between the capillary and the root zones is an *intermediate zone*, where water is held on the surface of the soil particles by electrostatic forces and is not easily removed by plants. Water is also held in this region in the form of water vapour within the air in the voids, and free water may also occur in the voids as water passes down from the surface. Where the water table is high, the intermediate zone may be missing.

As has been stated, soil contains water that may be accessible to plants even after the voids in the soil have been drained. The maximum water content (by volume) in the soil after the gravity water in the voids has drained is known as the *field capacity* of the soil. At any water content below this level, the difference between the actual soil moisture content and the field capacity is known as the *soil moisture deficit* (*SMD*). As the soil moisture is progressively reduced, plants find it increasingly more

Table 6.3   Soil water parameters (from Marshall and Holmes, 1979)

Soil type	Clay content: %	Saturation	Field capacity	Wilting point
Sand	3	0·40	0·06	0·02
Loam	22	0·50	0·29	0·05
Clay	47	0·60	0·41	0·20

difficult to extract moisture, and the minimum water content of the soil, below which plants will wilt and not recover, is termed the *wilting point*. Typical values of the field capacity and wilting points for the major soil types are given in Table 6.3.

The distribution of moisture with depth in a soil varies over a year according to the precipitation inputs and the evapotranspiration losses. The typical seasonal variation of non-saturated soil moisture content over a year is shown in Figure 6.3, together with notional precipitation and evaporation figures. In winter, the soil water content is likely to be at field capacity over the whole unsaturated zone. In the spring, with reduced rainfall and increased plant growth, an *SMD* will appear in the upper soil layers, extending to the full depth of the unsaturated zone in the summer. In autumn, the increased rainfall will remove the *SMD* in the upper layer, leaving a residual deficit in the lower layers.

6.1.3   Flow through soil

Flow in the saturated zone

When water reaches the saturated zone the water table begins to rise, resulting in a gradient in the free surface (phreatic surface), which causes the water to move horizontally. The velocity (v) at which the water moves is governed by the slope of the phreatic surface and the permeability of the rock, and the relationship generally used is that proposed by Darcy:

$$v = ks \tag{6.1}$$

where k is the intrinsic permeability of the soil or rock (m/day) and s is the gradient of the phreatic surface (the hydraulic gradient). The flow rate is

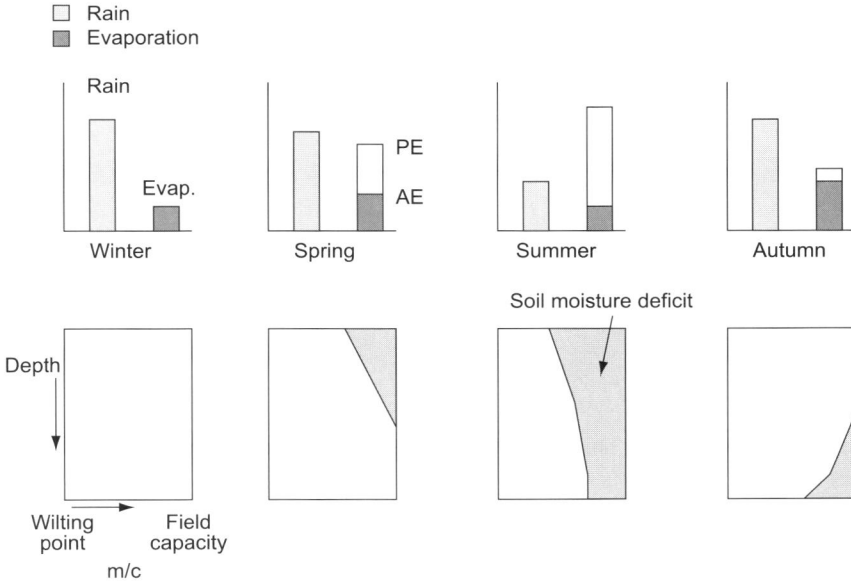

Figure 6.3   Typical seasonal variation of soil moisture

the product of the velocity and the effective area (A) available for flow, which is governed by the porosity (ρ) of the material, i.e.:

$$Q = k\rho As = KAs \tag{6.2}$$

where K is the overall coefficient of permeability.

For rocks with large fissures, the flow tends to be more turbulent and the linear relationship above may not apply. Where the upper water-surface in the aquifer is confined by being overlain by an impermeable material, the head used to calculate the hydraulic gradient is the piezometric head, which is the height to which the water would rise if were not confined (Figure 6.4(b)).

The solution of Equation (6.2) to determine a flow rate for a given soil permeability and given boundary conditions can be carried out either graphically or numerically. In the graphical approach, a flow net is constructed consisting of flow lines and equipotential lines (Figure 6.4). The equipotential lines connect points with the same potential (head), and flow lines are drawn perpendicular to these, i.e. parallel to the flow. A square grid is normally used, such that the increment of flow and head

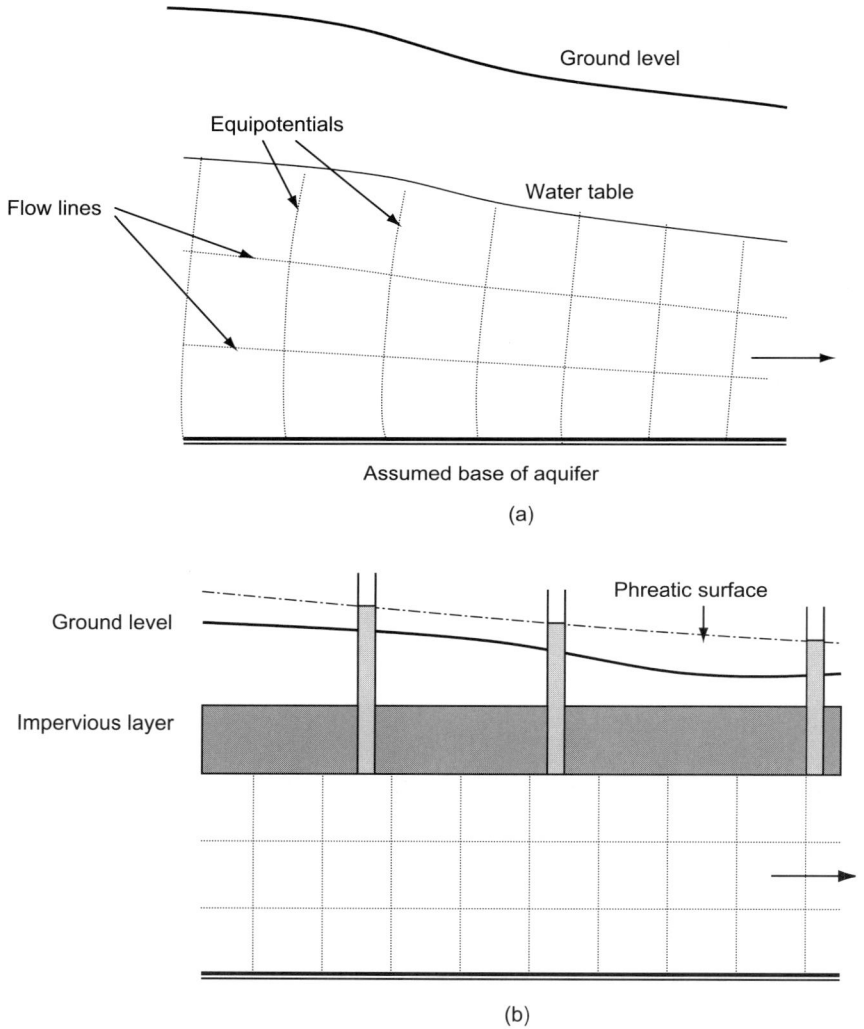

Figure 6.4   Flow nets in aquifers: (a) unconfined aquifer; and (b) confined aquifer

are kept the same. The intersection between flow lines and equipotential lines must always be perpendicular, but where there are sharp corners to the flow boundaries, the curvature of the lines may cause difficulties.

The problem can also be solved numerically by using the continuity principle. Assuming a steady state condition, a second order differential equation can be developed, i.e.:

$$q = \frac{Q}{y} = -K_z \frac{dh}{dz}$$

$$\frac{dq}{dx} = -K_z \frac{d^2 h}{dx^2} \tag{6.3}$$

where q is the flow per unit width and h is the height to the phreatic surface or the depth of aquifer. The negative sign follows from the decrease in head in the direction of flow.

For unsteady flow conditions, the principle of continuity applied to an element of sides x, y, z (Figure 6.5(a)) gives:

Inflow = Outflow + Change in storage

i.e., for flow in the x direction:

$$Q_x = \left(Q_x + \frac{Q}{x} x \right) + S_s xyz \frac{h}{t}$$

$$\therefore \frac{Q}{x} x + S_s xyz \frac{h}{t} = 0 \tag{6.4}$$

where S_s is the specific storage and $\partial h / \partial t$ is the rate of change of head with time. Applying the Darcy Equation (Equation (6.3)) gives:

$$xyz \left(-K \frac{{}^2 h}{x^2} \right) + S_s xyz \frac{h}{t} = 0$$

or:

$$K \frac{{}^2 h}{x^2} = S_s \frac{h}{t} \tag{6.5}$$

The analysis can easily be extended to two (or three) dimensions:

$$K_x \frac{^2 h}{x^2} + K_y \frac{^2 h}{y^2} = Ss \frac{h}{t} \tag{6.6}$$

A common numerical approach to the solutions of such equations is the finite difference method (see Section 7.7.2), where the differentials are replaced by the difference using values of head (h) at finite distances Δx (or Δt) apart (Figure 6.5(b)). The finite difference form of Equation (6.5) becomes:

$$K_x \frac{h_{i-1} - 2h_i + h_{i+1}}{(\Delta x)^2} = S_s \frac{h_i - h_{i-1}}{\Delta t} \tag{6.7}$$

Equations can be set up centred on each node in turn, which are solved either iteratively or as a series of simultaneous equations, provided that the boundary conditions are known. This approximation assumes that the

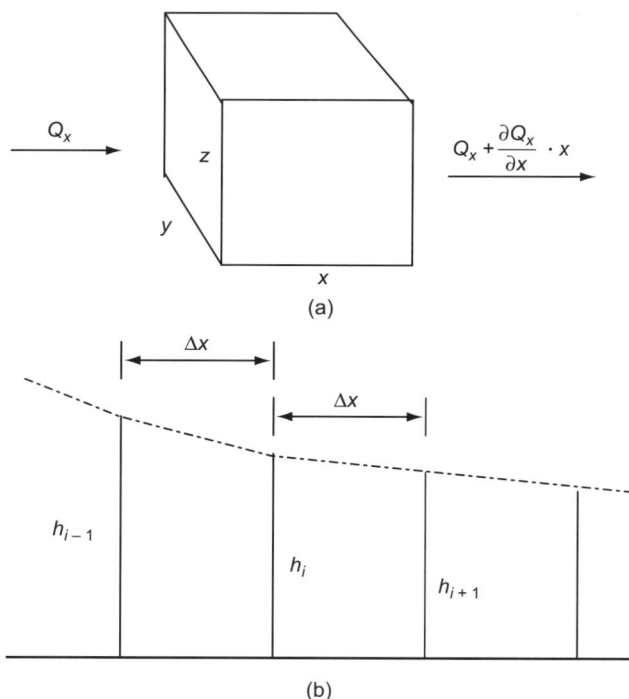

Figure 6.5   (a) Element of flow; and (b) finite difference representation of aquifer flow

assumes that the flow is horizontal and therefore care should be taken where there is significant non-horizontal flow, for example near a river.

Flow in the unsaturated zone

Whereas the flow of water in the saturated zone is controlled by the permeability of the soil (which is relatively stable) and the hydraulic gradient, water in the unsaturated zone is subject to sub-atmospheric pressure or suction. This results from the attraction of water to the soil particles and capillary pores, which leads to the formation of a thin film surrounding each particle. The thinner the film, the greater the suction pressure and water will tend to flow from areas of low-suction pressure to areas of high suction. Where water advances into a previously dry soil, there is a *wetting front* where the suction gradient can amount to several bars of pressure per centimetre of soil, which is many times greater than the normal hydraulic gradient.

Another difference between saturated and unsaturated flow is in the hydraulic conductivity. When the soil is saturated, all the pores are water-filled and are able to transmit a flow. As the soil desaturates, some of the pores become air-filled and the conductive portion of the soil's cross-sectional area, and hence the effective conductivity, decreases correspondingly. In granular soils, water sometimes remains only in capillary wedges at the contact points between the particles, thus forming discrete and discontinuous pockets of water (Figure 6.2). Therefore, the transition from saturated to unsaturated flow generally entails a sharp reduction in hydraulic conductivity, sometimes by several orders of magnitude.

The flow per unit area under a suction head gradient $\partial\varphi/\partial z$ is:

$$q = -K(\psi)\frac{\partial\psi}{\partial z} \tag{6.8}$$

where the permeability K is a function of the suction head (φ). This can be expressed in terms of the soil moisture content (θ):

$$-\frac{\partial\psi}{\partial z} = -\frac{\partial\psi}{\partial\theta} \times \frac{\partial\theta}{\partial z} = -\frac{1}{c} \times \frac{\partial\theta}{\partial z} \tag{6.9}$$

Where c is the ratio of the moisture content to the suction head. Hence:

$$q = -\frac{K}{c}\frac{\partial\theta}{\partial z} = -D\frac{\partial\theta}{\partial z} \tag{6.10}$$

where D is the effective diffusivity, which, like K, is a function of the soil moisture content and is introduced to enable the equation to be expressed in a similar form to the standard diffusion equation. The rate of change of moisture content can be determined using the continuity relationship, resulting in a second order equation analogous to the diffusion equation:

$$\frac{\partial \theta}{\partial t} = \frac{\partial q}{\partial z} = \frac{\partial}{\partial z}\left(D\frac{\partial \theta}{\partial z}\right) \qquad (6.11)$$

The vertical flow per unit area in an unsaturated soil under gravity only is (Equation 6.3):

$$q = -K\frac{dh}{dz} = -K \qquad (6.12)$$

since the hydraulic gradient in the vertical direction is unity (i.e. for every centimetre of vertical height the head changes by 1 cm). The rate of change of moisture content for both gravity and suction flow is therefore:

$$\frac{\partial \theta}{\partial t} = \frac{\partial}{\partial z}\left(D\frac{\partial \theta}{\partial z}\right) - \frac{\partial K}{\partial z} \qquad (6.13)$$

This equation, known as the *Richards Equation*, can, in principle, be solved using finite difference or other techniques. However, in practice, the relationship of diffusivity with depth and moisture content is difficult to establish and may vary according to whether the soil is wetting or drying, i.e. there is often a hysteresis effect. The relative importance of the suction and gravity components of the above equation depends on the initial boundary conditions. When the soil is dry, the suction term predominates and flow may be horizontal or vertical. As the soil becomes wetter, the suction term decreases and the gravity term becomes the main driving force.

The problem can be simplified by assuming an exponential (or similar) decrease in hydraulic conductivity with depth. This leads to the development of a notional 'perched water table', which is the upper limit of a saturated region within the unsaturated zone and it is the hydraulic conductivity at this depth which is assumed to control the vertical movement of water in the unsaturated zone. This is the basis of the *TOPMODEL*, which is described in Chapter 7.

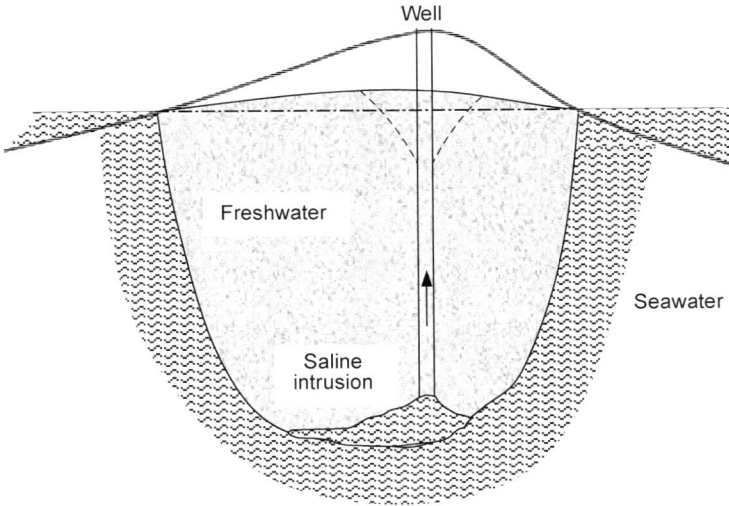

Figure 6.6   Saline intrusion under a small island

Saline intrusion

In coastal areas formed from permeable rock, groundwater can be present as either freshwater or seawater. Because of the different densities of seawater and freshwater, a lens of freshwater often exists in the aquifer on top of seawater. The thickness of the lens increases from the coast inland, and to maintain hydraulic equilibrium there is approximately 40 m of freshwater below mean sea level for every metre above sea level. Where a cone of depression is created by pumping from a well, an inverted cone in the boundary between saltwater and freshwater is also created, which can lead to saline intrusion into the well. Once this has happened, it is very difficult to reestablish the equilibrium. This problem is particularly acute for small islands with limited freshwater inflows (Figure 6.6), but it also occurs in parts of the United Kingdom. The risk of saline intrusion can be minimised by using a shallow well with horizontal galleries or by controlling the rate of abstraction from coastal wells during periods of low rainfall.

The effect of groundwater levels on river flows

Most rivers derive their flows from both surface runoff and groundwater. Surface runoff from impermeable ground occurs mainly in winter and tends to be intermittent and of short duration. For rivers draining permeable catchments and for most rivers in summer, flows are almost entirely derived from groundwater. Typically, such rivers have their sources in areas of wetland that are fed by water rising from an aquifer. In some cases, groundwater abstraction has reversed this process and river flow ceases completely in the summer months causing severe ecological problems. The situation can be mitigated by abstracting surface water rather than groundwater in summer months and by augmenting river flow from boreholes some distance away.

River bed aquifers

Ephemeral (non-permanent) rivers are major features of many arid or semi-arid areas. They typically exist in well-defined steep-sided channels, with flat floors, in-filled with alluvial sands. These rivers are recharged by sporadic storm events during a relatively short wet season, which results in a temporary surface flow in the channel. However, within the sand bed there is usually a more continuous, though very modest flow, which can be extracted relatively easily to provide a source of clean water for a significant part of the dry season. The amount of water in the so-called *sand rivers* can be enhanced by constructing a 'groundwater dam'. This can be either a barrier constructed below ground level, which obstructs the flow in the natural aquifer, or it can be a more conventional dam which extends above ground and traps the sediment, thereby enhancing the possible volume of storage (Nilsson, 1988).

Although the volume of water which can be stored within an aquifer is less than that in a conventional reservoir, there are many advantages of using groundwater dams. Evaporation losses are reduced, as are the losses caused by siltation and vegetation growth. The water extracted also requires very little treatment, and the risk of diseases caused by mosquitoes and snails is much less.

Flow measurement in aquifers

The velocity of flow in an aquifer can be measured by adding a salt solution to a borehole and measuring the decrease in salt concentration within the borehole over a period of time (Herbert *et al.*, 1997).

Assuming that, as the salt solution is added, the concentration in the borehole is instantaneously raised to C_I and that the water enters the borehole from an aquifer thickness equal to the screened length (L_{SCR}) of the borehole (Figure 6.7) at a uniform concentration of C_B, it can be shown (Appendix 6.2) that the mean velocity is given by:

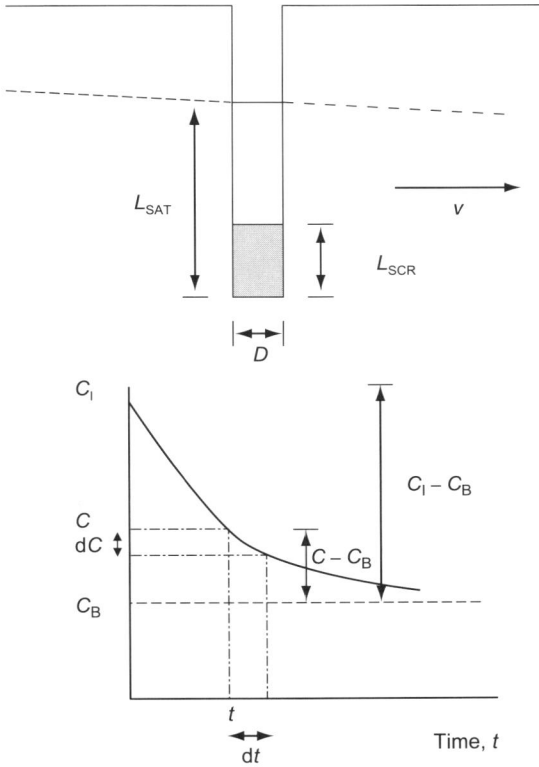

Figure 6.7   Measurement of flow in aquifers by salt dilution

$$v = -\frac{\pi D L_{SAT}}{4n L_{SCR} \alpha t} \ln\left(\frac{C - C_B}{C_I - C_B}\right)$$ (6.14)

where D and L_{SAT} are the diameter and saturated length of the borehole, respectively, and C is the salt concentration measured at time t. The parameters n and α are the porosity of the aquifer and the ratio of the width of flow intercepted by the borehole to the borehole diameter.

Therefore, plotting $\ln\{(C - C_B)/(C_I - C_B)\}$ against t should give a straight line graph with a slope:

$$m = -\frac{4n L_{SCR} \alpha v}{\pi D L_{SAT}}$$

i.e.:

$$v = \frac{\pi D L_{\text{SAT}}}{4 n L_{\text{SCR}} \alpha m}$$ (6.15)

6.2   Surface flow

Surface runoff occurs either when precipitation exceeds the infiltration capacity of the soil or when the water table reaches the surface. Runoff due to infiltration excess may happen immediately after a short period of very intense rainfall or it may be the result of a prolonged period of rainfall saturating the ground. Initially, the water may flow in the form of thin threads around individual obstructions, but as the runoff increases, the threads will merge to form sheet flow (Figure 6.8). On impervious surfaces, after the initial surface absorption, droplets begin to form and coalesce into small pools which, on flat slopes, begin to join together to form puddles in surface depressions. Once the depressions are filled, overland flow begins (Figure 6.9). On steep slopes this process starts much earlier than on flat slopes.

As the generated runoff moves downhill, its depth and velocity changes, depending on the slope and the nature of the surface. Although this can be represented mathematically, for example by the kinematic wave theory, the resulting equations are complex and time consuming to solve. Also, since the runoff behaviour of the catchment is an aggregation of many different types of surface, identifying appropriate physical parameters, such as roughness, as input for a mathematical model is very difficult.

The problem of overland flow can be simplified by assuming an empirical relationship between flow per unit width (q) and depth (y) of the form:

$$q = a y^b$$ (6.16)

A semi-empirical equation for channel flow can be used, for example the Manning Equation (Equation (6.27)), in which case $a = S^{1/2}/n$ and $b = 5/3$, where S is the surface slope and n is the Manning friction coefficient. For a constant rainfall input of i mm/h the flow will increase to an equilibrium when:

$$q = ix$$ (6.17)

(a)

(b)

(c)

(d)

Figure 6.8   Depression storage and evaporation on a paved surface: (a) initial wetting; (b) depression storage connected; (c) partial drying, some depression storage; and (d) almost complete drying

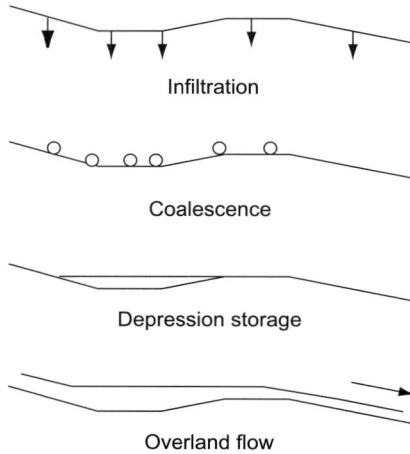

Figure 6.9   Initial surface processes on an impervious surface

At equilibrium, the depth of flow (D_e) will be constant and can be calculated from (Fleming, 1975):

$$D_e = \int_0^L y\,dx$$

$$= \frac{1}{a^{1/b}} \int_0^L q^{1/b}\,dx$$

$$= \frac{i^{1/b}}{a^{1/b}} \int_0^L x^{1/b}\,dx$$

$$= \frac{bi^{1/b} L^{1+1/b}}{a^{1/b}(b+1)}$$

$$= \frac{i^{0\cdot6} n^{0\cdot6} L^{1\cdot67}}{S^{0\cdot3}} \quad \text{(using the Manning Equation)} \qquad (6.18)$$

Crawford and Linsley (1966) proposed an empirical function relating the depth of overland flow rate to the surface detention which, when combined with the Manning Equation (Equation (6.27)), gives:

$$q = \frac{S^{1/2}}{n} \left(\frac{D}{L}\right)^{5/3} \left\{ 1 + 0.6 \left(\frac{D}{D_e}\right)^3 \right\}^{5/3}$$ (6.19)

The surface detention at time $j+1$ can be calculated from the continuity equation:

$$D_{j+1} = D_j + i\Delta t - F\Delta t - q\Delta t$$ (6.20)

where i, F and q are the unit rainfall, infiltration and outflow respectively, during time interval Δt.

6.3 Channel flow

The initial sheet-flow of surface runoff will tend to follow preferential paths and eventually small channels or rills will be formed which progressively erode to form larger channels. Channel flow is also supplied from emerging groundwater. As in the case of groundwater flow, flow in channels depends on the slope of the water surface. However, the flow is not constrained by the permeability of the ground and the velocity of flow in channels is several orders of magnitude greater than that of groundwater. Channels that are largely fed by groundwater, therefore, have a much slower response than those which are fed mainly by surface or near surface flow.

6.3.1 *Separation of channel flow components*

Hydrograph separation

As has been indicated above, there are various routes that water may follow between its initial contact with the ground and its arrival in a stream channel. The flow that passes through the ground before emerging in a channel, termed *baseflow*, varies only slowly with time and there is generally a lag of several days or weeks between changes in rainfall and changes in flow. In some cases, the baseflow may be negative in that there is flow from the channel into the ground. This occurs in many arid areas where rivers fed by mountain catchments pass over dry permeable soils. It also occurs temporarily in many channels when there is a significant change in water level. As the water level rises, some water is stored within the bank material and is released as the water level falls.

In contrast to baseflow, the *surface* or *direct flow* component varies rapidly with rainfall with a lag of a few hours or less. The direct flow component includes the flow that travels horizontally through the upper soil layers (often termed *interflow*), as well the runoff from relatively impervious areas and precipitation, which falls directly on the surface of channels and water bodies.

The exact boundary between baseflow and surface flow in a given hydrograph is difficult to define as it depends on the detailed hydrogeology of the catchment. The essential problem is to identify the start and end of the direct runoff component and the shape of the boundary between them. At the start of a storm, the sharp increase in flow usually provides an easily identifiable indication of the start of the direct runoff, but the end of the direct runoff is not as clearly defined (Figure 6.10(c)). However, there are several techniques to identify the end of the direct component, which, while not producing an exact solution, provide fairly practical and consistent methods.

One characteristic of the baseflow component is that in the recession phase the baseflow will generally follow an exponential decrease, i.e.

$$Q_t = Q_0 e^{-kt} \tag{6.21}$$

where Q_t is the baseflow at time t, Q_0 is the baseflow at the start of the recession and the coefficient k is a property of the catchment aquifer. Therefore, if the recession limb of the hydrograph is plotted on a log scale, the hydrograph of baseflow becomes linear as the flow becomes entirely baseflow (Figure 6.10(a)). The point at which a recession hydrograph becomes baseflow can be determined by plotting the recession limbs of several hydrographs to a log scale on tracing paper and moving them (at the correct flow rate) until they form a tangent to a common straight line, which is referred to as the master depletion curve. The master depletion curve is then plotted at a natural scale on tracing paper and superimposed on each recession hydrograph in turn. The end of the direct flow component is then taken as the point of separation of the hydrograph from the master depletion curve (Figure 6.10(b)).

Where an extensive flow record is not available, it is possible to identify the end of the baseflow separation by calculating the ratio $Q_t/Q_{t+\Delta t}$, where Δt is any convenient time interval, e.g. 1 h. From Equation (6.21), it can be seen that, for constant values of k and Δt, the ratio $Q_t/Q_{t+\Delta t}$ will be constant. In practice, the ratio generally decreases but if it is plotted against time, two straight lines are often apparent and the intersection of these lines can be used to identify the end of the baseflow separation (Figure 6.10(c)) (Example 6.1).

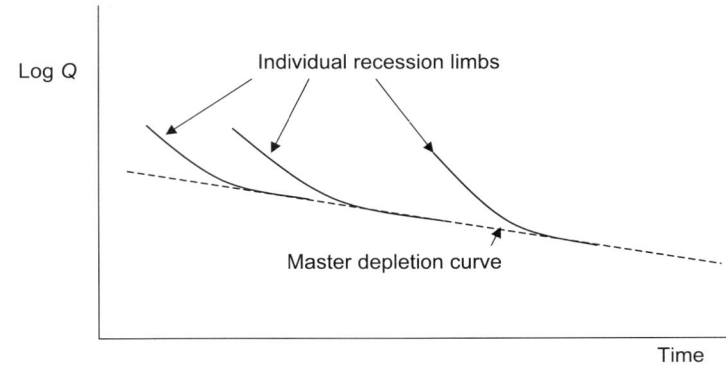

(a)

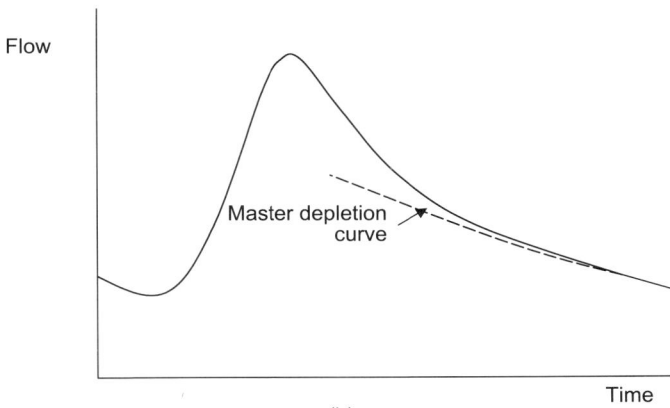

(b)

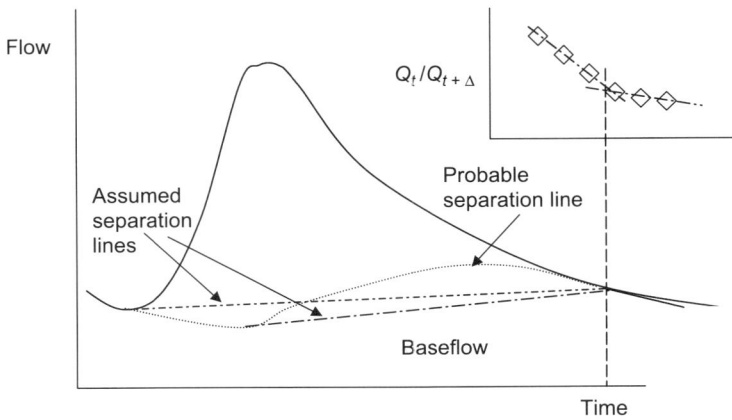

(c)

Figure 6.10   Separation of baseflow using a master depletion curve

Example 6.1

The flows shown below were recorded as part of the recession limb of a storm hydrograph. Estimate the end point of the separation between direct flow and baseflow.

Time (h)	25	26	27	28	29	30	31	32
Flow (m³/s)	12·5	9·7	7·8	6·5	5·5	4·8	4·2	3·7

Time	Flow, Q	$Q/Q_{t+\Delta t}$
25	12·5	1·29
26	9·7	1·25
27	7·8	1·2
28	6·5	1·18
29	5·5	1·15
30	4·8	1·14
31	4·2	1·14
32	3·7	

Plotting $Q/Q_{t+\Delta t}$ shows a change in gradient at about 28 h.

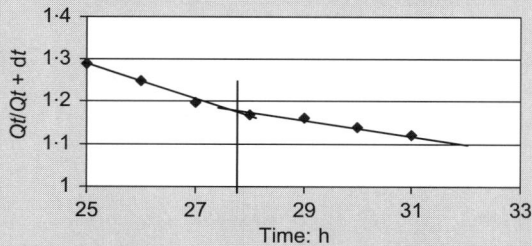

If very little flow data are available, several empirical formulae exist which give the approximate time (N) between the peak of the hydrograph and the end of the baseflow separation, e.g.:

$$N = 0{\cdot}8A^{0{\cdot}2}$$

(6.22)

where A is the catchment area in km^2 and N is in days.

Having established the end points of the baseflow separation line, the actual separation line can be drawn either as a straight line or as a curve. The actual separation line will probably decease as the baseflow recession continues before the rainfall has an impact on groundwater flows (Figure 6.10(c)). A common assumption is that the baseflow recession is a continuation of the flow hydrograph up to the peak of the flow hydrograph after which there is a linear increase up to the end of the separation.

Baseflow index

The above methods, while applicable to individual hydrographs, are rather time consuming for a long flow record. A more practical way of separating the baseflow in a flow record is described by Gustard *et al.* (1992) as follows.

(a) Divide the average daily flows into non-overlapping blocks of five days and calculate the minima for these blocks (Q_1, Q_2, ..., etc.).

(b) Considering consecutive sequences of three minima (Q_1, Q_2, Q_3), (Q_2, Q_3, Q_4), ... , etc., the central value is an ordinate of the baseflow line if it is less than $0{\cdot}9 \times$ the outer values. This is continued for the length of the record, generating a sequence of baseflow ordinates QB_1, QB_2, ... , etc., with varying time periods between them.

(c) Intermediate values of the baseflow line are calculated by interpolation with the condition that if $QB_i > Q_i$ then $QB_i = Q_i$.

(d) The volume of baseflow is then calculated as the integral under the baseflow line, bearing in mind that baseflow separation cannot start on the first day of the record and will finish before the end of the original hydrograph. It is recommended that the separation for a multi-year record is carried out for the entire record.

(e) The ratio of the area under the baseflow line to the total area under the hydrograph of daily flow is known as the *baseflow index*, which can be calculated for each year separately.

The relative proportion of baseflow depends mainly on the geology of the catchments. Many catchments in the south-east of England, which are underlain by chalk aquifers providing a high baseflow with little surface runoff, have baseflow indices of 0·7 or more (i.e. 70% of the total flow volume derives from the baseflow). However, catchments in the west of Scotland have a relatively impermeable geology with baseflow indices of 0·2 or less.

6.3.2   General features of open channel flow

The analysis of flow in an open channel is mainly concerned with the relationship between the depth (or stage) and flow rate (or discharge) in the channel. This is normally represented by a *stage–discharge curve*. Open channel flow is characterised by having a free surface where the pressure is atmospheric, which contrasts with flow in a full conduit, where there is no free surface. The cross-section of flow in a channel is therefore not completely defined by the shape of the channel since, if the flow rate changes, there is generally a change in the water level (i.e. the cross-section) as well as velocity. By comparison, the flow cross-section in a closed pipe is obviously completely defined by the boundary of the pipe (although the pressure in the pipe is not defined).

Flow in open channels can be broadly classified as uniform or non-uniform flow and as steady or unsteady flow. *Uniform* flow occurs where there is no change in flow conditions between one section of a channel and another location. In natural channels, uniform flow is never completely achieved because of the natural variations in bed material and obstructions such as trees, etc., although uniform flow may be approached if a constant slope, roughness and shape exist over a reasonable length of the channel. In long, straight, artificial channels with a uniform slope and cross-section, uniform flow is more likely. *Steady* flow occurs where flow conditions do not change at a specific point with respect to time. Again, in natural channels, flow conditions are continually changing but in practice, the rate of change is often low enough for the flow to be considered as steady.

In general, there is no unique relationship between flow rate and water level in open channels: a range of flows can exist for any given depth, depending on the water surface slope. However, in the case of uniform flow, or at specific locations known as *controls*, there is a unique relationship between flow rate and water level. Controls may also be considered as locations where the capacity of the channel is restricted and they determine the water level upstream and/or downstream of the

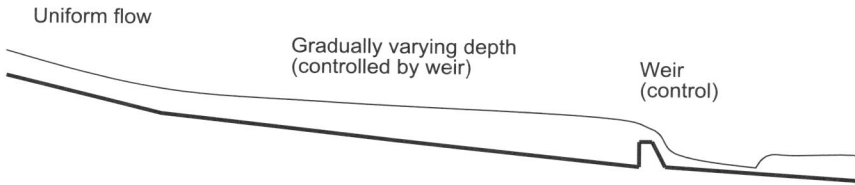

Figure 6.11   General water level profile in a channel

control. Figure 6.11 shows a typical longitudinal profile along a river channel. Upstream of the weir there is a region of gradually varying depth, where the water level is controlled by the weir. In the area immediately adjacent to and downstream of the weir, the water surface varies rapidly over a short distance and the water level in this region is also controlled by the weir. Above the region controlled by the weir, there may be a length of quasi-uniform flow.

6.3.3   Uniform flow

The essential feature of uniform flow is that the velocity of the water is such that the energy dissipated through friction and turbulence balances the loss in potential energy (or head) as the water moves down the slope of the channel. In other words, the friction head S_f equals the bed slope S_0. It follows that, for any given flow in a cross-section, there will be a unique depth, corresponding to uniform flow and this is referred to as the *normal depth*.

The first analysis of uniform open-channel flow was carried out by Chezy in 1768. He made the assumption that bed shear–stress was proportional to the square of the velocity. Thus, for a length L of channel the resistance shear force (R) is given by:

$$R = k \times P \times L \times V^2 \tag{6.23}$$

The term P in the above equation is known as the *wetted perimeter* and is the length of the cross-section that is contact with water (Figure 6.12).

The resistance force balances the force due to the component of the weight of the water (G) acting along the channel, which is:

$$G = \rho \times A \times L \times g \times i \tag{6.24}$$

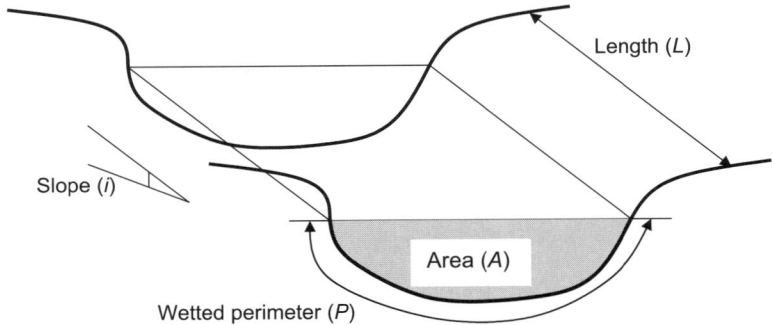

Figure 6.12   Analysis of an element of open-channel flow

where A is the cross-sectional area and i is the water surface slope (= bed slope, assuming $i = \sin(i)$).

Equating Equations (6.23) and (6.24) gives the Chezy Equation:

$$V^2 = \left(\frac{\rho g}{k}\right)\left(\frac{A}{P}\right)i$$

or:

$$V = C\sqrt{(mi)} \tag{6.25}$$

The term A/P is known as the *hydraulic radius* (m) and is a function of the geometric shape of the cross-section, while the constant C reflects the roughness of the channel bed. However, it has been found that the C coefficient is also a function of on the flow and cross-sectional shape, and the Chezy Equation is therefore of limited use where the flow is unknown.

Manning developed a more stable friction coefficient (n) and found an empirical relationship with the Chezy C coefficient:

$$n = \frac{m^{1/6}}{C} \tag{6.26}$$

which leads to the well-known Manning Equation:

$$V = \left(\frac{1}{n}\right)m^{2/3}i^{1/2} \tag{6.27}$$

Table 6.4   Manning friction coefficient for channels

Type of channel	Material	Manning's n: s/m$^{1/3}$
Natural watercourse	Earth	0·02–0·1
Natural watercourse	Gravel	0·03–0·07
Artificial channel	Earth	0·02–0·04
Artificial channel	Rock	0·02–0·05
Artificial channel	Concrete	0·012–0·017
Artificial channel	Dressed stone	0·013–0·02
Pipe	Clayware	0·011–0·016
Model	Perspex/glass	0·009–0·010

The n coefficient is a relatively stable property of the bed material and channel roughness, and Table 6.4 gives typical values of the Manning coefficient for various bed conditions.

In combination with the continuity equation, the Manning or Chezy equations can be used to find the flow rate corresponding to a given water level in a given channel. To carry out the reverse calculation, i.e. to find the water level for a given flow rate, an iterative procedure is necessary. Alternatively, a curve of depth against flow rate (stage–discharge curve) can be constructed using the Manning or Chezy equations. It sometimes helps to separate the physical characteristics of a channel from the hydraulic parameters. For example, using the Manning Equation:

$$\frac{Q}{i^{1/2}} = \frac{A}{n} m^{2/3} \qquad (6.28)$$

The right-hand side of the equation can be plotted as a function of depth from survey data, together with an estimate of the Manning friction coefficient. For a given flow rate and water surface slope, the left-hand side (often termed *conveyance*) can be calculated and hence the depth can be estimated (Examples 6.2 and 6.3).

Example 6.2

A rectangular channel 3 m wide carries a flow of 4 m³/s. The longitudinal slope of the channel is 1/1000 and the Manning roughness coefficient is 0·03 s/m$^{1/3}$. Estimate the depth of flow.

Try depth d = 1·5 m

Cross-sectional area, $A = 1.5 \times 3 = 4.5$ m^2

Wetted perimeter, $P = 2 \times 1.5 + 3 = 6$ m

Hydraulic radius, m $= \dfrac{4.5}{6} = 0.75$ m

$$Q = \frac{4.5 \times 0.75^{2/3}}{0.03 \times \sqrt{1000}} = 3.92 \text{ m}^3/\text{s}$$

Try depth d = 1·55 m

Cross-sectional area, $A = 1.55 \times 3 = 4.65$ m^2

Wetted perimeter, $P = 2 \times 1.55 + 3 = 6.1$ m

Hydraulic radius, m $= \dfrac{4.65}{6.1} = 0.76$ m

$$Q = \frac{4.65 \times 0.76^{2/3}}{0.03 \times \sqrt{1000}} = 4.09 \text{ m}^3/\text{s}$$

Take depth = 1·52 m

Example 6.3

A trapezoidal channel has a base width of 5 m and side slopes of 2 horizontal:1 vertical, and a Manning n coefficient of 0·02 s/m$^{1/3}$. Construct a conveyance–depth curve and estimate the depth of flow for a flow rate of 30 m³/s and a water surface slope of 0·004.

Conveyance is calculated as:

$$k = \frac{A}{n} m^{2/3}$$

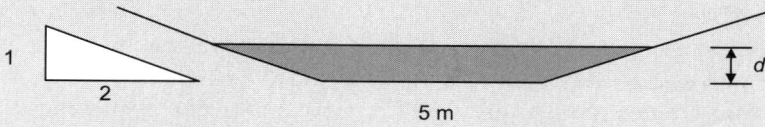

5 m

Tabulating values for depths:

Depth	Area	Wetted perimeter	m	k
0·5	3	7·24	0·41	83·4
1	7	9·47	0·74	286·1
1·5	12	11·71	1·02	609·9
2	18	13·94	1·29	1067·0

Plotting:

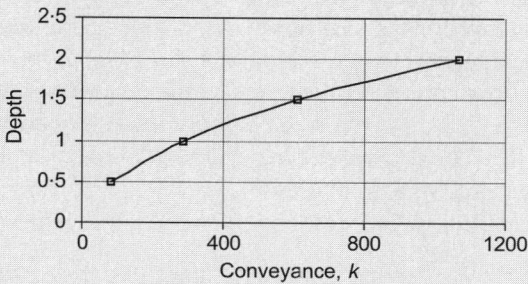

For $Q = 30$ m³/s and $i = 0.004$:

$$k = \frac{Q}{i^{1/2}} = \frac{30}{0.004^{1/2}} = 474$$

From graph:

$d = 1·3$ m

6.3.4 Non-uniform flow

As has been mentioned, the depth of flow in non-uniform flow conditions is largely determined by the nature of various controls, which may be natural features, such as rock sills, or artificial structures, such as weirs. Controls either present a restriction to flow or define the water level, for example, where a channel flows into a large water-body. As seen in Figure 6.11, a control can influence the water depth for some distance upstream and downstream of the structure. Control structures are characterised by a unique relationship between flow rate (Q) and water level, which, in general, can be expressed in terms of an equation of the form:

$$Q = CBH^n \tag{6.29}$$

where H is the head, which can be taken, approximately, as the vertical height of the upstream water surface above the crest of the structure, and B is the width of the flow over the structure. The coefficient C reflects energy losses due to friction and turbulence, as well as gravitational and other constants, while the exponent n is normally between 1·5 and 2·5. Examples of artificial control structures are given in Section 6.4.1.

Equations such as Equation (6.29) allow us to determine the water level at a control structure for a given flow rate in a river channel. In order to determine the water level at other locations in a river channel (apart from where uniform flow conditions can be assumed), a water surface profile must be established. Using the basic Saint-Venant Equation for flow in open channels, it can be shown that the differential equation of depth with respect to distance along the channel is:

$$\frac{\Delta d}{\Delta L} = \frac{S_O - i}{1 - \dfrac{Q^2 B}{g A^3}} \tag{6.30}$$

where S_O is the bed slope, i is the slope of the water surface, Q is the flow rate, B is the channel width, and A is the flow cross-sectional area.

Such profiles can be determined for a given length of channel, provided that the water depth is known at both ends of the channel, or that it includes a control of some form at one end. Starting at one end and working either upstream or downstream, the overall difference in depth over the channel length is divided into a number of intervals of depth (Δd), over which uniform flow is assumed (Figure 6.13).

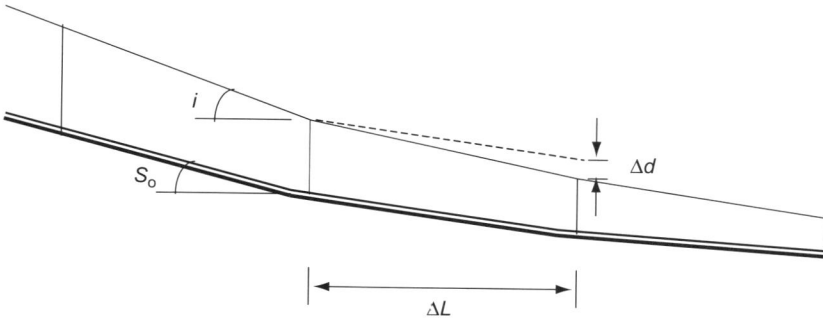

Figure 6.13   Determination of channel profiles

The slope of the water surface over each section is estimated from the Manning Equation using the mean cross-sectional area and hydraulic radius for the section, i.e.:

$$i = \left(\frac{vn}{m^{2/3}} \right)^2 \qquad (6.31)$$

Equation (6.30) can then be used to calculate the length (ΔL) of the section corresponding to the change in depth of Δd. This is repeated for each section to allow a profile of depth along the channel length to be constructed (Example 6.4).

Example 6.4

A rectangular channel 3 m wide carries a flow of 2 m³/s. The slope of the channel is 1/800 and the Manning n coefficient is 0·015 s/m$^{1/3}$. At the lower end of the channel is a 1·9 m wide broad-crested weir with a coefficient of discharge of 0·95. The crest of the weir is 0·475 m above the bed level. Determine the water level just upstream of the weir and, using three depth intervals, the profile of the water surface upstream to a point where the depth is 0·6 m.

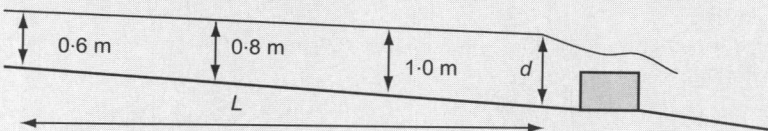

For a broad-crested weir:

$$Q = 1.705CdBH^{3/2}$$

$$2.0 = 1.705 \times 0.95 \times 1.9 \times H^{3/2}$$

$$H^{3/2} = \frac{2.0}{1.705 \times 0.95 \times 1.9}$$

$$H = 0.750 \text{ m}$$

$$d = 0.75 + 0.475 \text{ m} = 1.225 \text{ m}$$

Use intervals of depth of 0.6, 0.8, 1.0, 1.225 m.

For section $d = 0.6$ to 0.8 m:

Mean depth $d = \dfrac{0.6 + 0.8}{2} = 0.7$ m

Mean area $A = 0.7 \times 3.0 = 2.1 \text{ m}^2$

Wetted perimeter $P = 2 \times 0.7 + 3.0 = 4.4$ m

Hydraulic radius $m = \dfrac{2.1}{4.4} = 0.477$ m

Velocity $v = \dfrac{2.0}{2.1} = 0.952$ m/s

ws slope $i = \left(\dfrac{vn}{m^{2/3}}\right)^2 = \left\{\dfrac{0.952 \times 0.015}{0.477^{2/3}}\right\}^2 = 0.000547$

$$s - i = 0.00125 - 0.000547 = 0.000703$$

$$\frac{Q^2B}{gA^3} = \frac{2.0^2 \times 3.0}{9.81 \times 2.1^3} = 0.1321$$

$$\Delta d = 0.8 - 0.6 = 0.2 \text{ m}$$

$$\Delta L = \Delta d \frac{\left(\dfrac{1 - Q^2B}{gA^3}\right)}{s - i} = \frac{0.2 \times (1 - 0.1321)}{0.000703} = 247.0 \text{ m}$$

The results are tabulated opposite together with those for the other depth intervals.

d_1	d_2	Mean depth	Area	m	Mean vel.	wl slope, i	$s - i$	$Q^2 B/gA^3$	Δd	ΔL	L
0·600	0·800	0·700	2·100	0·477	0·952	0·000547	0·000703	0·132085	0·200	247·0	247·0
0·800	1·000	0·900	2·700	0·563	0·741	0·000266	0·000984	0·062147	0·200	190·6	437·6
1·000	1·225	1·113	3·338	0·639	0·599	0·000147	0·001103	0·032904	0·225	197·3	634·8

6.3.5 Unsteady flow

Where flow varies with time at a given location, it is referred to as unsteady flow. The variation can be over a matter of minutes, for example, during a surge wave, in which case forces due to the acceleration of the water need to be included in the analysis. However, the time-scale of interest to hydrologists is normally hours or days, as in, for example, the hydrograph resulting from a storm event. In such cases, flow can be considered as steady for each time-step. Of particular concern to hydrologists is the way the peak of such a hydrograph is reduced and delayed as it passes down a channel or through a reservoir, a process known as *flood routing*. Flow routing occurs in both channels and reservoirs. In both cases it results from the storage of water within the channel or reservoir, although the analysis is slightly different in the two cases.

Reservoir routing

The attenuation of the hydrograph as flow passes through a reservoir is due to the fact that, as the flow into the reservoir changes, the amount of storage also changes. This can be expressed as:

Inflow volume = Outflow volume + Change in storage volume

Over a small time interval dt, this can be expressed as:

$$P \times dt = Q \times dt + dS \tag{6.32}$$

Over a finite time interval Δt, this can be written as:

$$P_{av} = Q_{av} + \frac{\Delta S}{\Delta t} \tag{6.33}$$

where P_{av} and Q_{av} are the mean inflow and outflow respectively over a time interval Δt.

The mean inflow and outflow can be considered as the average of the values at the start and end of the time interval. Thus:

$$\frac{P_n + P_{n+1}}{2} = \frac{Q_n + Q_{n+1}}{2} + \frac{S_{n+1} - S_n}{\Delta t} \tag{6.34}$$

where the subscripts n and $n+1$ refer to the flows and storage at the beginning and end of a time interval Δt respectively. It is usually required

236

to find the values of the outflow hydrograph (Q) for given values of the inflow hydrograph (P). However, since the storage is also not known, a further relationship is required. This can be derived from a consideration of the hydraulic characteristics of the outflow structure and the topography of the reservoir. The outflow from most reservoirs passes over a weir or similar spillway structure, where there is a relationship between water level or head above the spillway crest and flow, which can be expressed as an equation of the form of:

$$Q = K \times H^n \tag{6.35}$$

where H is the height of the water level above the crest and K is a constant, incorporating the width and the discharge coefficient of the weir and the coefficient n is normally about 1·5.

The topography of a reservoir basin can be defined in terms of its plan area at different levels above the spillway crest. (Since there is assumed to be no outflow if the water level is below the crest, the shape of the reservoir at these levels is not relevant). The volume relationship can be found by numerical integration between known discrete values (Figure 6.14(a)). A method which is commonly used, assumes that the volume between two levels ΔH apart is given by:

$$\Delta S = 1/2(A_n + A_{n+1})\Delta H \tag{6.36}$$

Sometimes, the plan area of a reservoir at different levels can be represented by a continuous function, for example of the form:

$$A = C_1 \times H^m \tag{6.37}$$

where C_1 and m are empirical constants.

If this is the case, the volume of storage between the spillway crest and various levels is computed by integrating the plan area with respect to height:

$$S = \int A dH = \int C_1 H^m dH = \frac{C_1 H^{m+1}}{(m+1)} \tag{6.38}$$

The volume relationship (Equation (6.36) or (6.38)) can be combined with the spillway characteristics (Equation (6.35)) to give a relationship between outflow and storage. In other words, for any given water level, both the volume of storage and the outflow can be computed. The outflow–storage relationship is used together with the basic storage equation to determine the outflow hydrograph. It should be noted that a fundamental assumption of reservoir routing is that the water surface in the reservoir is horizontal.

Equation (6.34) can be rearranged as:

$$Q_{n+1} + \frac{2S_{n+1}}{\Delta t} = P_n + P_{n+1} - Q_n + \frac{2S_n}{\Delta t} \tag{6.39}$$

The term $Q + 2S/\Delta t$ is calculated *a priori* for various water levels using the combination of Equations (6.36) (or (6.38)) and (6.35), and can be plotted as a function of Q (Example 6.5).

Example 6.5

The flow into a reservoir is given below. The reservoir has a surface area of 15 ha at the crest level of the outflow weir and 40 ha at a level 0·5 m above the crest level. The width of the outflow weir is 12 m and its coefficient of discharge is 0·9. The initial outflow is 3 m³/s. Determine the outflow hydrograph.

Time (h)	0	1	2	3	4	5
Inflow (m³/s)	3	15	9	3	3	3

Reservoir characteristics
For water level at the crest:
    Storage $(S) = 0$
    Outflow $(Q) = 0$

$$\therefore \frac{Q + 2S}{\Delta t} = 0$$

For water level at 0·5 m above crest:
    Volume $S = 1/2(15 + 40) \times 10^4 \times 0\cdot5 = 137\,500$ m³
    Outflow $Q = 1\cdot705 \times 0\cdot9 \times 12 \times 0\cdot5^{3/2} = 18\cdot414 \times 0\cdot5^{3/2} = 6\cdot53$ m³/s

$$\frac{2S}{\Delta t} = \frac{2 \times 137\,500}{3600} = 76\cdot39 \text{ m}^3/\text{s}$$

$$Q + \frac{2S}{\Delta t} = 6\cdot53 + 76\cdot39 = 82\cdot92 \text{ m}^3/\text{s}$$

Outflow hydrograph

First time-step:

Inflow $P_n = 3.0 \text{ m}^3/\text{s}$

Inflow $P_{n+1} = 15.0 \text{ m}^3/\text{s}$

Outflow $Q_n = 3.0 \text{ m}^3/\text{s}$

$$\frac{2S_n}{\Delta t} = \frac{3}{6.53} \times 76.39 = 35.10 \text{ m}^3/\text{s}$$

$$Q_{n+1} + \frac{2S_{n+1}}{\Delta t} = P_n + P_{n+1} - Q_n + \frac{2S_n}{\Delta t}$$

$$= 3.0 + 15.0 - 3.0 + 35.10 = 50.10 \text{ m}^3/\text{s}$$

$$Q_{n+1} = \frac{50.10}{82.92} \times 6.53 = 3.94 \text{ m}^3/\text{s}$$

$$\frac{2S_{n+1}}{\Delta t} = 50.10 - 3.94 = 46.15 \text{ m}^3/\text{s}$$

Second time-step:

Inflow $P_n = 15.0 \text{ m}^3/\text{s}$

Inflow $P_{n+1} = 9.0 \text{ m}^3/\text{s}$

Outflow $Q_n = 3.94 \text{ m}^3/\text{s}$

$$\frac{2S_n}{\Delta t} = \frac{3.94}{6.53} \times 76.39 = 46.15 \text{ m}^3/\text{s}$$

$$Q_{n+1} + \frac{2S_{n+1}}{\Delta t} = P_n + P_{n+1} - Q_n + \frac{2S_n}{\Delta t}$$

$$= 15.0 + 9.0 - 3.94 + 46.15 = 66.21 \text{ m}^3/\text{s}$$

$$Q_{n+1} = \frac{66.21}{82.92} \times 6.53 = 5.21 \text{ m}^3/\text{s}$$

$$\frac{2S_{n+1}}{\Delta t} = 66.21 - 5.21 = 61.00 \text{ m}^3/\text{s}$$

The remaining calculations are tabulated overleaf.

Time: h	P_n: m³/s	P_{n+1}: m³/s	$-Q_n$: m³/s	$2S_n/\Delta t$: m³/s	$Q_{n+1}+2S_{n+1}/\Delta t$: m³/s	Q_{n+1}: m³/s	$2S_{n+1}/\Delta t$: m³/s	Q_n: m³/s	H: m
0	3·00	15·00	−3·00	35·10	50·10	3·94	46·15	3·00	0·30
1	15·00	9·00	−3·94	46·15	66·21	5·21	60·99	3·94	0·36
2	9·00	3·00	−5·21	60·99	67·78	5·34	62·44	5·21	0·43
3	3·00	3·00	−5·34	62·44	63·11	4·97	58·14	5·34	0·44
4	3·00	3·00	−4·97	58·14	59·17	4·66	54·51	4·97	0·42
5	3·00	1·50	−4·66	54·51	54·35	4·28	50·07	4·66	0·40

For each time-step in turn, the terms on the right-hand side of the equation are known (assuming that the initial outflow and storage are known), and the right-hand side is equated to the term $Q + 2S/\Delta t$ for the next time-step. From the relationship between $Q + 2S/\Delta t$ and Q, the corresponding value of Q for this time-step can be determined and hence the term $2S/\Delta t$. This allows the right-hand side of Equation (6.39) to be determined for this time-step and the process is repeated for each time-step. An example of a typical inflow and outflow hydrograph for a reservoir (Figure 6.14(b)) shows how reservoir routing results in a reduction or attenuation in the peak flow and a delay in the time of the peak flow. The amount of attenuation depends on the size of the reservoir and the width of the spillway: maximum attenuation occurs with a narrow spillway and a large surface area of the reservoir.

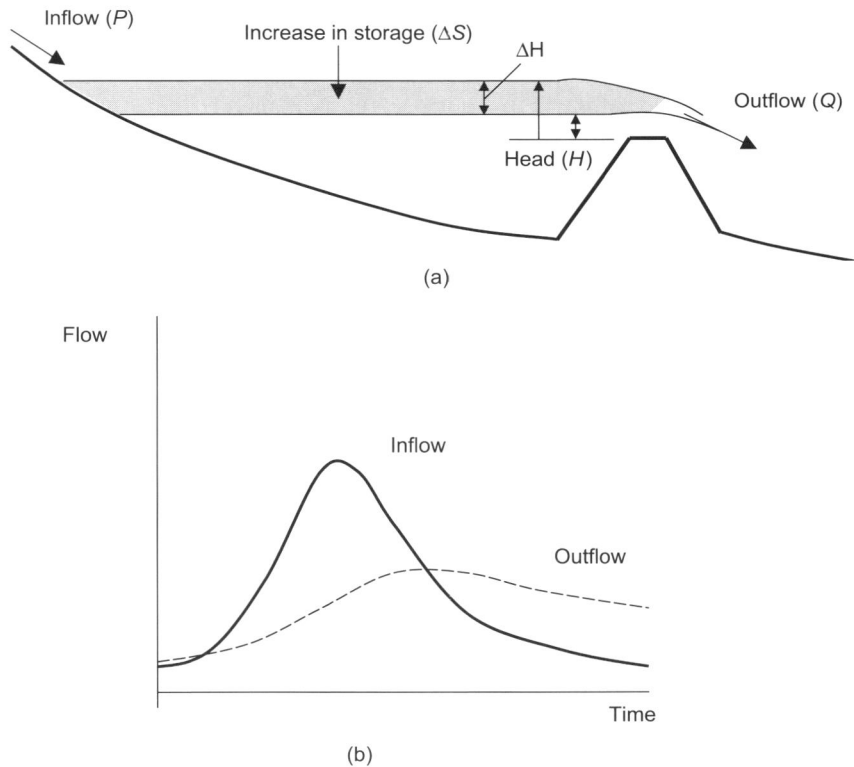

Figure 6.14 Reservoir routing: (a) terms used in reservoir routing; (b) typical inflow and outflow hydrographs

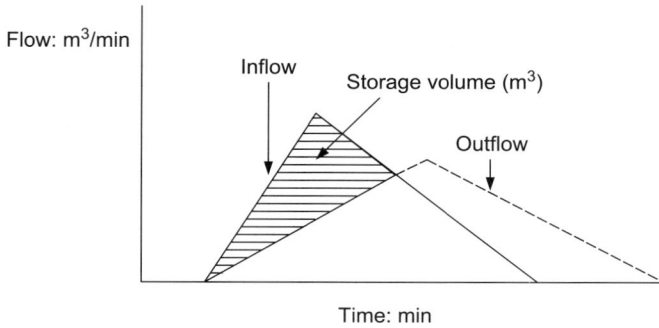

Figure 6.15   Direct estimation of storage volume

It is also possible to derive an inflow hydrograph from a specified outflow hydrograph using the same procedure with a slight modification to the storage equation. Furthermore, it is possible to estimate the required storage to achieve a given reduction in peak flow by using this procedure iteratively with the help of a spreadsheet or simple programme. Alternatively, it can be seen from the basic storage relationship (Equation (6.32)) that the overall storage volume is the difference between the integrals of inflow and outflow with time, in other words, the area between the inflow and outflow hydrographs. Therefore, if simple geometric shapes (e.g. triangles) are assumed for the inflow and outflow hydrographs, it is possible to estimate the storage volume directly (Figure 6.15).

Channel routing

As in the case of reservoirs, the routing in river channels is due to the change in storage volume within the channel. The shape of the outflow hydrograph in relation to the inflow hydrograph is similar to that for a reservoir (Figure 6.14(b)). The main difference in the two processes is that, in a reservoir, the water surface is essentially horizontal and, therefore, there is a unique relationship between storage and outflow, whereas in river channels, the water surface slope is not defined and the volume of storage in the channel is therefore a function of both the inflow and outflow. In addition to the storage of water in the channel, a certain amount of water is also stored within the banks of the channel. When the water level rises due to an increase in flow, some of the water is stored in the voids of the bank material. Conversely, when the flow is receding, water comes out of the bank storage and the water level is enhanced.

The storage in a channel can be represented by a linear function of the inflow (P) and outflow (Q), for example:

$$S = k[xP + (1-x)Q] \tag{6.40}$$

where k and x are the *storage constant* and *weighting factor*, respectively.

The storage constant (k) represents the ratio of storage volume (usually m^3) and flow (usually m^3/s) and thus has the units of time. It depends on the geometric shape of the channel and the length of the reach, and can be considered as approximately equal to the mean travel-time of the flow in the reach. It is important that the storage constant has the same units as the flow time units, but these do not have to be the same as the time interval used in the calculations. The weighting factor (x) reflects the relative importance of inflow and outflow in determining the storage. For a reservoir, the weighting factor would be zero but for a channel it is usually between zero and 0·5, most often between 0·2 and 0·3.

By substituting equation (6.40) into the basic storage equation (6.34), the routing equation becomes (Appendix 6.3):

$$Q_{n+1} = C_0 P_{n+1} + C_1 P_n + C_2 Q_n \tag{6.41}$$

where:

$$C_0 = \frac{-(kx - 0·5\Delta t)}{(k - kx + 0·5\Delta t)}$$

$$C_1 = \frac{(kx + 0·5\Delta t)}{(k - kx + 0·5\Delta t)}$$

$$C_2 = \frac{(k - kx - 0·5\Delta t)}{(k - kx + 0·5\Delta t)}$$

The constants C_0, C_1 and C_2 are readily calculated from the storage constant and weighting factor, and, for a given inflow hydrograph and initial outflow value, the outflow at the next time-step can be determined from Equation (6.41) (Example 6.6).

Again, it is possible to determine the inflow hydrograph if the outflow hydrograph is given, using the same formulation. It is also possible to estimate the storage constant and weighting factor for a channel, if both the inflow and outflow hydrographs are known. Referring to the storage

Example 6.6

The inflow hydrograph to a length of channel is given below:

Time (h)	0	1	2	3	4
Inflow (m³/s)	1	2	4	1	1

The storage coefficient of the channel (k) = 5 h and the weighting factor (x) = 0·2. Determine the outflow of the channel assuming the initial outflow is 1 m³/s.

$$C_0 = \frac{-(kx - 0.5\Delta t)}{(k - kx + 0.5\Delta t)} = \frac{-(5 \times 0.2 - 0.5 \times 1)}{(5 - 5 \times 0.2 + 0.5 \times 1)} = \frac{-0.5}{4.5} = -0.111$$

$$C_1 = \frac{(kx + 0.5\Delta t)}{(k - kx + 0.5\Delta t)} = \frac{-(5 \times 0.2 + 0.5 \times 1)}{4.5} = \frac{1.5}{4.5} = 0.333$$

$$C_2 = \frac{(k - kx - 0.5\Delta t)}{(k - kx + 0.5\Delta t)} = \frac{(5 - 5 \times 0.2 - 0.5 \times 1)}{4.5} = \frac{3.5}{4.5} = 0.778$$

Check $-0.111 + 0.333 + 0.778 = 0.1000$ *OK*

First time-step:

$$Q_{n+1} = C_0 P_{n+1} + C_1 P_n + C_2 Q_n$$

$$P_1 = 1.0, \quad P_2 = 2.0$$

$$Q_2 = -0.111 \times 2.0 + 0.333 \times 1.0 + 0.778 \times 1.0 = 0.89 \text{ m}^3/\text{s}$$

Second time-step:

$$P_1 = 2.0, \quad P_2 = 4.0$$

$$Q_2 = -0.111 \times 4.0 + 0.333 \times 2.0 + 0.778 \times 0.89 = 0.91 \text{ m}^3/\text{s}$$

The remaining calculations are shown in the table opposite.

Time: h	P_n:m³/s	$C_0 \times P_{n+1}$: m³/s	$C_1 \times P_n$: m³/s	$C_2 \times Q_n$: m³/s	Q: m³/s
0	1·00	−0·22	0·33	0·78	1·00
1	2·00	−0·44	0·67	0·69	0·89
2	4·00	−0·11	1·33	0·71	0·91
3	1·00	−0·11	0·33	1·50	1·93
4	1·00	−0·11	0·33	1·34	1·73
5	1·00	−0·11	0·33	1·22	1·56
6	1·00	−0·11	0·33	1·12	1·44
7	1·00	—	—	—	1·34

equation (6.40), it can be seen that the storage S is a linear function of $[xP + (1-x)Q]$ and, therefore, if S is plotted against $[xP + (1-x)Q]$, a straight line graph results, with a slope equal to the storage constant (k), provided that the correct value of x is used. The procedure involves estimating an initial value of x and then plotting and iterating values of x until a straight line is achieved (Figure 6.16).

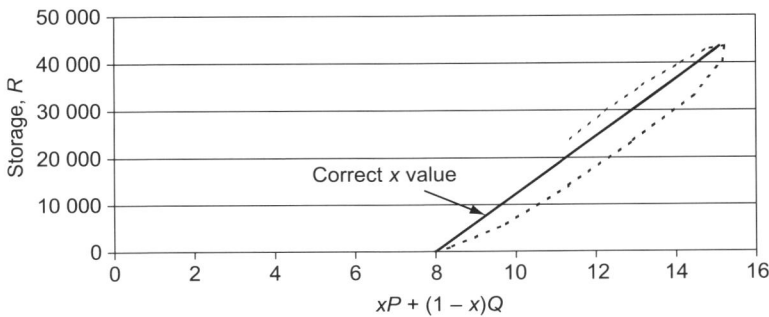

Figure 6.16   Estimation of channel routing constants

Example 6.7

Using the inflow and outflow hydrograph from Example 6.6, show that the storage coefficient (k) = 5 h and the weighting factor (x) = 0·2.

Try x = 0·25
For $P = 1$, $Q = 1$:

$$xP + (1 - x)Q = 0·25 \times 1 + 0·75 \times 1 = 1·0 \text{ m}^3/\text{s}$$

Storage $(S) = 0 \text{ m}^3$ (assume initial strorage is zero)

For $P = 2$, $Q = 0·89$:

$$xP + (1 - x)Q = 0·25 \times 2 + 0·75 \times 0·89 = 1·17 \text{ m}^3/\text{s}$$

Storage $(S) = [\text{inflow} - \text{outflow}] \times \Delta t$

$$= 0 + \left[1/2(1·0 + 2·0) - 1/2(1·0 + 0·89) \right] \times 3600$$

$$= 2000 \text{ m}^3$$

The remainder of the calculation is given in the table below.
Plotting $xP + (1 - x)Q$ against S does not give a coincident line (see graph (a)).

Try x = 0·20
For $P = 1$, $Q = 1$:

$$xP + (1 - x)Q = 0·2 \times 1 + 0·8 \times 1 = 1·0 \text{ m}^3/\text{s}$$

Storage $(S) = 0 \text{ m}^3$ (as before)

For $P = 2$, $Q = 0·89$:

$$xP + (1 - x)Q = 0·2 \times 2 + 0·8 \times 0·89 = 1·11 \text{ m}^3/\text{s}$$

Storage $(S) = 2000 \text{ m}^3$ (as before)

The remainder of the calculation is given in the table below.
Plotting $xP + (1 - x)Q$ against S does not give a coincident line (see graph).

Storage constant $(k) = \dfrac{S}{(xP + (1-x)Q)} = $ slope

$$= \dfrac{13\,432}{[(1\cdot75 - 1\cdot0) \times 3600]} = 4\cdot98 \text{ h}$$

Time: h	P: m³/s	Q: m³/s	$xP+(1-x)Q$: m³/s	S: m³/s	$xP+(1-x)Q$: m³/s	S: m³/s
			$x = 0\cdot25$		$x = 0\cdot20$	
0	1·00	1·00	1·00	0	1·00	0
1	2·00	0·89	1·17	2000	1·11	2000
2	4·00	0·91	1·69	9556	1·53	9556
3	1·00	1·93	1·70	13 432	1·75	13 432
4	1·00	1·73	1·54	10 447	1·58	10 447
5	1·00	1·56	1·42	8126	1·45	8126
6	1·00	1·44	1·33	6320	1·35	6320
7	1·00	1·34	1·26	4915	1·27	4915

(a)

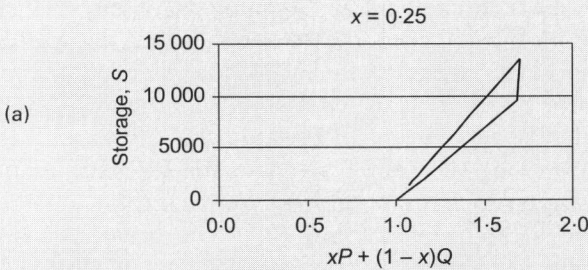

$x = 0\cdot25$

(b)

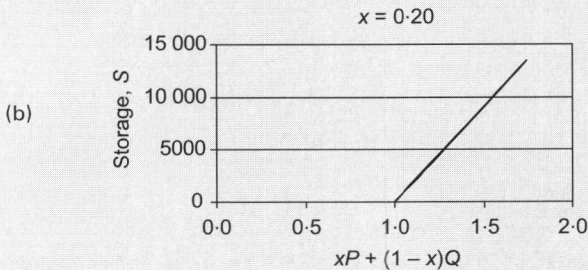

$x = 0\cdot20$

6.4 Channel flow measurement

In open channels, flow is normally measured by recording the water level (stage) and then establishing a relationship between water level and flow (discharge). This stage-discharge relationship, or rating curve, can be established either by taking a series of flow measurements at various water levels or by constructing a permanent structure, such as a weir or a flume, which acts as a control.

6.4.1 Flow measuring structures

As noted earlier, control structures are characterised by a unique relationship between water level (or head) and flow, which can generally be represented by an equation of the form of:

$$Q = C \times BH^n \tag{6.42}$$

Such a relationship can be used to convert a continuous record of water level into a flow hydrograph. In the past, the water level was usually recorded using a paper chart on a rotating drum with a pen connected to a float. More recently, water level is recorded digitally using pressure transducers and data loggers. The remote data loggers can be interrogated from a central control and the data can be transmitted by land-line or radio. Care needs to be taken that the stage–discharge relationship used for a particular structure is valid, as changes in the bed morphology or debris can distort the relationship. It should also be remembered that a stage–discharge characteristic derived for normal flows may not apply to extreme flows, for example, where bank overtopping occurs. Ideally, any structure should be located in a straight, unobstructed length of channel with a low velocity of approach and where flow is controlled by the channel resistance (normal flow conditions). The accuracy of the flow readings for most structures is within ±5% at a 95% confidence level.

The principle types of flow-measuring structures are broad-crested weirs, thin-plate weirs and flumes (Figure 6.17).

Broad-crested weirs

A broad-crested weir has a crest with a substantial crest dimension in the direction of flow, so that the stream lines become parallel with the crest

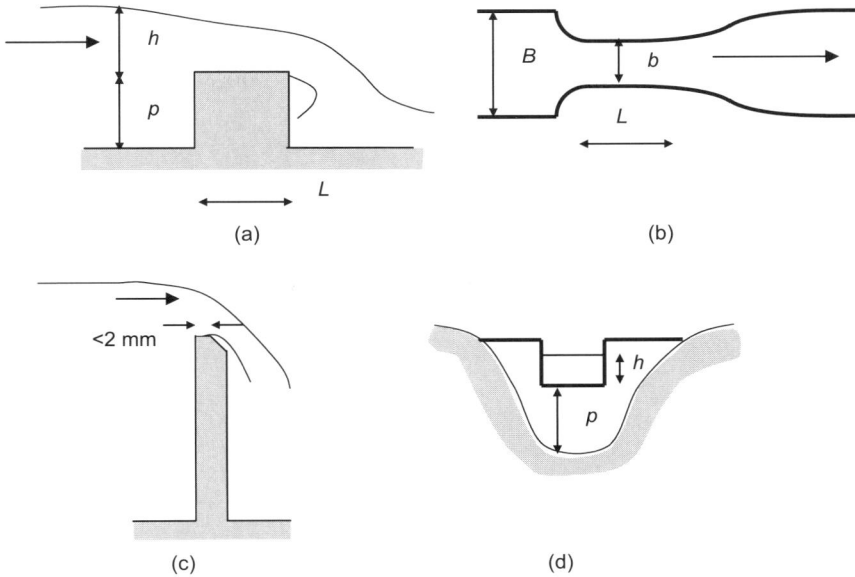

Figure 6.17   Examples of flow measurement structures: (a) broad-crested weir; (b) flume; (c) thin-plate weir — section; (d) rectangular thin-plate weir — elevation

(Figures 6.17(a) and 6.18). Where the channel is approximately rectangular in cross-section, the weir should extend over the full width of the channel. The site for a broad-crested weir should be selected so that the channel is straight for a sufficient length to ensure a regular velocity distribution and that the structure is not drowned at high flows. The upstream water-level should be recorded at a position three to four times the maximum head from the upstream face of the weir.

Thin-plate weirs

Thin-plate weirs usually consist of a triangular or rectangular notch extending only over part of the width of the channels with a crest width of 1–2 mm (Figure 6.17 (c) and (d)). This enables the overflowing water (the *nappe*) to spring clear of the crest, but sufficient difference between upstream and downstream water-levels must be provided to allow this to happen. Because of the thinness of the plate, they are also less robust and unsuitable for large channels or where there is a lot of debris. However, they do not require a rectangular cross-section and are more precise at small flows. In order to combine the precision at low flows with a capacity for high flows,

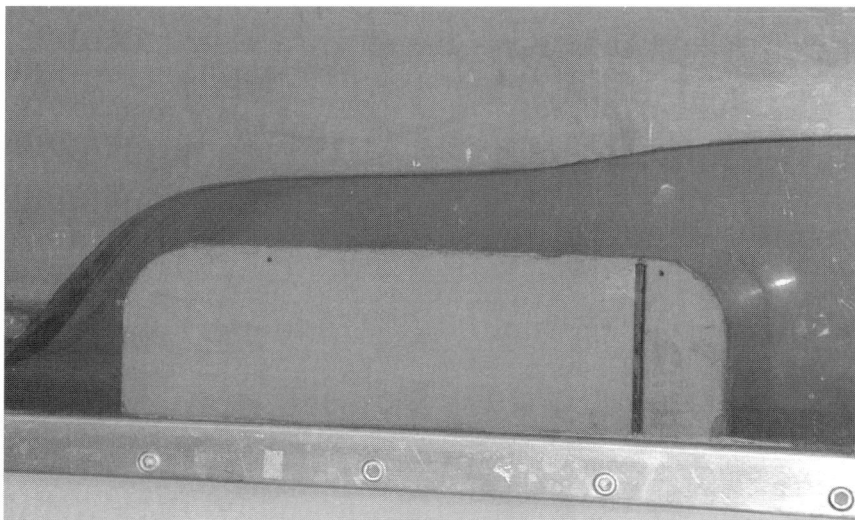

Figure 6.18   Flow over a broad-crested weir

Figure 6.19   Compound rectangular weir at a gauging station (note thin-crest thickness and gravel accretion on upstream)

compound weirs are sometimes used. These consist of a narrow rectangular or triangular notch set in a wider opening (Figure 6.19).

Flumes

A flume is a gradual constriction in the width of the channel, sometimes also incorporating a rise in the bed level (Figures 6.17(b) and 6.20). The converging section is quite sharp and is followed by a prismatic (constant cross-section) and a gradual expansion zone. Flumes are therefore well suited to channels carrying debris and silt, and also have the advantage of requiring a smaller difference between upstream and downstream water levels than weirs. However, care is required in designing flumes to ensure that they do not become 'drowned' at high flows due to high tailwater levels.

The values of the respective coefficients in Equation (6.42) and the limitations of the devices are given in Table 6.5. More information on the characteristics of each device and their applications can be found in British Standard 3680 (BS 3680).

Figure 6.20   Laboratory flume looking upstream

Table 6.5 Characteristics of weirs and flumes (BS 3680)

Type	C	n	Coefficient of discharge	Limitations
Flume	$1 \cdot 75 \, Cd$	1·5	$(1 - 0 \cdot 006L/b) \times$ $(1 - 0 \cdot 003L/h)^{3/2}$	$h > \max(0 \cdot 05L, 0 \cdot 05 \text{ m})$ $bh/(B(h + p)) = 0 \cdot 7$ $b \geq 0 \cdot 1 \text{ m}$ $h_1/b \leq 3$ $h \leq 2 \text{ m}$
Broad-crested weir	$1 \cdot 705 \, Cd$	1·5	0·85 to 1·245 (see BS 3680)	$h \geq 0 \cdot 06 \text{ m}$ $p > 0 \cdot 15 \text{ m}$ $0 \cdot 1 < h/L < 1 \cdot 6$
Thin-plate rectangular	$2/3 \, \sqrt{2g} \, Cd$	1·5	$0 \cdot 596 + 0 \cdot 091 \, h/P$	$h/p > 2 \cdot 5$ $h > 0 \cdot 03 \text{ m}$ $b \geq 0 \cdot 20 \text{ m}$ $p \geq 0 \cdot 10 \text{ m}$
Thin-plate triangular	$8/15 \, \text{Tan}\vartheta \, Cd\sqrt{2g}$	2·5	0·585 to 0·60 (see BS 3680)	$h/P \leq 0 \cdot 4$ $h/b \leq 0 \cdot 2$ $0 \cdot 05 > h > 0 \cdot 38 \text{ m}$ $p \geq 0 \cdot 45 \text{ m}$ $b \geq 1 \cdot 0 \text{ m}$

Key: L = length of a broad-crested weir or the prismatic throat of a venturi flume in the direction of flow, b = width of a broad-crested weir or the prismatic throat of a venturi flume or the surface width of a thin-plate weir perpendicular to the direction of flow, p = crest height above channel bed, h = height of upstream water surface above the structure crest, B = channel width

6.4.2 Flow measurement techniques in channels

Where it is not economic to install a permanent structure, it is possible to establish a stage–discharge or rating curve for a natural channel, provided it is reasonably straight and free from major obstructions. In such a case, it can be assumed that the flow rate increases as some function of water level (stage). Therefore, if a series of discrete discharge measurements are made in a channel, together with the corresponding stage, a stage–discharge curve can be constructed. Individual flow measurements are also useful to check the calibration of weirs, etc., especially if they become submerged.

Flow can be measured by integrating measurements of velocity across a channel cross-section, using a current meter or various ultrasonic devices or by measuring the change in concentration of a marker introduced into the flow

Current metering

A current meter is a small propeller either mounted on a rod or suspended from a cable, and which is placed at various locations in the channel (Figure 6.22). The speed of rotation of the propeller is recorded using a data logger and is converted into a water velocity using calibration data provided. Because of the variation of velocity vertically and horizontally in a channel, it is necessary to measure the velocity at several locations. Normally, at least ten locations across the channel are used, depending on the size of the channel, and the locations should be spaced so as to include approximately equal flow areas. For relatively shallow depths, a single reading can be taken at each location at a depth of 0·6 times the water depth from the surface, which is where the velocity equates to the mean velocity with depth. For deeper channels, a more accurate value can be obtained by taking the mean of velocity readings at 0·2 times depth and 0·8 times depth for each section (Figures 6.21 and 6.22).

To obtain the flow rate, the individual velocity measurements need to be integrated over the flow cross-section, which can be done in various ways. One of the most straightforward is to assume that the width of each section extends between the mid-points between the neighbouring stations and that the depth measured at the velocity station represents the average depth (Figure 6.21(b)), in which case the flow through a section is given by:

$$q = \frac{1}{2}(x_{n+1} - x_{n-1})\mathrm{d}_n v_n \tag{6.43}$$

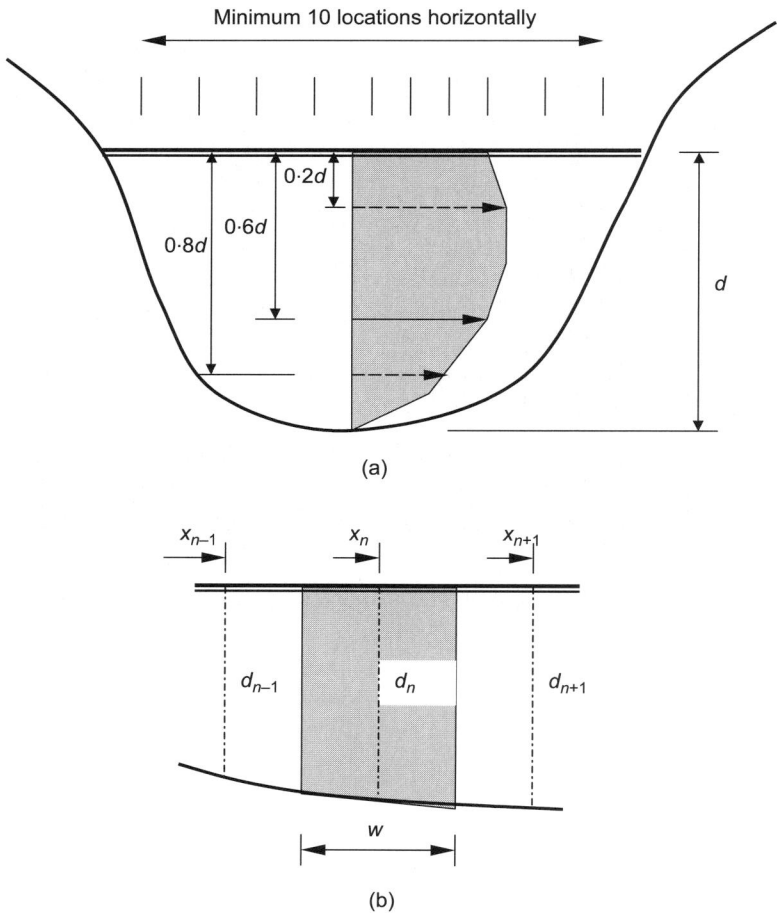

Figure 6.21   Current metering: (a) velocity measurement; and (b) computation method

Adjustments may be needed at the initial and final stations, which are assumed to be at the bank where the depth may, or may not, be zero.

Current metering can only be carried out where the flow is reasonably uniform with no bends or obstructions and little turbulence.

Dilution method

For steep, turbulent flow, an alternative technique, known as dilution measurement, can be used. This relies on the principle of mass balance,

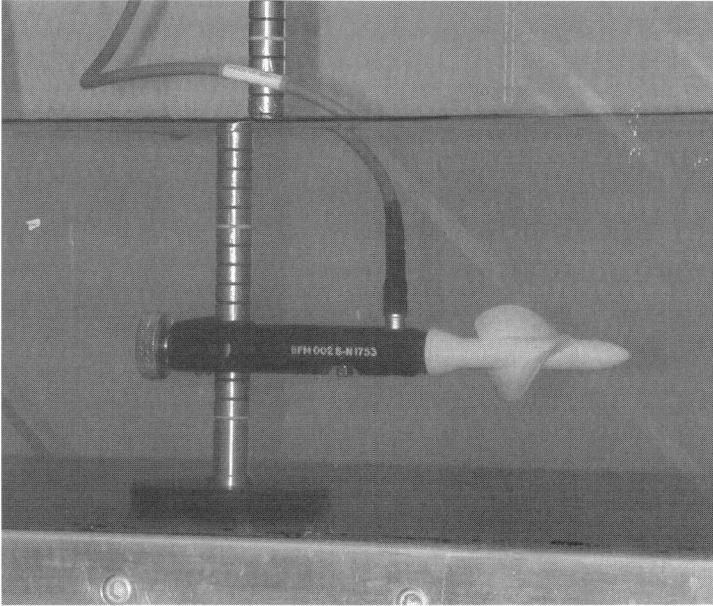

Figure 6.22   Current meter

under which the mass flow of a solute (defined as the product of flow multiplied by concentration) is assumed to remain constant over a short length of channel. This principle can be used in two ways. In one approach, a known mass (*M*) of solute (normally salt) is added to a stream, either directly or dissolved in a small quantity of water. The concentration of salt in the stream at a distance downstream is then monitored at regular intervals of, say, one minute. From the above definition:

Mass flow = Flow × Concentration

$$m = Q \times C \qquad\qquad (6.44)$$

Over a short period d*t*:

$$\int m \times \mathrm{d}t = Q \int C \mathrm{d}t \qquad\qquad (6.45)$$

or:

$$Q = \frac{M}{\int C dt} \tag{6.46}$$

where C is the observed concentration due to the added salt and M is the total mass of salt. There is normally some background concentration (C_0), which can be measured by taking readings before the salt is added, and so C can be expressed as:

$$C = C_1 - C_0 \tag{6.47}$$

Hence:

$$Q = \frac{M}{\int (C_1 - C_0) dt} \tag{6.48}$$

where C_1 is the actual observed salt concentration.

The integral can be evaluated by plotting the concentrations and using a graphical or numerical integration technique (Figure 6.23).

An alternative approach uses a continuous flow of a salt solution of known concentration (C_2), which is added to the main flow (Q) at a known rate (q). The mean concentration in the stream then increases from the initial background level (C_0) to a new level (C_1) (Figure 6.24). The mass balance equation then becomes:

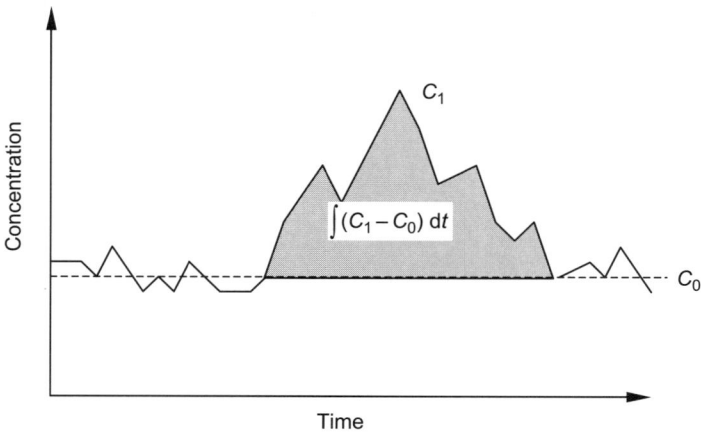

Figure 6.23   Constant volume dilution method

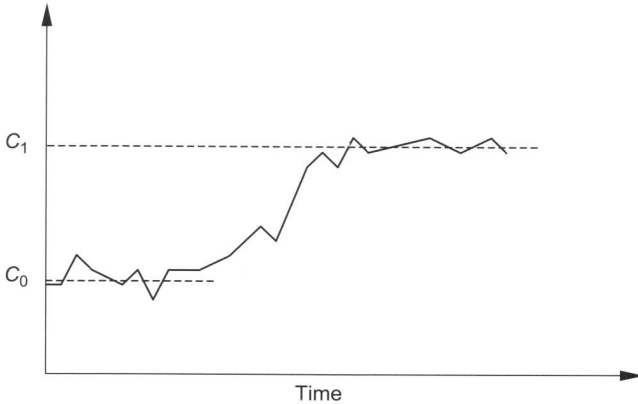

Figure 6.24   Constant flow dilution method

$$qC_2 + QC_0 = (Q+q)C_1 \qquad (6.49)$$

Rearranging this gives:

$$Q = \frac{C_2 - C_1}{C_1 - C_0} q$$

$$\approx \frac{C_2}{C_1 - C_0} q \qquad (6.50)$$

if $C_1 \ll C_2$.

For both dilution methods, the observation point needs to be located sufficiently far downstream for complete mixing to occur (at least 25 times the width of the channel) and this technique is thus well suited to turbulent mountain streams. Nevertheless, care should be taken that the addition of the solute does not have any serious harmful effect on the ecology of the watercourse. The volume of salt should be chosen such that concentration of salt in the water course does not normally exceed 200 mg/l with an absolute maximum of 700 mg/l. Flow estimation by dilution gauging should generally give results within about $\pm 5\%$ of flow readings from conventional weirs (Butterworth *et al.*, 2000). Portable meters with integrated software to perform the discharge calculations are now available.

Flow measurement using air bubbles

This method of flow measurement involves the release of air bubbles from a perforated pipe laid on the bed of a channel (Toop *et al.*, 1997). It is assumed that the terminal velocity of the bubbles is constant and is given by the equation:

$$V_T = \sqrt{\frac{8rg}{3C_D}} \qquad (6.51)$$

where r is the bubble radius and C_D is the drag coefficient ($=2.72$).

The flow can then be found by integrating the horizontal area (A) between the line of the pipe and the line of the first bubbles, and is given by:

$$Q = V_T \times A \qquad (6.52)$$

The assumption of uniform and constant bubble velocity implies a steady horizontal uniform flow without turbulence and secondary currents. Tests have shown the method to be quick and accurate for single measurements, but for continuous flow recording, complex image processing equipment would be required (Figure 6.25).

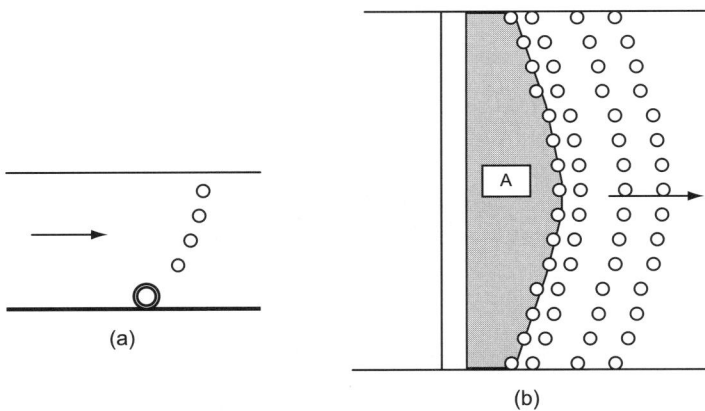

Figure 6.25   Air bubble method of flow measurement: (a) section; and (b) plan

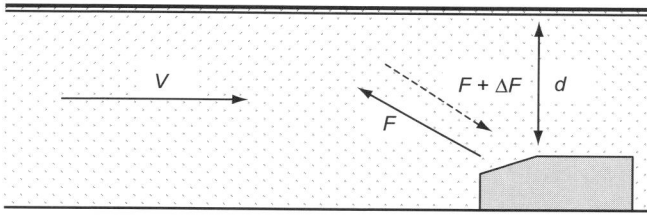

Figure 6.26   Ultrasonic velocity measurement using the Doppler principle

Ultrasonic methods

A technique that is particularly suited to small artificial channels or partly full pipes uses ultrasonic sound waves. In one system, a signal of known frequency is transmitted from a transducer on the base of the channel (Figure 6.26). The frequency of the return signal, which is reflected from sediment and air bubbles in the flow, is compared with that of the original signal. The difference in frequency (ΔF) is due to the Doppler Effect, resulting from the motion of the particles and can be used to estimate the velocity of the flow. A pressure transducer is also used to estimate the depth, and from the dimensions of the channel or pipe the flow rate can be calculated. This method has the advantage that there is very little disturbance to the flow and it is particularly suited to channels which contain sediment or debris. However, since only one velocity reading is used, it is not suitable for natural channels where there is a wide variation in velocity.

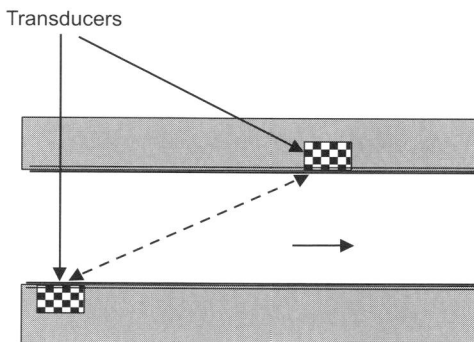

Figure 6.27   Ultrasonic velocity measurement

An alternative ultrasonic method involves the measurement of the transit times of sound pulses transmitted alternately in downstream and upstream directions (Figure 6.27). Two transducers are mounted in staggered positions on opposite banks. Each transducer acts alternately as a transmitter and receiver, sending and receiving pulses of sound. The sound will naturally travel faster in the downstream direction than in the upstream direction, and the difference in the upstream and downstream transit times can be used to estimate the velocity of the water. Allowance needs to be made for the temperature gradients, which can cause refraction of the sound waves and also for the attenuation of the signal due to air bubbles and sediment.

6.4.3 *Rating curves for channels*

Most measurements of discharge are carried out at low or moderate flows, and it is often required to estimate discharges corresponding to large floods where only high water marks are available. It is therefore necessary to extend a rating curve beyond the range of observations.

A simple graphical extension of the stage-discharge curve using natural scales will work if the extension is small and the rating curve is well defined over its upper range. Extending the rating curve is easier if the relationship is linearised by, for example, plotting on logarithmic axes. The logarithmic extension method assumes a relationship of the form:

$$Q = K(S - a)^b \qquad (6.53)$$

where Q and S are the discharge and stage respectively, and K, a and b are constants. The constant a represents the intercept of the stage–discharge curve on the stage axis and can be estimated initially by a simple plot. If a is non-zero, the stage–discharge curve will be non-linear when plotted on a log–log scale. Various values of a can be tried until the data plots as a straight line on a log–log scale. The constants b and K can then be estimated from the slope and the intercept of the plot.

An alternative approach is based on the Chezy Equation (Equation (6.25)). For a wide shallow rectangular channel the hydraulic radius (m) tends to the depth (d), and the Chezy Equation can thus be written as:

$$Q = AC\sqrt{d}\sqrt{i} \qquad (6.54)$$

For a given location, this can be expressed as:

$$Q = KA\sqrt{d} \qquad\qquad (6.55)$$

Thus, if the discharge is plotted against $A\sqrt{d}$ for each of the observed measurements, a linear relationship should result.

Another similar method assumes that the cross-section can be represented by:

$$A = Bd^{m} \qquad\qquad (6.56)$$

in which case the Chezy Equation can be written as:

$$Q = \text{constant} \times d^{n} \qquad\qquad (6.57)$$

which can also be linearised by plotting on a log scale.

Although a channel rating curve can be extended using the above techniques, when the water level is above bankfull level, there is generally a significant change in the cross-sectional shape and friction coefficient due to vegetation and other obstructions, and a simple linear relationship may not be valid.

It is also often found that, even at normal flows, the relationship between stage and discharge is not unique. In other words, for the same stage there may be more than one discharge. This is due to the effect of channel and bank storage, and to the variation in the slope of the water surface. Furthermore, when the flow is increasing, the slope of the water surface for a given depth is greater than for the same depth when the flow is receding (Figure 6.28). Since, for varying flow, the discharge depends on the slope of the water surface rather than the slope of the channel, the flow on the rising part of a hydrograph (Q_1) will be greater then the flow at the same depth on the falling part (Q_2).

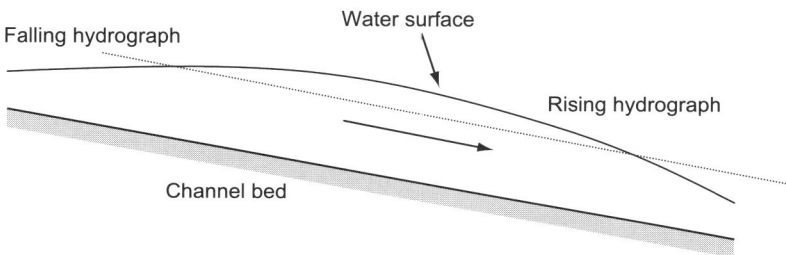

Figure 6.28   Variation in water surface slope

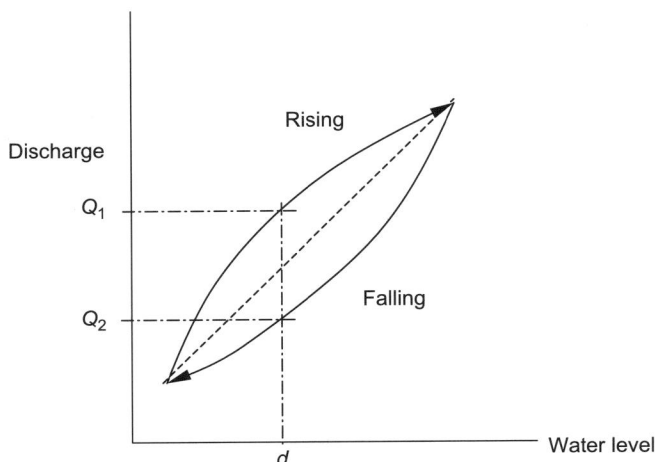

Figure 6.29   Water level discharge curves

As a result of these phenomena, it is possible, in principle, to define separate rating curves for the rising and falling limbs of a hydrograph. Sometimes, the discharge readings during a flood are corrected for a theoretical steady state condition, but, in the long term, the rising and falling stages will tend to cancel (Figure 6.29).

6.5   Summary

There are several flow pathways which water can follow from the point at which it falls on the ground to where it is observed at the outflow of a catchment. They may involve flow over the surface of the ground or through the body of the soil or rock. The essential hydrological process which is involved with all these modes of flow is attenuation, that is the delay in the peak flow and the reduction in the peak flow level. This can vary from a few minutes in the case of surface flow, to several days or longer where the flow is mainly through the ground. In open channels, the natural attenuation of the channel may be enhanced by the construction of artificial reservoirs.

6.6 References

Boorman, D. B., Hollis, J. M. *et al.* (1995). *Hydrology of Soil Types: a hydrologically-based classification of the soil types of the United Kingdom.* Centre for Ecology and Hydrology, Wallingford.

British Standards Institution. *BS-3680: Methods of Measurement of Liquid Flow in Open Channels.* BSI, London.

Butterworth, J. A., Hewitt, E. J. *et al.* (2000). Discharge Measurement Using Portable Dilution Gauging Flowmeters. *Journal of CIWEM*, **14**, 436–441.

Crawford, N. H. and Linsley, R. K. (1966). *Digital Simulation in Hydrology. The Stanford Watershed Model IV.* Stanford University.

Fleming, G. (1966). *Computer Simulation Techniques in Hydrology.* Elsevier, New York.

Gustard, A., Bullock, A. *et al.* (1992). *Low Flow Estimation in the United Kingdom.* Institute of Hydrology, Wallingford.

Herbert, R., Barker, J. A. *et al.* (1997). Exploiting Groundwater from Sand Rivers in Botswana using Collector Wells. *Proceedings of the 30th International Geology Congress.*

Marshall, T. J. and Holmes, J. W. (1979). *Soil Physics.* Cambridge University Press, Cambridge.

Nilsson, A. (1988). *Groundwater Dams for Small Scale Water Supply.* IT Publications, London.

Toop, C. J., Webster, P. *et al.* (1997). Improved Guidelines for the Use of the Rising Air Float Technique for River Gauging. *Journal of CIWEM*, **11**, 61–66.

Appendix 6.1
The Hydrology of Soil Type (HOST) classification scheme (Boorman *et al.*, 1995)

Notes to table on following two pages.

Gleyed layers refer to layers of soft clay.

Also unclassified (urban) areas and lakes.

No extensive UK soil types exist outside the table or within the bold portions of the table.

IAC used to index lateral saturated hydraulic conductivity.

IAC# used to index soil water storage capacity.

Substrate hydrogeology	Mineral soils				Peat soils
	Groundwater or aquifer	No impermeable or gleyed layer within 100 cm	Impermeable layer within 100 cm or gleyed layer at 40–100 cm	Gleyed layer within 40 cm	
Weakly consolidated, microporous, by-pass flow uncommon (chalk)	Normally present and at >2 m	1			
Weakly consolidated, microporous, by-pass flow uncommon (limestone)		29			
Weakly consolidated, macroporous, by-pass flow uncommon		2	12	13	14
Strongly consolidated non or slightly porous, by-pass flow common		3			
Unconsolidated, macroporous, by-pass flow very uncommon		4			
Unconsolidated, microporous, by-pass flow common		5			

Substrate hydrogeology	Groundwater or aquifer	Mineral soils — No impermeable or gleyed layer within 100 cm	Impermeable layer within 100 cm or gleyed layer at 40–100 cm (IAC#>7.5)	(IAC#<7.5)	Gleyed layer within 40 cm (IAC<12.5 [1 m day^{-1}])	(IAC>12.5 [1 m day^{-1}])	Peat soils — Drained	Undrained
Unconsolidated, macroporous, by-pass flow very uncommon	Normally present and at >2 m		6					
Unconsolidated, microporous, by-pass flow common			7		8	9	10	11
Slowly permeable	No significant groundwater or aquifer	15	17	20		23		25
Impermeable (hard)		16	18	21				26
Impermeable (soft)			19	22	24			
Eroded peat								27
Raw peat								28 (8·06)

Appendix 6.2
Estimation of velocity in an aquifer by salt dilution method (from Herbert *et al.*, 1997)

The velocity in the aquifer is measured by adding a salt solution to the borehole water and measuring the decrease in salt concentration over a period of time.

Referring to Figure 6.7, the following assumptions are made.

- The concentration within the borehole remains uniform and equal to the concentration leaving the borehole.
- At time zero the concentration in the borehole is instantaneously raised to C_I.
- Water enters the borehole from an aquifer thickness equal to the screened length of the borehole (i.e. there is no vertical flow in the aquifer).
- Water upstream of the borehole is at a uniform concentration of C_B.
- The flow is steady state.

The volume of water in the borehole is:

$$V = \frac{\pi D^2 L_{SAT}}{4}$$

After a time t a volume ΔV has entered (and left) the borehole, where:

$$\Delta V = v \times n \times L_{SCR} \times D \times \alpha \times t$$

where n is the porosity of the aquifer and α is the ratio of the width of the aquifer contributing flow to the borehole to the borehole diameter.

The flow of water removes the salt and reduces the concentration exponentially.

During a time t to $t + dt$, the concentration reduces from C' to $C' - dC$, where:

$$dC' = \frac{-C' \times dV}{V}$$

and:

$$C' = C - C_B$$

dV is the volume of water added in time dt, integrating:

$$\int_{C_I-C_B}^{C-C_B} \frac{1}{C'}dC' = -\frac{\Delta V}{V}$$

$$\ln\left(\frac{C-C_B}{C_I-C_B}\right) = -\frac{\Delta V}{V} = -\frac{vnL_{SCR}D\alpha t}{\dfrac{\pi D^2 L_{SAT}}{4}}$$

$$v = -\frac{\pi\Delta L_{SAT}}{4nL_{SCR}\alpha t}\ln\left(\frac{C-C_B}{C_I-C_B}\right)$$

Appendix 6.3
Derivation of channel routing formula

From the basic storage equation (Equation (6.34)):

$$\frac{P_n + P_{n+1}}{2} = \frac{Q_n + Q_{n+1}}{2} + \frac{S_{n+1} - S_n}{\Delta t}$$

rearranging:

$$Q_{n+1} + \frac{2S_{n+1}}{\Delta t} = P_n + P_{n+1} - Q_n + \frac{2S_n}{\Delta t}$$

Assuming a storage relationship for a channel:

$$S_n = k[xP_n + (1-x)Q_n]$$

Therefore:

$$Q_{n+1} + \frac{2k}{\Delta t}(xP_{n+1} + Q_{n+1} - xQ_{n+1}) = P_n + P_{n+1} - Q_n + \frac{2k}{\Delta t}(xP_n + Q_n - xQ_n)$$

Multiplying by $\Delta t/2$:

$$\frac{\Delta t}{2}Q_{n+1} + kxP_{n+1} + kQ_{n+1} - kQ_{n+1}$$

$$= \frac{\Delta t}{2}P_n + \frac{\Delta t}{2}P_{n+1} - \frac{\Delta t}{2}Q_n + kxP_n + kQ_n - kQ_n - kxQ_n$$

$$Q_{n+1}(k - kx + 0.5\Delta t) = P_{n+1}(-kx + 0.5\Delta t) + P_n(kx + 0.5\Delta t) + Q_n(k - kx - 0.5\Delta t)$$

Hence:

$$Q_{n+1} = C_0 P_{n+1} + C_1 P_n + C_2 Q_n$$

where:

$$C_0 = \frac{-(kx - 0.5\Delta t)}{(k - kx + 0.5\Delta t)}$$

$$C_1 = \frac{(kx + 0.5\Delta t)}{(k - kx + 0.5\Delta t)}$$

$$C_2 = \frac{(k - kx - 0.5\Delta t)}{(k - kx + 0.5\Delta t)}$$

Chapter 7

Hydrological models

This chapter considers the various methods which may be used to model the relationship between rainfall and the consequent runoff. The two basic processes involved in the transformation of rainfall to runoff are the losses due to evaporation, interception, etc., and the attenuation due to the storage on the surface or in the ground. Some models consider only the surface or near surface runoff, treating any flow into the ground as a loss with an allowance for baseflow being made at a later stage.

7.1   General principles of hydrological modelling

The transformation between rainfall and runoff depends on a number of factors, including:

- the space–time distribution of precipitation and evaporation
- the topography of the basin in terms of the direction and magnitude of slopes
- the conductivity, porosity and storage capacity of the soil layers
- the hydro-geological characteristics of the underlying rock strata
- the vegetation, land use and agricultural practice.

Ideally, these data should be available for all points on the surface of the catchment and for some depth below the surface over the time-scale of the model. This is clearly impossible and so all models involve approximations of data. Some models, referred to as *lumped* models, assume that the whole catchment can be represented by a single set of parameters, i.e. that there is no variation across the catchment. In

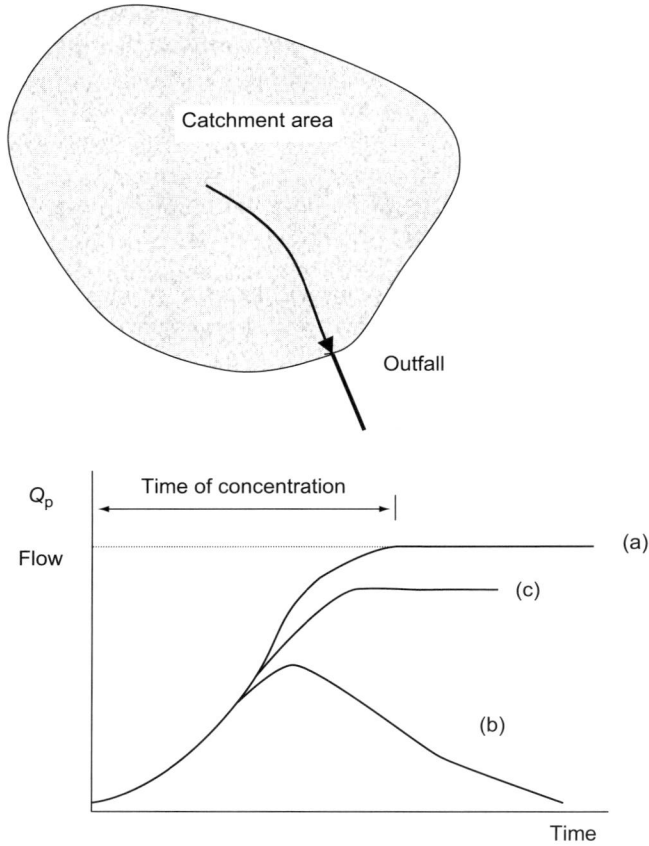

Figure 7.1   Typical drainage area and hydrograph

which is known as the *Modified Rational Formula*. The runoff coefficient (C) allows for losses due to infiltration, evaporation and storage in surface depressions, etc. In urban areas most of the runoff is generated by paved areas and the area (A) is sometimes equated to the area of paved surfaces connected to the drainage system on the basis that there is no runoff from unpaved areas. The runoff coefficient then represents the proportion of runoff from the paved areas only. In practice, there is usually some runoff from unpaved areas, just as there are losses associated with infiltration and depression storage, etc., on paved areas. The overall proportion of runoff therefore depends on the proportion and permeability of paved

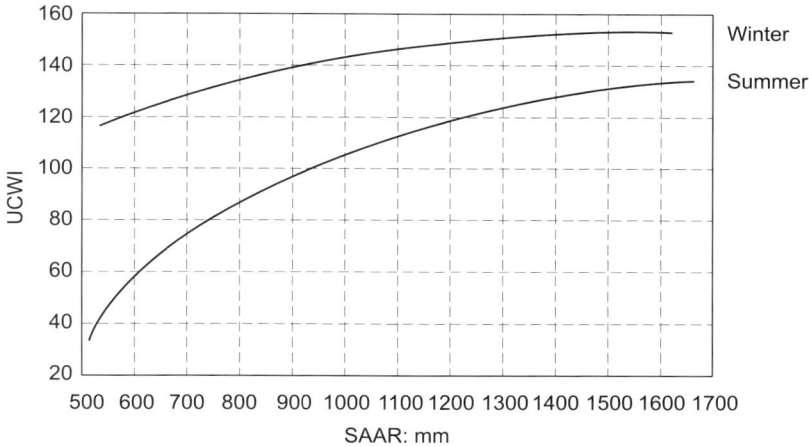

Figure 7.2 Relationship between UCWI and SAAR (reproduced with the permission of the Centre for Ecology and Hydrology, Wallingford)

areas, the wetness of the ground, the catchment slope and the soil type, as well as the rainfall depth and duration. The proportion of runoff may be estimated on the basis of local experience or an empirical equation such as Equation (7.4) may be used (Hydraulics Research, 1983):

$$PR = 0.829 \times PIMP + 25.0 \times SOIL + 0.078 \times UCWI - 20.7 \tag{7.4}$$

where PR is the proportion of runoff (from the total catchment area), $PIMP$ is the percentage impermeable area, $SOIL$ is the soil index, and $UCWI$ is the Urban Catchment Wetness Index.

$SOIL$ is a parameter that varies between 0·15 for high-infiltration soils, such as sands and gravels, and 0·5 for low-infiltration soils, such as clays. The Urban Catchment Wetness Index ($UCWI$) is a parameter based on the Standard Average Annual Rainfall ($SAAR$) and can be estimated from a map (Figure 7.2). It should be stressed that the above equation is based solely on empirical data and it is not reliable if used outside of a certain range for each variable. It is therefore unwise to use the equation for small sub-catchments which may, for example, have 100% impervious areas.

The main difficulty in the use of the Rational Method is in the choice of the rainfall intensity. As described in Chapter 5, for a given location the design storm intensity depends on the designated return period and the duration of the rainfall event. The choice of a return period is essentially an economic or social decision and is related to the consequences of any

Table 7.1   Return periods for urban drainage in Scotland

Location	Design condition	Return period: years
Average slope greater than 1%	Pipe full — no surcharge	1
Average slope 1% or less	Pipe full — no surcharge	2
Sites with severe consequences of flooding	Surcharging	5

flooding in terms of economic and social disruption. Table 7.1 gives examples of recommended return periods for urban drainage schemes in Scotland (WRc, 2001).

The duration of the design rainfall event is important because there is generally an inverse relationship between average intensity and duration. The design storm duration for the Rational Method is usually taken to be equal to the time of concentration for a particular sewer length. If the rainfall duration were less than the time of concentration, the peak flow at the outfall would not be reached (Figure 7.1 curve (b)). However, if the duration of the storm were longer than the time of concentration, the corresponding intensity would be less and, therefore, the runoff would be less (curve(c)). It follows that, in an urban catchment consisting of many sub-catchments draining to various pipe lengths, the design storm will be different for each pipe length and will generally increase towards the outfall. The assumption of the storm duration being equal to the time of concentration implies that the contributing area of the sub-catchment increases uniformly with time, i.e. that the time-area diagram is linear (Section 7.3). In practice, the time-area diagram will generally be non-linear, in which case the duration of the critical design storm will be less than the time of concentration. A more accurate design storm duration can be estimated by drawing a tangent to the time area diagram.

The Rational Method also simplifies the flow process by assuming that the time of concentration for a point in a drainage system is the sum of the *time of entry* (the time for overland flow from the furthest part of a sub-catchment to the nearest entry point of the pipe system) and the *time of flow* (the time to flow from the entry point to the point in question). The time of entry depends on the length, slope and nature of the sub-catchment, as well as the rainfall intensity. Typical values recommended in the *Wallingford Procedure* range from three minutes for a five-year return period storm on a small catchment to eight minutes for a one-year return period storm on a large flat catchment. The time of flow is estimated from hydraulic tables using mean velocities, assuming the pipe is flowing full. Since the velocity is a function of the pipe diameter, an iterative process is required when designing a new system (Example 7.1).

The Modified Rational Method has been incorporated in the *Wallingford Procedure* for stormwater drainage. However, there are a number of limitations to the method, which make it unsuitable for catchments greater than about 1500 hectares in area or for pipes greater than about 600 mm in diameter. In particular, it only gives a peak flow and it does not allow for storage in the pipes, which will attenuate the flow in large pipes. Also, the Rational Method assumes a uniform rainfall intensity and may underestimate flows where the rainfall pattern is highly peaked, i.e. where the peak intensity is significantly greater than the mean value. It has also been found to be unreliable for irregularly shaped catchments where there is not a uniform increase in area with distance from the outfall.

Example 7.1

An urban catchment is drained by a system consisting of four pipes, as shown in the diagram overleaf. The details of the sub-catchments and the pipes are given in the table below. Using the Modified Rational Method with the rainfall intensity duration data given below, suggest suitable pipe sizes. The runoff coefficient is 0·8 and the time of entry is 5 minutes.

Duration: min	Intensity: mm/h	Run-pipe	1–1	1–2	2–1	1–3
5	51·8	Length (m)	70	95	120	75
6	44·6	Gradient (1 in.)	80	100	120	100
7	39·5	Area (ha)	0·74	1·65	0·42	0·81
8	35·4					

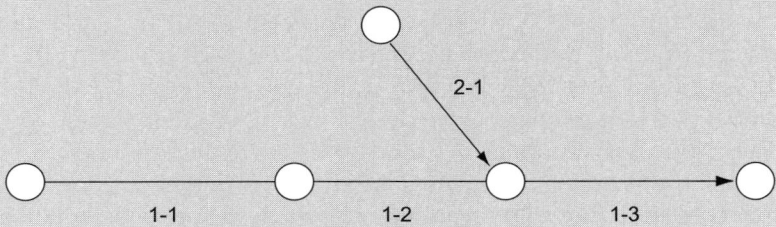

For pipe-run1 length 1 (sub area 1–1):
Gradient = 1 in 80, assume diameter = 150 mm (minimum allowable)
Pipe full velocity = 1·12 m/s (from Colebrook-White formula or pipe tables)

$$\text{Time of flow} = \frac{70}{1 \cdot 12 \times 60} = 1 \cdot 04 \text{ min}$$

Time of concentration = 5·0 + 1·04 = 6·04 min
Design rainfall = 44·4 mm/h (by interpolation from intensity duration data)
Contributing area = 0·74 ha
Design flow = 3·61 × 0·8 × 44·4 × 0·74 = 94·9 l/s (Modified Rational Formula)
Pipe capacity = 19·9 l/s (from Colebrook-White formula or pipe tables)

This is inadequate, therefore increase diameter to, say, 300 mm. The calculations are repeated in column 3 of the table below.

(1)	(2)	(3)	(4)	(5)	(6)
Run-pipe		1–1	1–2	2–1	1–3
Length	(m)	70	95	120	75
Gradient	(1 in.)	80	100	120	100
Diameter	(mm)	300	450	300	600
Velocity	(m/s)	1·76	2·03	1·43	2·43
Time of flow	(min)	0·66	0·78	1·40	0·52
Time of conc.	(min)	5·66	6·44	6·40	6·96
Rainfall	(mm/h)	47·0	42·3	42·6	39·7
Area	(ha)	0·74	1·65	0·42	0·81
Cum. area	(ha)	0·74	2·39	0·42	3·62
Flow	(l/s)	100·5	292·3	51·6	415·3
Capacity	(l/s)	124·3	323·2	101·3	688·2

For pipe 1–2, the contributing area will be the cumulative area (i.e. 0·74 + 1·65 = 2·39 ha), and the time of concentration will be the time of concentration for the preceding pipe + time of flow. *For pipe 1–3*, the contributing area will the total area and the time of concentration will be the longer of the times of concentration for the upstream pipes (6·44 min) + time of flow (0·52 min).

7.3   The time-area method

The time-area concept was first proposed in the context of flood routing and, later, it was developed for urban drainage design by the Transport and Road Research Laboratory. The basic principle is that a catchment can be divided into zones of equal travel time, i.e. the time taken by a water particle to travel from its point of impact to a specified outfall. For example, in Figure 7.3 the lines represent travel time increments.

The average flow recorded at the outfall during the first time increment of a storm will be

$$Q_1 = a_1 \times R_1 \tag{7.5}$$

where a_1 is the area of the first time zone and R_1 is the effective rainfall over the first time increment. Over the next interval the average flow will be:

$$Q_2 = a_1 \times R_2 + a_2 \times R_1 \tag{7.6}$$

where a_2 is the area of the second time zone and R_2 is the rainfall during the second period.

In general discrete terms, the flow at time t can be expressed as:

$$Q(t) = \sum_{n=0}^{n_{max}-1} (A_{n+1} - A_n)R(t-n) \tag{7.7}$$

where A is the cumulative time area up to a time t.

The variation of the cumulative time area with distance from the outlet obviously depends on the shape and surface characteristics of the catchment. The time-area diagram can be estimated for small urban catchments from a study of drainage maps, but for rural catchments it is

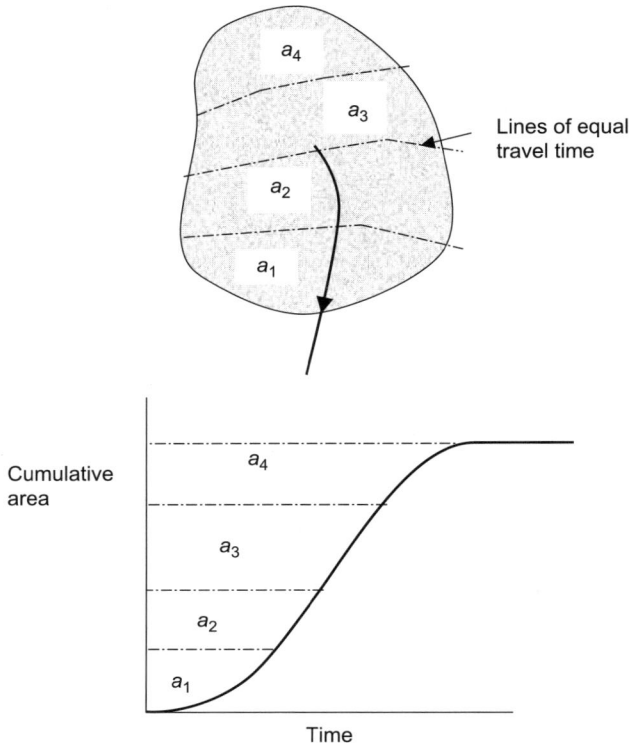

Figure 7.3   Division of catchment into time areas

more difficult. Calver (1996) proposed a general topographical relationship between the cumulative area and the distance from the outlet of the form:

$$(1-x') = (1-A')^{1/c} \tag{7.8}$$

where x' and A' are the normalised distance from the outfall (x/x_{max}) and the normalised area (A/A_{max}) respectively, and the constant c is about 3 depending on the stream density. The variation of cumulative area with time also depends on the velocity at which the water moves across the catchment. It has been found that this velocity tends to be greater in the wetter areas near the channels. Thus, the areas bounded by travel time zones will tend to increase towards the boundaries of the catchment. In most cases, the travel time in the channels is much less than the over-ground response time and can be neglected.

Although it may be considered as a lumped model, the time-area formulation has the potential to perform as a distributed model by incorporating non-uniform rainfall and a spatially varying catchment (Saghafian, 1998), and, in any case, it can reflect the shape and drainage pattern of the catchment, although the determination of the time-area boundaries can be very difficult (Example 7.2).

7.4   The unit hydrograph method

As has been seen above, one of the problems with the time area concept is identifying which areas of the catchment belong to which time zones. The unit hydrograph concept avoids this problem as it represents the various time delays by a time distribution of runoff, which is not directly related to specific areas.

7.4.1   *General principles of the unit hydrograph*

The shape of the hydrograph of surface (or near surface) runoff from a period of uniform rainfall for a given catchment is assumed to depend on:

- the duration of rainfall
- the depth of rainfall
- the losses due to evaporation and flow to groundwater, etc.
- the physical characteristics of the catchment.

The first three of these factors obviously vary with time, whereas the physical characteristics of the catchment, such as the shape, slope, vegetation, soil type, etc., can generally be assumed to remain constant with time, at least over the short term. In other words, for given values of rainfall duration and depth, and allowing for any losses, the hydrograph of direct runoff will be a property of the catchment only. The unit hydrograph is defined as 'the hydrograph resulting from a unit depth and duration of effective rainfall'. A unit hydrograph must therefore be associated with specified units of depth and duration. In the United Kingdom, the depth units are normally 1 mm or 1 cm, and the duration is normally 1 minute or 1 h. The term *effective rainfall* means the rainfall after

Example 7.2

An urban catchment can be divided into three sub-catchments with equal drainage times of 5 minutes. The distances from the outlet to the boundaries of the sub-catchments are 0·3 km, 0·7 km and 1·2 km respectively. If the total area of the catchment is 3·6 km², estimate the areas of the sub-catchments. Assuming a runoff coefficient of 0·2, calculate the runoff hydrograph at 5-minute intervals for the storm given.

Time: min	Rainfall intensity: mm/h
0–5	15·0
5–10	19·0

Using the relationship:

$$1 - \frac{A}{A_{max}} = \left(1 - \frac{x}{x_{max}}\right)^3$$

the contributing areas are estimated as shown in the table.

Distance, x	x/x_{max}	$1 - x/x_{max}$	$1 - A/A_{max}$	A/A_{max}	A	ΔA
0·3	0·25	0·75	0·42	0·58	2·09	2·09
0·7	0·58	0·42	0·07	0·93	3·35	1·26
1·2	1·0	0·0	0·0	1·0	3·6	0·25

$t = 5$ min

$$Q = \frac{0.2 \times 2.09 \times 10^6 \times 15.0}{1000 \times 3600} \ \text{m}^3/\text{s} = 0.056 \times 2.09 \times 15.0 = 1.74 \ \text{m}^3/\text{s}$$

$t = 10$ min

$$Q = 0.056 \times (2.09 \times 19.0 + 1.26 \times 15.0) = 3.28 \ \text{m}^3/\text{s}$$

$t = 15$ min

$$Q = 0.056 \times (1.26 \times 19.0 + 0.25 \times 15.0) = 1.55 \ \text{m}^3/\text{s}$$

$t = 20$ min

$$Q = 0.056 \times (0.25 \times 19.0) = 0.27 \ \text{m}^3/\text{s}$$

all losses resulting from evaporation, infiltration, interception, etc., have been deducted.

There are therefore three stages in the application of the unit hydrograph:

(a) the estimation of effective rainfall (which is covered in Chapter 5)
(b) the construction of the unit hydrograph and its convolution with the effective rainfall hyetograph
(c) the addition of the baseflow component.

7.4.2 *The convolution of unit hydrographs with rainfall*

Although the rainfall intensity is assumed to be constant over the specified interval, the unit hydrograph method can be used to construct hydrographs for complex rainfall patterns, provided that they can be broken down into periods of uniform rainfall with a duration equal to the duration associated with the unit hydrograph. The method of convolution depends on two important principles.

The principle of proportionality

This assumes that the ordinates of the hydrograph are proportional to the rainfall. For example, if the depth (or intensity) of rainfall is doubled, the ordinates of the hydrograph will be doubled (Figure 7.4).

The principle of superposition

This assumes that the hydrograph from two consecutive periods of rainfall can be found by adding the hydrograph of the first period of rainfall to that from the second period of rainfall, after displacing the second hydrograph by a time equal to the time of the start of the second period of rainfall. This principle can also be used for two contiguous periods of rainfall (Figure 7.5).

Therefore, it can be seen that any pattern of rainfall can be modelled by a combination of vertical scaling and horizontal superposition. This is illustrated in Table 7.2 where a 1 cm, 1 h unit hydrograph is given in column two. Q_{15} and Q_{20} are the flows from the separate 1 h periods of rainfall of 1·5 cm and 2 cm respectively (note that the flow from the second hour of rainfall is delayed by 1 h). The final column represents the aggregate flow.

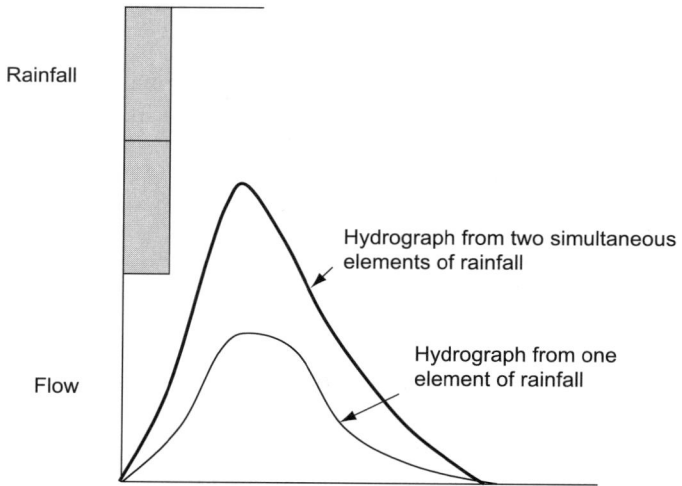

Figure 7.4   Proportionality of unit hydrograph

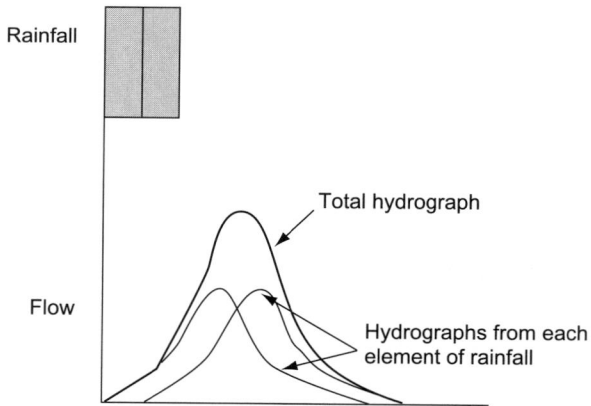

Figure 7.5   Superposition of unit hydrographs

Table 7.2   Application of unit hydrograph

Time: h	Unit hydrograph	Q_{15}	Q_{20}	Q_{tot}
0	0	0	—	0
1	6	9	0	9
2	3	4·5	12	16·5
3	0	—	6	6

7.4.3   Changing the duration of the unit hydrograph

Although a unit hydrograph is associated with a particular duration of rainfall, a unit hydrograph for one duration can be changed into a unit hydrograph of another duration, which may be necessary to suit the time interval of the rainfall record. Two cases may be considered.

Changing a unit hydrograph to a longer duration

If, for example, it is required to convert a five-minute, 1 cm unit hydrograph into a 10-minute, 1 cm unit hydrograph, the hydrograph for two consecutive periods of five minutes is first produced by adding the two unit hydrographs (with the second hydrograph displaced by five minutes), following the principle of superposition outlined above. This will result in the hydrograph for 2 cm of rainfall over 10 minutes and the ordinates must therefore be divided by two to give the unit hydrograph (for 1 cm of rainfall). It should be noted that the area under the unit hydrograph (which represents the depth of rainfall) does not change with the duration of the rainfall: the longer duration unit hydrograph has a longer time-base and lower flow ordinates.

Changing a unit hydrograph to a shorter duration

To change a unit hydrograph to a shorter period it is necessary firstly to produce an S-curve. This is the hydrograph resulting from a series of contiguous rainfall events. It can be produced by adding a series of unit hydrographs, each displaced by the original duration (t_1) associated with the unit hydrograph (Figure 7.6(a)).

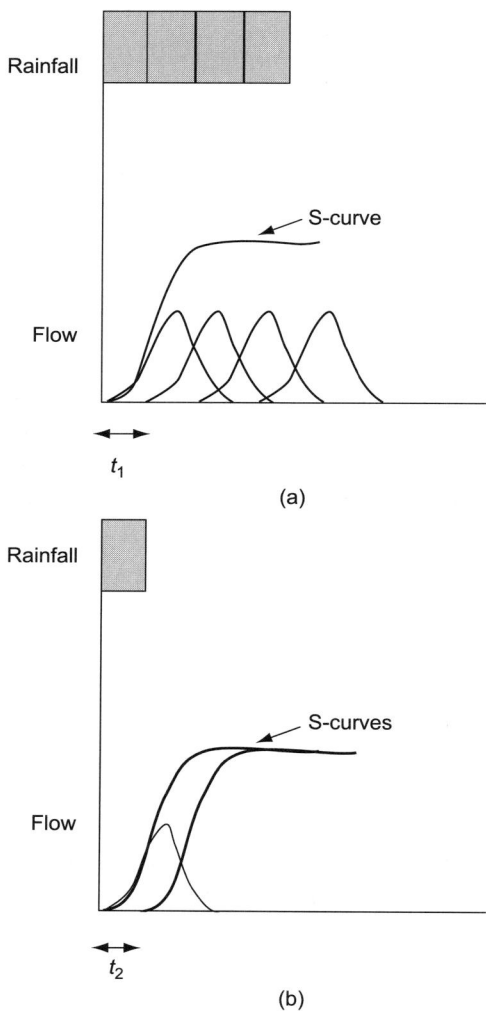

Figure 7.6 (a) Construction of an S-curve; and (b) construction of unit hydrograph from S-curves

Therefore, the S-curve is the hydrograph from a long continuous period of rainfall with an intensity of $1/t_1$. The maximum value of the S-curve represents the flow when the whole catchment is contributing and can be calculated from the product of the catchment area and the rainfall intensity (allowing for appropriate units). To produce the required unit hydrograph for a different period (t_2), the opposite procedure is then carried out: that is, the ordinates of the S-curve are displaced by a period t_2 and subtracted from the original ordinates (Figure 7.6(b)). Since this represents the hydrograph for rainfall with an intensity of $1/t_1$, the ordinates must then be multiplied by t_1/t_2 to give the required unit hydrograph.

Table 7.3 shows how a three-minute unit hydrograph is derived from a five-minute unit hydrograph. It can be seen that it is not necessary to repeat the unit hydrograph indefinitely. Column four is the S-curve displaced by three minutes and since, in the first five minutes, the unit hydrograph is the same as the sum of the S-curve, column three can be filled in progressively.

Although a unit hydrograph refers to a particular duration of rainfall, the principle of a unit hydrograph can be extended to the concept of an *instantaneous unit hydrograph*, which is the runoff from an infinitely short period of effective rainfall. It was seen above that the unit hydrograph of period t_2 can be found by subtracting two S-curves t_2 apart, i.e.:

$$U_{t_2} = (S - S_{-t_2})^{t_1/t_2} \tag{7.9}$$

As t_2 is reduced, the difference in the S-curves tends to dS and t_2 tends to dt and, in the limit, the instantaneous unit hydrograph ordinate at time t becomes:

$$U_t = \frac{\mathrm{d}S}{\mathrm{d}t} \tag{7.10}$$

The instantaneous unit hydrograph is therefore the slope of the S-curve and is a representation of the physical properties of a catchment which affect its response to rainfall.

7.4.4   *Derivation of unit hydrographs from existing records*

Unit hydrographs can be constructed from observed hydrographs with associated rainfall records either by analysis of individual flood events or by consideration of the lag between the rainfall and the resulting hydrograph peak.

Unit hydrographs from observed flood events

Before flow records can be used to construct a unit hydrograph, the baseflow element has to be removed (Section 6.3.1), as a unit hydrograph

Table 7.3   S-curve method

(1)	(2)	(3)	(4)	(5)	(6)
Time: min	5-min unit hydrograph	S-curve	Displaced S-curve	ΔS	3-min unit hydrograph
0	0·0	0·0	—	—	—
1	1·0	1·0	—	—	—
2	2·0	2·0	—	—	0·0
3	4·0	4·0	0·0	4·0	6·7
4	8·0	8·0	1·0	7·0	11·7
5	11·0	11·0	2·0	9·0	15·0
6	13·0	14·0	4·0	10·0	16·7
7	14·0	16·0	8·0	8·0	13·3
8	13·0	17·0	11·0	6·0	10·0
9	11·0	19·0	14·0	5·0	8·3
10	9·5	20·5	16·0	4·5	7·5
11	8·5	22·5	17·0	5·5	9·2
12	7·0	23·0	19·0	4·0	6·7
13	6·8	23·8	20·5	3·3	5·5
14	5·5	24·5	22·5	2·0	3·3
15	4·7	25·2	23·0	2·2	3·7
16	4·0	26·5	23·8	2·7	4·5

only describes surface runoff. The records are then examined to locate isolated storms of almost uniform intensity. If a number of such storms are available, a series of unit hydrographs can be derived. The total depth of rainfall for each storm is established by comparing the area under the hydrograph with the measured rainfall depth and adjusting either or both to obtain agreement. The flow ordinates are then normalised by dividing by the rainfall depth, and the time-base of the hydrograph is adjusted using one of the methods outlined above. Finally, the unit hydrographs obtained in this way from a number of storm events are compared to obtain an average curve.

In most cases it is very difficult to find suitable, isolated, uniform intensity storms, and other more complex methods may be used to obtain unit hydrographs from multiple storms. These are described in various sources, e.g. Linsley *et al.* (1988).

Unit hydrographs from the hydrograph lag

The time to peak of the unit hydrograph can be estimated from flow and rainfall records using the time lag between the rainfall event and the corresponding flood. In the *Flood Estimation Handbook* the following regression equation is proposed to estimate the time to peak for the instantaneous unit hydrograph:

$$Tp(0) = 0.879 LAG^{0.951} \tag{7.11}$$

The *LAG* is defined as the time between the centroid of the rainfall event and the peak runoff (or the centroid of multiple runoff peaks). The lag can be estimated manually from a chart record of water level (stage), but at least six months of records are recommended and the value of the lag should be averaged from several storm events.

7.4.5 Derivation of unit hydrograph from catchment characteristics

If no flow records are available, it is possible to derive a unit hydrograph from the physical characteristics of a catchment. However, it should be recognised that this will be much less reliable than using observed flow records.

Nash (1959) suggested that the transformation of effective rainfall into runoff was analogous to the attenuation of a series of n linear reservoirs where the inflow (P) and outflow (Q) were related by:

$$P = K \times Q \tag{7.12}$$

If a unit impulse of rainfall is applied to the first reservoir, the resulting outflow from the nth reservoir can be regarded as the instantaneous unit hydrograph given by:

$$U(0,t) = \frac{1}{K\Gamma(n)} \left(\frac{t}{K} \right)^{n-1} e^{-t/k} \tag{7.13}$$

where $\Gamma(n)$ is the gamma function.

For catchments with flow and rainfall records, the values of n and K are related to the moments of the flow hydrograph and rainfall hyetograph:

$$K = m_1 m_2$$
$$n = \frac{1}{m_2} \tag{7.14}$$

where:

$$m_1 = Q_1' - R_1'$$

and:

$$m_2 = \frac{Q_2 - R_2}{(Q_1' - R_1')^2}$$

Q'_1 and R'_1 are the first moments of the hydrograph and hyetograph, respectively, about the origin, and Q_2 and R_2 are the corresponding second moments about the centroid. Where there are no flow records, the following regression equations are suggested:

$$m_1 = 36.7 A^{0.3} OLS^{-0.3} = 23.1 L^{0.3} EA^{-0.33}$$
$$m_2 = 0.39 L^{-0.1} \tag{7.15}$$

The Nash model is limited to small or medium-sized catchments in the UK unless more specific regression equations are obtained using gauged catchments.

Some workers have related the peak flow and the time to peak of the instantaneous unit hydrograph to various geomorphological features of the catchment, such as the stream lengths and density leading to the concept of the *geomorphological instantaneous unit hydrograph* (GIUH). Bhaskar *et al.* (1997) developed a regression relationship between the peak flow and time to peak of the instantaneous unit hydrograph, and the ratio of the stream lengths and catchment areas for different orders of streams using a topographical ordering classification. It was also shown that these could be related to the parameters n and K in Equations (7.13) and (7.14)

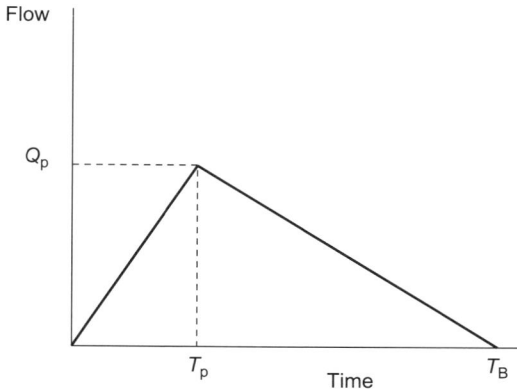

Figure 7.7   Simplified unit hydrograph

and hence the shape of the instantaneous unit hydrograph could be found.

The *Flood Studies Report* (NERC, 1975) simplifies the shape of the unit hydrograph to a triangle which is defined by three properties: the time to peak (Tp), the base time (T_B) and the peak flow (Qp). It uses empirical regression coefficients between the time to peak and various catchment descriptors representing the catchment area, the length and slope of the main stream channel, and the proportion of the catchment which drained through a lake or reservoir. These descriptors are usually derived manually from maps and this procedure is rather time consuming and involves an element of skill and subjectivity (Figure 7.7).

The *Flood Estimation Handbook* (Houghton-Carr, 1999), which supersedes the *Flood Studies Report*, uses the same triangular model for the unit hydrograph but uses descriptors which are based on digital terrain models and are calculated automatically. This means that the parameters can be more complex and subtle, and can therefore represent the hydrological characteristics of the catchments more reliably. The *Flood Estimation Handbook* uses the time to peak of the instantaneous unit hydrograph ($Tp(0)$) as the basis for estimation and the proposed regression equation is:

$$Tp(0) = 4 \cdot 270 DPSBAR^{-0 \cdot 35} PROPWET^{-0 \cdot 80} DPLBAR^{0 \cdot 54} (1 + URBEXT)^{-5 \cdot 77}$$

$$(7.16)$$

The four catchment descriptors used in the above equation describe the topography of the catchment, the soil wetness and the extent of urban development, and are normally computed from a *Digital Elevation Model* (DEM) grid. The parameter *DPSBAR* measures the catchment *slope* and is calculated from the mean of all the slopes (m/km) between adjacent nodes on the DEM grid. The parameter *DPLBAR* which describes the *shape* of the catchment is the mean distance between each node in the catchment and the outlet. The parameter *PROPWET* describes the wetness of the catchment, reflecting the antecedent weather conditions, which have a major effect on the proportion of runoff from a storm event. The soil wetness is indexed by the *soil moisture deficit* (SMD), which is determined as the cumulative sum of daily rainfall, less actual evaporation, calculated on a monthly basis. The parameter *PROPWET* for a catchment is defined as the proportion of time that the *SMD* was less than 6 mm over the period 1961–90. *URBEXT* is a measure of the proportion of urban area in the catchment, and is described in Section 5.7.4.

The time to peak of the instantaneous unit hydrograph needs to be adjusted so that the period of the unit hydrograph coincides with the data interval. It is recommended that the time interval should be about 10–20% of $Tp(0)$. The time to peak for a unit hydrograph of period ΔT can be calculated using:

$$Tp = Tp(0) + \frac{\Delta T}{2} \tag{7.17}$$

or using the S-curve technique outlined above.

The peak flow of the unit hydrograph (Up) and the base time (T_B) are estimated by regression and by continuity (since the area under the unit hydrograph must equal the volume of effective rainfall) as:

$$Up = \left(\frac{2 \cdot 2}{Tp}\right) \times AREA$$

$$T_B = 2 \cdot 52 Tp \tag{7.18}$$

Finally, the ordinates of the unit hydrograph at intervals of ΔT are estimated using the linear proportionality of the triangular shape.

7.4.6 *The use of the* Flood Estimation Handbook *unit hydrograph for reservoir spillway capacity*

Reservoirs in the UK with a capacity of more than 25 000 m³ are subject to the Reservoirs Act 1975 which requires, *inter alia*, regular inspection by a qualified Panel Engineer, including an assessment of the spillway capacity. The spillway capacity for a dam is based on the *probable maximum flood* (PMF), which can be regarded as the largest flood which might ever occur. It is normally estimated using the unit hydrograph method by applying a notional *probable maximum precipitation* (PMP), as described in Chapter 4, to the catchment, using the most extreme combination of characteristics. The inflow hydrograph to the reservoir is found by convoluting the rainfall hyetograph with the unit hydrograph, as described in the previous section. However, to allow for the more rapid runoff in extreme flood conditions, the time to peak of the unit hydrograph is reduced by a third. Since the area of the unit hydrograph is fixed, the effect of this is to increase the flow ordinates. This adjustment is carried out before the correction to the time to peak for the data interval. The resulting hydrograph is then routed through the reservoir using a standard reservoir routing procedure described in Section 6.3.5. In addition, the effects of wind and wave surcharge are added to the level predicted by the routing exercise.

The severity of the assumptions made in the estimation of the maximum flood level depends on the potential consequences of overtopping for a given dam. The recommendations for UK dams (ICE, 1996) divide dams into four categories, as shown in Table 7.4.

7.4.7 *Summary of the* Flood Estimation Handbook *unit hydrograph method*

The *Flood Estimation Handbook* unit hydrograph model is summarised in Figure 7.8. As has been noted, the unit hydrograph is a linear transformation between effective rainfall and direct runoff and is assumed to be a property of a given catchment. It can be derived either by analysis of observed flood hydrographs or by a regression with the hydrograph lag. Using a triangular approximation, the shape of the unit hydrograph can be defined in terms of three parameters. If no flow records are available, these parameters can be estimated by regression with certain catchment characteristics. The unit hydrograph is then convolved with the effective rainfall hyetograph to produce the hydrograph of direct runoff. A component of baseflow then needs to be added to give the complete hydrograph (Example 7.3).

$$Tp = Tp(0) + \frac{\Delta T}{2} = 3\cdot76 + 0\cdot5 = 4\cdot26 \text{ h}$$

$$Up = 2\cdot2 \times \frac{AREA}{Tp} = 2\cdot2 \times \frac{15\cdot5}{4\cdot26} = 8\cdot00 \text{ m}^3/\text{s}$$

$$T_B = 2\cdot52 Tp = 10\cdot74 \text{ h}$$

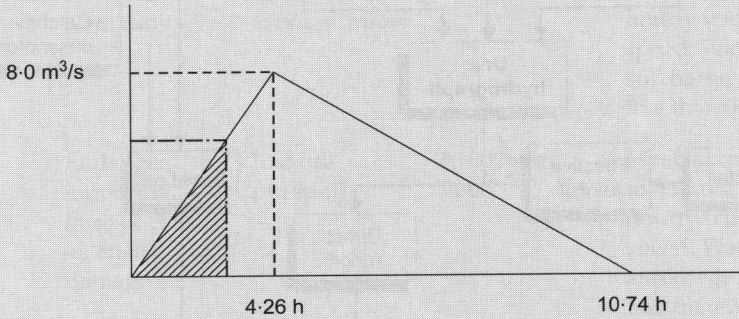

Ordinates of the unit hydrograph at hourly intervals are obtained by interpolation.

Time: h	Ordinates of unit hydrograph	Runoff from 1·25 cm	Runoff from 0·93 cm	Total runoff: m³/s
0	0·00	0·00	—	0·0
1	1·88	2·347	0·00	2·3
2	3·76	4·695	1·747	6·4
3	5·63	7·042	3·493	10·5
4	7·51	9·390	5·240	14·6
5	7·09	8·862	6·986	15·8
6	5·85	7·318	6·593	13·9
7	4·62	5·773	5·444	11·2
8	3·38	4·229	4·295	8·5
9	2·15	2·684	3·146	5·8
10	0·91	1·140	1·997	3·1
11	0·00	0·00	0·848	0·8

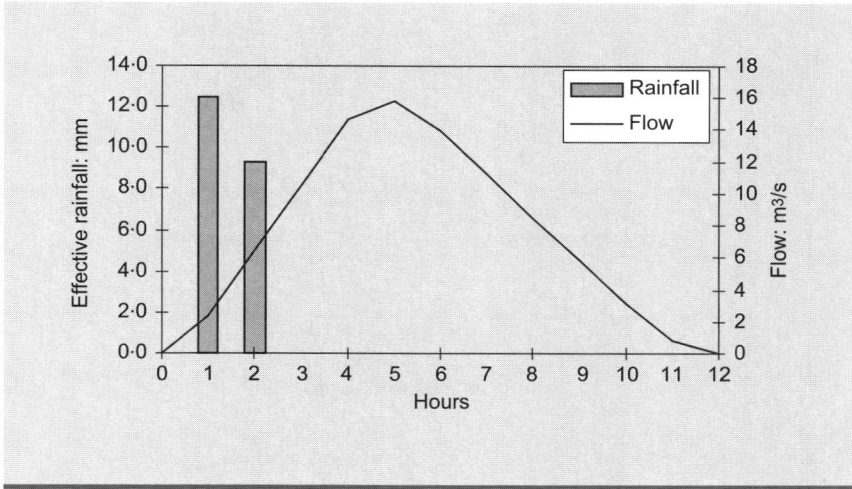

The structure of a typical physically based runoff model is shown in Figure 7.9. The models usually use a water balance equation to allocate the applied rainfall into surface runoff, evaporation, infiltration and other losses. There is then a storage accounting procedure which calculates the volume in each store at each time-step, and a routing procedure which translates the output from each store into the river channel and which usually involves one or more notional storage tanks. In the more complex models, the flow through both saturated and unsaturated soil is modelled using fundamental physical relationships and the model is often linked to channel flow models.

7.5.1 Water balance principle

Physically based models are founded on the fundamental water–balance relationship:

$$Q = P - E - D - \Delta S \qquad (7.19)$$

where Q is the runoff, P is precipitation, E is evapotranspiration, D is the drainage to the groundwater and ΔS is the change in water storage in the soil. This is used together with a simple continuity equation relating the inflows and outflows to the soil store to calculate the runoff. In some cases, the outflows from the soil store may be simple overflows, which

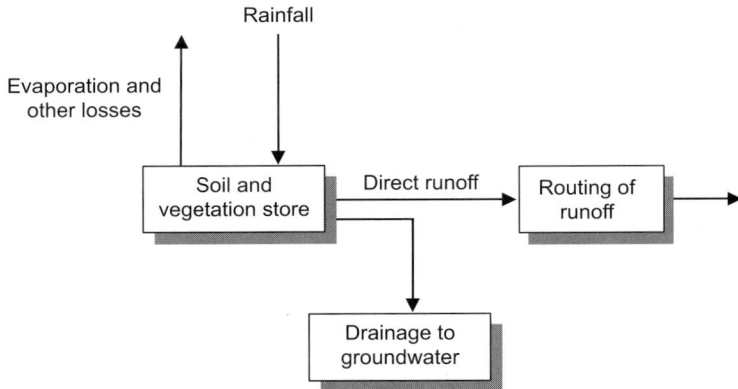

Figure 7.9   The structure of a typical rainfall–runoff model

carry all the excess flow when the soil store is full, otherwise they may be functions of storage (*S*) of the form:

$$Q = CS^n \tag{7.20}$$

As described in Chapter 5, the actual evaporation depends on the potential evaporation, which is largely a function of meteorological conditions, such as radiation and wind speed. Values of potential evaporation are available from the UK Meteorological Office and other sources, and the ratio of actual to potential evaporation is usually taken to be a function of the volume of water in one or more of the storage tanks representing storage within the soil or on the surface. Some models use a linear function so that evaporation decreases linearly as the contents of the store decrease below a maximum capacity. Others use a negative exponential function, which means the evaporation falls more rapidly as the store becomes more empty. The depression storage may also be deducted from the gross rainfall at this stage.

7.5.2   The modelling of snowmelt

The contribution of snowmelt to runoff can be significant in many high-latitude countries. A typical snowmelt model (Arnell, 1996) assumes that precipitation falls as snow when the temperature falls below a critical threshold and that the laying snow begins to melt when temperatures exceed another threshold value. It is also assumed that the drainage from the snowpack is a function of the liquid content of the snow. As the snow begins to melt there is initially little runoff until the liquid content has reached a critical value. The total drainage from a snowfield therefore consists of a 'slow' and a 'fast' component. The structure of a typical snowmelt model is given in Figure 7.10. The parameters for the above snowmelt model are given in Table 7.5.

7.5.3 Routing procedures

The routing of the runoff is normally achieved by using a number of notional reservoirs with a simple linear storage–outflow relationship given by:

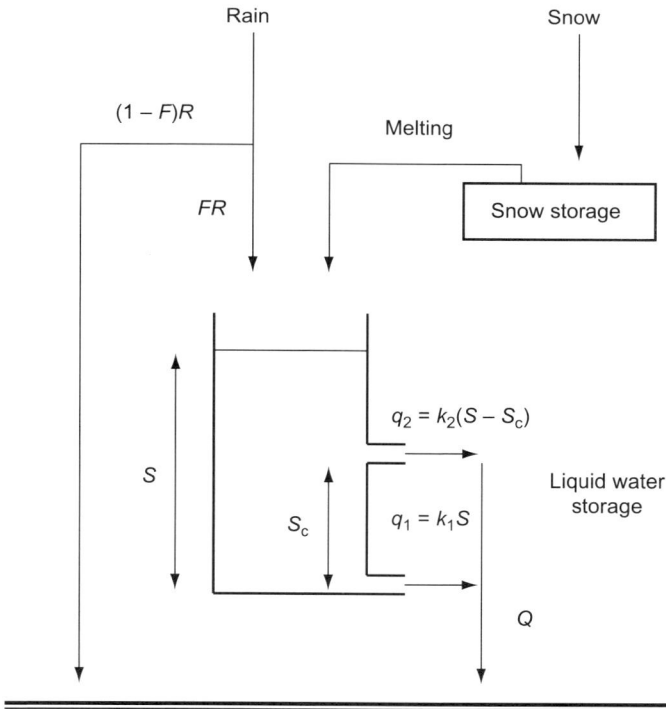

Figure 7.10   Structure of a typical snowmelt model

Table 7.5   Parameters for snowmelt model

Parameter	Description	Value
T_{crit}	Threshold temperature for snowfall	1°C
T_{melt}	Threshold temperature for snowmelt	0°C
K_{melt}	Melt rate	4 mm/°C/day
S_c	Volume of snowpack at which entire catchment is covered by snow	100 mm
S'	Proportion of liquid water in snow pack above which drainage increases	0·04
k_1	Slow drainage rate	0·15
k_2	Fast drainage rate	0·85

$$S = KQ$$

$$\text{and } I = Q + \frac{dS}{dt} \tag{7.21}$$

The inflow-outflow relationship for the two reservoirs becomes:

$$I_1 = Q_1 + k\frac{dQ_1}{dt}$$

$$Q_1 = I_2 = Q_2 + k\frac{dQ_2}{dt}$$

$$I_1 = Q_2 + 2k\frac{dQ_2}{dt} + k^2\frac{d^2Q_2}{dt^2} \tag{7.22}$$

where I and Q are the inflows and outflows, respectively (the suffixes referring to the two reservoirs), and S is the storage (Figure 7.11).

7.5.4   The Wallingford Hydrograph Method

This is a computer-based conceptual model specifically intended for urban catchments. It is incorporated in the *Wallingford Procedure* and in

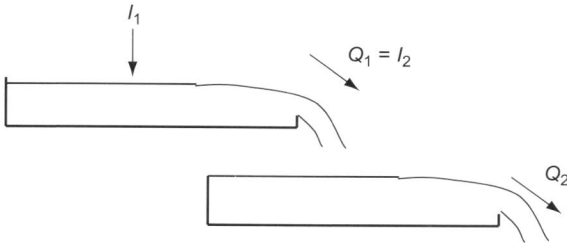

Figure 7.11   Double reservoir model

more recent software such as *InfoWorks*. In contrast to the Rational Method, it can incorporate observed rainfall hyetographs or synthetic storm profiles and produces a complete hydrograph of flow. The model uses the double reservoir concept, representing the attenuation in the surface runoff and pipe-flow phases by separate reservoirs. The modelling can be divided into three elements, namely the rainfall generation (described in Chapter 4), the surface runoff and the pipe flow.

In generating the surface runoff, the Wallingford Model allows for initial depression storage by using an empirical relationship with the average ground slope given in Equation (5.1). This depression storage is subtracted from the rainfall depths for each time period of the rainfall profile until the total depression storage has been filled, giving the net rainfall profile (Figure 5.2).

The storage constants for the reservoirs depend on the rainfall intensity, contributing area, slope and the nature of the surface. It is recommended that the storage constant k is found from:

$$k = C \times i*^{-0.39} \tag{7.23}$$

where $i* = 0.5(1 + i_{10})$, i_{10} is the 10 minute moving average rainfall intensity, $C = 0.117 S^{-0.13} A^{0.24}$, S is the slope (m/m), and A is the contributing area.

The procedure calculates 10 standard hydrographs using combinations of three sizes of catchment and three nominal ground slopes with an additional hydrograph for pitched roofs. Hydrographs for sub-catchments are produced from these standard hydrographs according to the size and slope of the sub-catchment. These hydrographs are then routed through the pipe system using standard hydraulic pipe formulae, incorporating weirs, orifices and overflow structures as required. The effect of surface flooding through manhole surcharging can also be included.

7.5.5 The SHE model

The SHE (Système Hydrologique Européen) model is an example of a comprehensive three-dimensional grid-based model. The rectangular grid used can vary from 50 m to 2000 m, and each element has a specified value of elevation and parameters which allow for the calculation of interception, evapotranspiration and snowmelt, as well as vertical flow into the unsaturated soil layer and from the unsaturated layer into the saturated zone. The grid elements are linked by two-dimensional surface runoff and groundwater components, and are also connected to channel flow elements.

It can be seen that, even for a moderate-sized catchment, the amount of data and computational resources required for such a model are considerable. However, in some cases it is possible to specify distributed precipitation and other meteorological data across the catchment. Some care is required in assembling the data, since each grid element represents a significant area and the parameter values may not be those measured at spot locations.

7.6 Physically based partially distributed models

It was mentioned above that the application of fully distributed models is limited by the amount of data and computing resources required. It was also noted that certain meteorological parameters can be assumed to be fairly regularly distributed across the catchment. It is also likely that the hydrological parameters of the catchment, in terms of the water balance and runoff characteristics, are evenly distributed across the catchment. If this is the case, then the hydrological response of the catchment could be represented by a statistical distribution of certain parameters rather than individual values at discrete locations. In their simplest form, partially distributed models do not make any assumptions about the characteristics of specific areas of the catchment. Such models are, in effect, a transfer function between the rainfall input and the runoff. Other models assume certain areas of the catchment have similar hydrological responses based on soil type, aspect, slope and vegetation, etc. It is now easier to identify such areas if catchment data are available in digital form using standard statistical classification procedures. An alternative approach is to assume that the hydrological response of a catchment depends essentially on the topography and soil type, which has been shown to be valid, at least for catchments with moderate to steep slopes and shallow soils.

7.6.1 Variable storage model

Moore's Probability Distributed Model (Moore, 1985; Arnell, 1996) is based on the assumption that there is a variation of soil moisture capacity across a catchment, which can be represented as a series of reservoirs with varying capacities. Those with the lowest capacity will naturally fill first after rainfall, and the overflow is then routed to the channel as runoff (either as quick storm runoff or slow runoff). The proportion of the catchment producing runoff therefore increases as the storm progresses. Likewise, the smaller stores will also empty first at the end of the rain event and the area producing runoff will progressively decrease. There is no attempt to relate the distribution of storage to any physical characteristic, although such a concept has been found to represent the general performance of many catchments.

One form of the storage distribution function is proposed as:

$$F(c) = 1 - \left[\frac{1-c}{c_{max}} \right]^b \tag{7.24}$$

where c is the soil moisture capacity and c_{max} is the maximum capacity. The exponent b represents the degree of spatial variability: a value of zero indicates a constant capacity and 1 implies a uniform variation across the catchment (Figure 7.12).

The maximum amount of water that can be held in storage is the area under the distribution curve, i.e.:

$$S_{max} = \int_0^{c_{max}} (1 - F(c)) \mathrm{d}c \tag{7.25}$$

Therefore, the combination of S_{max} and c_{max} can be used to determine the value of the exponent b.

If all the stores with a capacity of, say, c^* are full, then the volume of water stored is:

$$S = \int_0^{c^*} (1 - F(c)) \mathrm{d}c \tag{7.26}$$

Hence, the capacity c^*, below which all of the stores are filled for a given value of total soil moisture storage S, can be evaluated as a function of S_{max} and b. The model uses a negative exponential function to calculate the

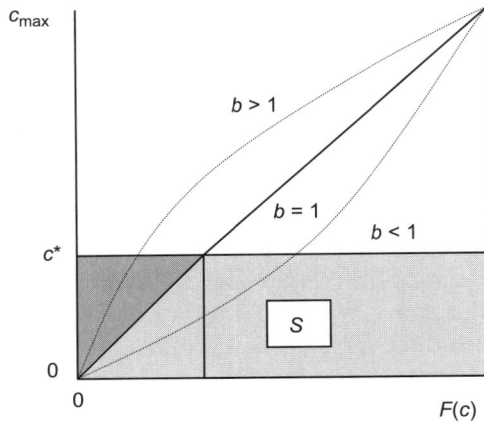

Figure 7.12   Distribution of soil moisture storage

ratio of actual to potential evaporation from the storage volume. The volume of effective rainfall in excess of that needed to saturate the soil becomes direct runoff which occurs over that part of the catchment where the soil is filled to capacity and the runoff is routed through a double reservoir as described above. The model requires only the five parameters shown in Table 7.6.

Table 7.6   Parameters of the variable storage model

Parameter	Description
c_{max}	Maximum soil moisture capacity in the basin (mm)
S_{max}	Maximum amount of soil moisture over the basin (mm)
K	Soil drainage coefficient (mm/day)
G_{rout}	Baseflow routing coefficient
S_{rout}	Channel routing coefficient

7.6.2 TOPMODEL

TOPMODEL is another variable contributing area conceptual model in which the main features are the use of the topography in determining runoff and a negative exponential law relating the hydraulic conductivity of the soil to the vertical distance from the ground level. In this model the total flow comprises the surface runoff and the flow in the saturated soil zone. The surface runoff is generated either where the saturated soil zone appears at the surface (saturation excess) or where rainfall intensity exceeds the infiltration capacity of the soil (infiltration excess). The infiltration flow from the unsaturated zone to the saturated zone is assumed to be determined by the soil conductivity at the top of the water table (depth z_i) (Figure 7.13) under the unit hydraulic gradient.

The hydraulic conductivity of the soil at depth z is assumed to follow a relationship:

$$K_Z = K_0 \exp(-fz_i) \tag{7.27}$$

where K_0 is the transmissivity of the soil at the surface and f is a decay factor that is assumed to be constant over the catchment. It is assumed that the water table is parallel to the ground surface (with a slope $\tan(\beta_i)$), so that the flow in the saturated zone at location i is given by:

$$q_i = T_i(z_i)\tan(\beta_i) \tag{7.28}$$

T_i is the transmissivity and is given by:

$$T_i(z) = \int_{zi}^{Z} K_Z(x)\mathrm{d}x$$

$$= \frac{K_0}{f}[\exp(-fz_i) - \exp(-fZ)]$$

$$= \frac{1}{f}[K_{Zi} - K_Z] \tag{7.29}$$

where Z is the height of the bottom of the saturated layer above a datum.

Substituting in Equations (7.27) and (7.28), and neglecting the conductivity at depth Z (K_Z):

Figure 7.13 Slope tan β, Water table, z_i, Z, q_i, Datum, (a)

Area drained per unit contour length, (b)

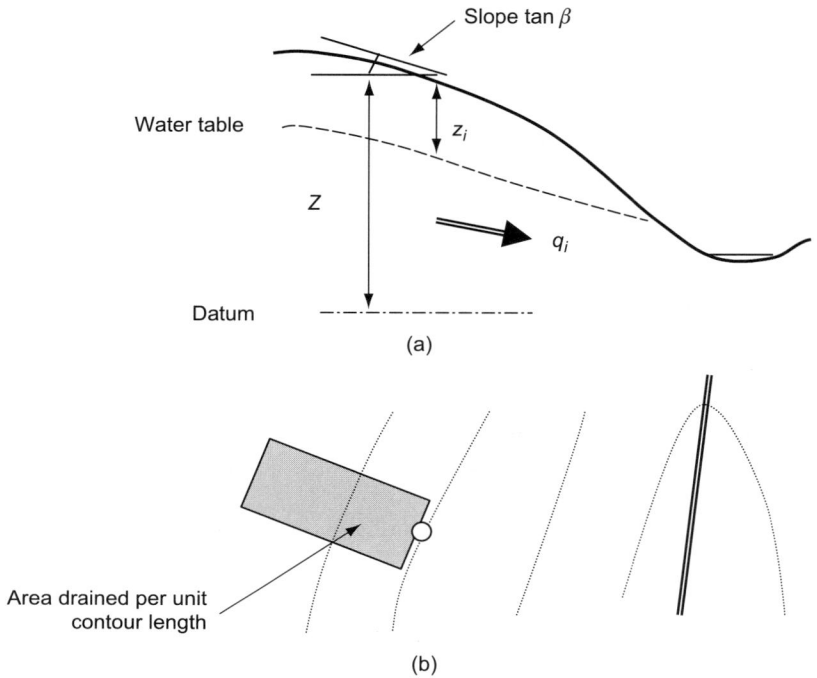

Figure 7.13   Notation used in TOPMODEL: (a) section; and (b) plan

$$q_{si} = \left(\frac{K_0}{f}\right) \tan(\beta_i)\exp(-fz_i) = T_0 \tan(\beta_i)\exp(-fz_i) \qquad (7.30)$$

where T_0 is the transmissivity of the saturated soil which, like K_0 and f, may be assumed to be constant over the whole catchment unless some kind of distribution is appropriate.

The change in the overall level of the water table over a time interval Δt can be found from continuity principles:

$$z_{t+\Delta t} = z_t - (Q_{ui} - Q_s)\frac{\Delta t}{A} \qquad (7.31)$$

where Q_{ui} and Q_s are the integrals of q_{ui} and q_s with respect to the area. The depth of the water table (z) over the catchment is assumed to be related to the decay parameter f and a *topographic index* $\gamma_i = \ln(\alpha_i/\tan\beta_i)$ (where α_i is the area per contour length draining through location i) by:

$$z_i = \overline{Z} - \frac{1}{f}\left[\ln\left(\frac{\alpha_i}{\tan\beta_i}\right) - \lambda + \ln T_0\right] \tag{7.32}$$

where:

$$\lambda = \frac{1}{A}\int_0^A \ln\left(\frac{\alpha_i}{T_0 \tan\beta_i}\right)dA$$

The topographic index is a measure of the propensity to develop saturated conditions: high values are associated with a long upstream slope and low slope angle. The distribution of the index can be calculated from an analysis of a digital terrain model.

The model has been found to provide a good simulation of discharges especially where the groundwater is not deep, and it also provides a reasonable estimate of the varying contributing area. Areas with deeper groundwater are more difficult to model because of the complex wetting up period at the start of the wet season and the effect of possible perched water tables.

7.7 Hydraulic modelling

The above hydrological models can be used to generate runoff for particular rainfall events. In order to asses the extent of flooding for such events, some form of hydraulic model is required which relates flow rate and water level. In the early days of modelling, *physical* models of river channels were necessary but *numerical* models have become more available with the advent of cheap, powerful personal computers. Numerical models have the advantage that they do not need expensive laboratory facilities, but they can be misleading where there are complex flow patterns, such as eddy currents. In both cases, the model requires estimates of friction coefficients and other data, which may be subject to considerable errors. Therefore, some kind of validation process, where water levels and velocities are compared against independently observed values, is essential. In all modelling, considerable skill and experience is required to set up the models and to interpret the results.

7.7.1   Physical models

Physical models rely on the principle of hydraulic similarity, which includes geometric, kinematic and dynamic similarity. *Geometric* similarity requires that the ratio of any linear dimension on the model to the corresponding dimension at full scale (prototype) is always the same. *Kinematic* similarity refers to ratios of velocity and flow, while *dynamic* similarity refers to ratios of forces. The forces involved may include pressure, gravity, viscosity or surface tension, and it is not possible to ensure dynamic similarity for all forces. In the case of hydraulic models, the predominant force is usually gravity and it is normally a requirement that the ratio of gravity to inertia forces is the same for the model as the prototype. This ratio is the *Froude Number* ($v/\sqrt{(gd)}$) and the use of scaling such as Froude scaling allows the ratios of velocity, time, etc., to be determined in terms of the basic model scale. For example:

$$\frac{V_p}{\sqrt{gd_p}} = \frac{V_m}{\sqrt{gd_m}}$$

$$\frac{V_p}{V_m} = \sqrt{\frac{d_p}{d_m}} = x^{1/2} \tag{7.33}$$

where x is the geometric scale ratio and the suffixes p and m refer to the prototype (full scale) and model parameters respectively. A summary of the Froude scale ratios for various quantities is given in Table 7.7.

One problem with physical models of river systems is that the horizontal extent of the area to be modelled may require a small model-scale to ensure that the model size is not impractically large. However, if the resulting model water-depths are less than about 6 mm, surface tension effects become significant and it is then difficult to measure the water levels accurately. These problems can be overcome by the use of distorted scale models with a larger scale for the vertical dimension than the horizontal. This also ensures that the velocities are large enough to represent the sediment transport accurately and that the natural turbulent flow does not become laminar. However, there are disadvantages in distorted scale-models in that they do not model transverse currents and eddies accurately. The Froude scale ratios for distorted models are also shown in Table 7.7.

Another problem with physical models of river systems is that, to represent the roughness of the prototype channel, some form of artificial roughness, for example vertical rods, is often needed. However, it should

Table 7.7   Froude scale ratios

Quantity	Natural scale	Distorted scale	
	1:x	1:x horizontal	1:y vertical
Length	x	x	y
Area	x^2	x^2	xy
Volume	x^3	x^2y	x^2y
Time	$x^{1/2}$	$x/y^{1/2}$	$x/y^{1/2}$
Velocity	$x^{1/2}$	$y^{1/2}$	$y^{3/2}/x$
Flow rate	$x^{5/2}$	$xy^{3/2}$	$xy^{3/2}$

be recognised that much of the energy lost in natural channels results from eddies at bends and sharp changes in section rather than frictional losses. It is therefore not necessary to model the frictional characteristics of the bed exactly, although the accurate representation of the head loss at structures is important. The accurate modelling of sediment transport in mobile bed models is also a problem and may necessitate the use of special lightweight material.

7.7.2   *Numerical models*

A numerical open-channel model is a system of equations that link water level to flow rate.

The basic data required are:

- the geometry of the river cross-sections and channel network
- the roughness values for the channel and floodplain
- the hydraulic characteristics of any control structures
- a flow and/or water level time-series at one or more boundaries.

The output is a time-series of water level at various locations in the channel.

The full equations for flow in open channels were developed by Saint-Venant and consist of a dynamic equation and a continuity equation:

$$S_f = i - \frac{\partial y}{\partial x} - \frac{\partial}{g\partial x}\left(\frac{Q^2}{A^2}\right) - \frac{\partial Q}{gA\partial t} \qquad (7.34a)$$

$$\frac{\partial Q}{\partial x} + B\frac{\partial y}{\partial t} = 0 \qquad (7.34b)$$

where S_f is the friction slope, i is the bed slope, B is the channel width, y is the depth, Q is the flow rate, and A is the cross-sectional area.

If only the first term on the right-hand side of Equation (7.34a) is used and it is combined with Equation (7.34b), the result is the *kinematic* wave equation, which only considers frictional and gravitational forces:

$$\frac{\partial Q}{\partial t} + c\frac{\partial Q}{\partial t} = 0 \qquad (7.35)$$

The kinematic equation describes a wave which travels at a velocity c but does not attenuate and is appropriate where uniform or nearly uniform flow conditions apply. If the first two terms on the right-hand side of Equation (7.34a) are used together with Equation (7.34b), the result is the *diffusive* wave equation:

$$\frac{\partial Q}{\partial t} + c\frac{\partial Q}{\partial t} = D\frac{\partial^2 Q}{\partial x^2} \qquad (7.36)$$

This formulation would be used where flow varies relatively slowly, such as channel routing and where backwater effects or reverse flow may occur. Where flow or water level varies rapidly with time, for example where a sluice gate is suddenly opened, the full equation should be used.

In the case of uniform flow, the equations can be easily solved separately using the Manning and continuity equations, and for gradually varied flow the equations can be solved approximately (Equation (6.24)). However, most models generalise the problem by solving the full equations. The above equations are partial differential equations, where Q is a function of distance (x) and time (t). The most common method of solution of this type of equation uses finite differences, where the partial differentials $\partial Q/\partial t$, etc., are replaced by finite differences:

$$\frac{Q_{j+1} - Q_j}{\Delta t} \quad \text{or}$$

$$\frac{Q_j - Q_{j-1}}{\Delta t} \quad \text{or}$$

$$\frac{Q_{j+1} - Q_{j-1}}{2\Delta t} \tag{7.37}$$

where Q_{j+1}, Q_j and Q_{j-1} are consecutive values of Q with a time interval of Δt. The three alternative expressions are known as *forward*, *backward* and *central difference* representations, respectively. The second order differential is given by:

$$\frac{\partial^2 Q}{\partial x^2} = \frac{\dfrac{Q_{i+1,j} - Q_{i,j}}{\Delta x} - \dfrac{Q_{i,j} - Q_{i-1,j}}{\Delta x}}{\Delta x}$$

$$= \frac{Q_{i+1,j} - 2Q_{i,j} + Q_{i-1,j}}{\Delta x^2} \tag{7.38}$$

By substituting these expressions for the partial differentials and rearranging the equations, the unknown value $Q_{i+1,j}$ can be isolated and the equation solved. Such as method is termed *explicit*. However, the explicit formulation is only valid for $\Delta t < \frac{1}{2}(\Delta x)^2$, which imposes a severe limitation in terms of the grid spacing in relation to the time-step. A more general, finite difference formulation can be obtained by replacing the second order differential by the mean of its finite difference representations on the $(j + 1)$th and the jth time rows. Hence, the differential equation:

$$\frac{\partial Q}{\partial t} = \frac{\partial^2 Q}{\partial x^2} \tag{7.39}$$

can be represented as:

$$\frac{Q_{i,j+1} - Q_{i,j}}{\Delta t} = \frac{1}{2}\left\{ \frac{Q_{i+1,j+1} - 2Q_{i,j+1} + Q_{i-1,j+1}}{(\Delta x)^2} + \frac{Q_{i+1,j} - 2Q_{i,j} + Q_{i-1,j}}{(\Delta x)^2} \right\} \tag{7.40}$$

giving:

$$-rQ_{i-1,j+1} + (2+2r)Q_{i,j+1} - rQ_{i+1,j+1} = rQ_{i-1,j} + (2-2r)Q_{i,j} + rQ_{i+1,j} \quad (7.41)$$

where:

$$r = \frac{\Delta t}{(\Delta x)^2}$$

In general, the left-hand side of Equation (7.41) contains three unknown values of Q and the right-hand side contains three known values. If there are N internal grid points in the x direction, then, for the first time-step ($j = 0$) and $i = 1, 2, ..., N$, the equation will give N simultaneous equations in terms of the initial boundary conditions. Similarly, at the next time-step, there will be N simultaneous equations involving the values at the first time-step. Such a method, in which the calculation involves the solution of sets of simultaneous equations, is known as an *implicit* method and applies to all finite values of r (Smith, 1978).

7.7.3 Modelling of sewer systems

The modelling of a typical sewer system can involve a complex representation of both free surface and pressurised flow, including weirs, overflows and pumps. In practice, models vary from very coarse planning models to more detailed models for drainage area studies and very fine models for detailed design, where each manhole is represented by a node. The level of data required for each model therefore depends on the scale of the model. There are particular problems associated with urban drainage models in relation to the estimation of the extent of the area contributing to each sewer length and the degree of impermeability of the surface. In the modelling of existing drainage systems, there are also problems in estimating the roughness and condition of the pipes, both of which have a considerable effect on their hydraulic performance. (WaPUG, 1998). The design rainfall input to drainage models is normally an intense summer storm, although there is evidence that prolonged periods of wet weather followed by lesser events may be more significant in terms of flood risk as they appear to be a feature of changes in the climate of the UK (Balmforth, 2002).

7.8   Model calibration and validation

The reliability of any model depends on the amount and reliability of the data and the degree to which the algorithms used in the model represent the actual processes involved. Since many of the parameters can only be estimated approximately, there is considerable justification for adjusting their values to give the best fit with observed data, a process known as *calibration*. Successful calibration is a combination of science, experience and intuition. Because of the limitations of the data and the model complexity, no model will give an exact fit for any length of data. However, there will be almost infinite combinations of parameter values that will give a reasonably close fit.

One of the problems in calibration is in measuring the closeness of fit between the model data and the corresponding observed data. A popular measure is based on the sum of the squares of the differences between the two time-series:

$$\sigma^2 = \frac{1}{N}\sum(y_o - y_m)^2 \tag{7.42}$$

where y_o and y_m are the observed time-series and the model output respectively. This may be transformed into a measure of efficiency as:

$$E = \left[\frac{1-\sigma^2}{\sigma_o^{\,2}}\right] \tag{7.43}$$

where $\sigma_o^{\,2}$ is the variance of the observed time-series.

In a simple case where there are, say, only two parameters, efficiencies can be calculated for various random combinations of parameter values. Thus, a surface is generated which, in an ideal case, will have a single peak representing the optimum combination of parameter values (Figure 7.14(a)).

Techniques exist, e.g. Nelder and Mead (1965), where, starting with an arbitrary combination of parameter values, the peak or optimum solution can be located iteratively without a large amount of computation. Ideally, the method should converge on the same peak regardless of the starting point. However, in many cases the surface is irregular, consisting of many local peaks and the method may not locate the overall optimum (Figure 7.14(b)). In other words, different solutions are obtained from different starting points. This problem is particularly common where some of the parameters are relatively insensitive, i.e. where the parameter has little or no effect on the model result or where some parameters interact with different pairs of parameter values, giving very similar goodness of fit. The concept can easily be extended to three or more parameters, although the results are not easy to represent visually.

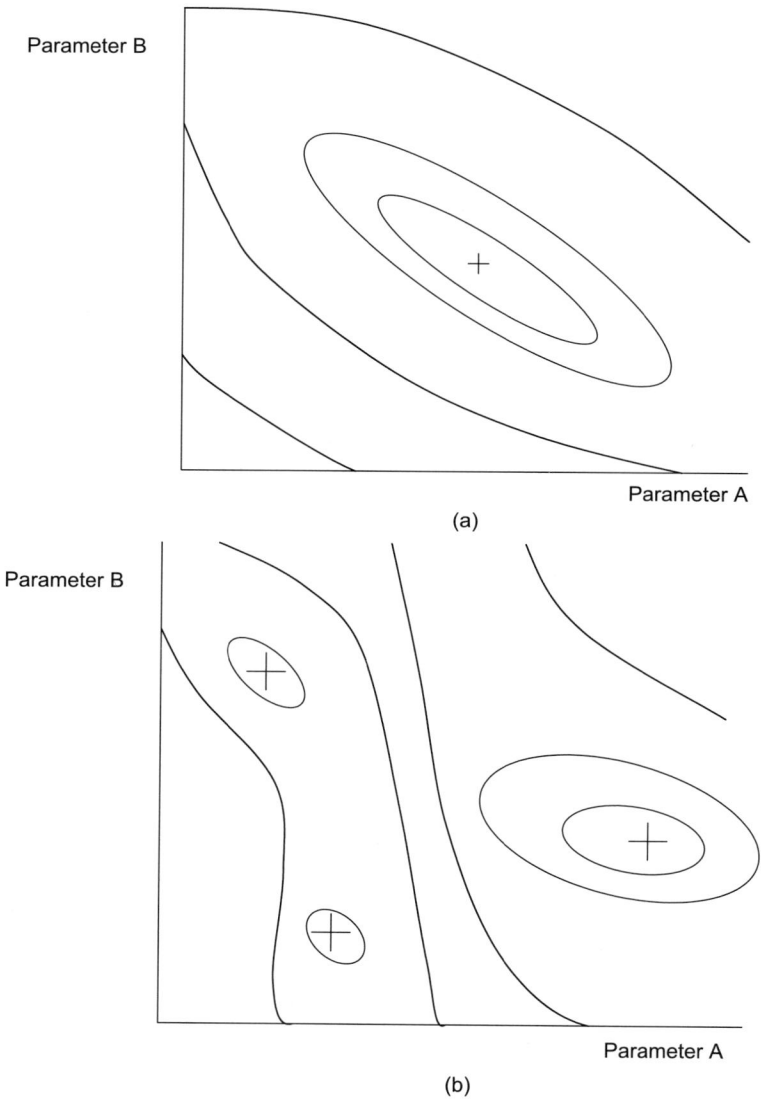

Figure 7.14   *Response surfaces for parameter optimisation: (a) single peak; and (b) multiple peaks*

There are other problems with the optimisation of model parameters based on a measure of goodness of fit. Firstly, the greatest absolute deviations are likely to be near the peak flows and therefore greater weight will be given to peak flows compared to periods of low flow. This may be desirable where the prediction of flood events is the main objective, but it can give a poor fit for other cases. Secondly, such a measure is very sensitive to errors in the timing of flood peaks. Large errors may result from relatively small differences in the peak times. Another problem is that the use of squared errors assumes that the errors are randomly distributed, whereas in most hydrological series the residual may be auto-correlated in time or may be proportional to the flow.

There are also more fundamental problems with parameter estimation based on goodness of fit, as it implies that there is a single combination of parameters that will give the best result. In practice, there will be, as has been stated, a wide range of parameter combinations which could give equally valid results. A more reliable way of estimating model parameters is to estimate the performance associated with each of a large number of possible parameter combinations and then to weight each combination according to its performance. The result is a distribution of values of model output at each time-step.

7.9   Probabilistic models

In many cases the required output from rainfall–runoff models are the flows and levels corresponding to different frequencies or probabilities of occurrence. It is generally assumed that a storm of a given probability will result in a peak flow with the same probability but this is not strictly true. A more accurate estimation of the probabilities of occurrence of different flow rates can be obtained from repeated runs of a standard deterministic model to transform given probability distributions of rainfall into distributions of runoff volume and intensity. Alternatively, it can be achieved using a analytical probabilistic model incorporating a simple rainfall–runoff model. A typical simple rainfall–runoff model is:

$$V_r = 0 \qquad V < V_d$$

$$= \phi(V - V_d) \qquad V \geq V_d \tag{7.44}$$

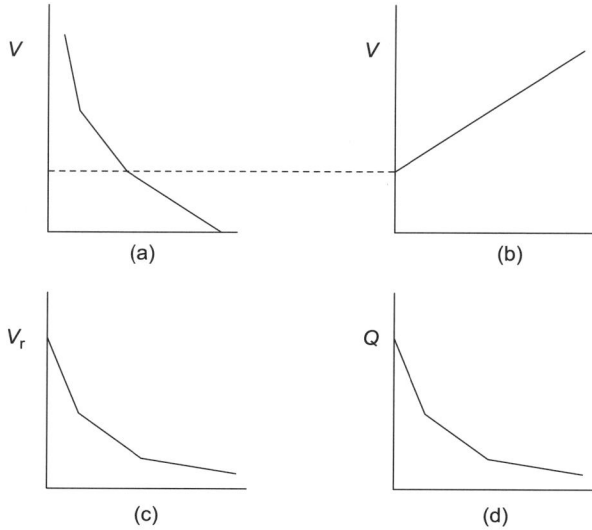

Figure 7.15   Probabilty runoff model: (a) probability; (b) V_r; (c) probability; and (d) probability

where V_r is the volume of runoff from a rainfall volume V and V_d is the depression storage, and ϕ is the runoff coefficient (Figure 7.15). Figure 7.15(a) shows the specified rainfall PDF and Figure 7.15(b) represents the transformation between rainfall volume and runoff volume. Figure 7.15(c) is the cumulative probability function for runoff volume showing a zero probability where the rainfall volume is less than the depression storage (V_d). The runoff can also be expressed in terms of the intensity or flow rate (Figure 7.15(d)).

7.10   Expert systems

As has been mentioned, most hydraulic models are limited by the range of conditions used for calibration and a large number of parameters does not always result in a robust model. The application of expert systems techniques allows knowledge to be incorporated in the forecasting systems with a simpler modelling technique. Expert systems are designed to hold the knowledge of an expert in a given field and to apply that knowledge to data in order to make decisions or to provide 'decision support'. As new data arrive, they are compared with the knowledge base rules and decisions are made. Expert systems are analogous to, and have been derived from, the human thought processes. Some specialised artificial intelligence languages have been developed for expert systems, although any conventional programming language can be used.

Expert systems have been developed as decision support in areas such as drought alleviation and reservoir operation, as well as in calibrating parameters in hydrological models. It has also been used in the real-time 'tuning' of the parameters of a runoff model (Wedgwood, 1995).

7.11   Summary

This chapter has considered the various methods that can be used to model the relationship between rainfall and runoff. The Rational Method is a simple transform method, suitable for small urban catchments which can be divided into sub-catchments. The time-area method requires assumptions to be made about the proportion of a catchment which is contributing to flow at any time, which implies assumptions about overall flow velocities. The unit hydrograph method assumes a unique time distribution of flow, which is normalised with respect to the depth and duration of effective rainfall. All these methods allow for losses by removing a certain proportion of rainfall, which may vary according to the catchment wetness. The physically based models use the concept of tanks representing soil storage providing a routing mechanism. In the Wallingford Model, losses are incorporated by using notional percentages of impermeable area and an empirical formula for depression storage. Other, more general models, use evapotranspiration algorithms and a storage accounting procedure.

The problem of calibration applies to all models. At the most basic level, it involves comparing the results from the model with observed data using some objective measure of goodness of fit. However, the combination of model parameters which result in the 'best fit' will not be the only valid set of parameters and more extensive methods of assessing the reliability of a model may be necessary.

7.12 References

Arnell, N. (1996). *Global warming, river flows and water resources*. Wiley, Chichester.

Balmforth, D. (2002). *Climate Change and SUDS. Sustainable Urban Drainage Systems*. Scottish Hydraulics Study Group, Glasgow.

Bevan, K. J. (2000). *Rainfall Runoff Modelling*. Wiley, Chichester.

Bhaskar, N. R., Parida, B. P. and Nayak, A. K. (1997). Flood Estimation for Ungauged Catchments Using the GIUH. *Journal of Water Resources Planning and Management*, **123**, No. 4, 228–238.

Calver, A. (1996). Development and Experience of the TATE Rainfall Runoff Model. *Proceedings of the Institution of Civil Engineers, Water, Maritime and Energy*, **118**, 168–176.

Houghton-Carr, H. (1999). *Flood Estimation Handbook Vol. 4 — Restatement and Application of the FSR rainfall–runoff Method*. Institute of Hydrology, Wallingford.

Hydraulics Research (1983). *Design and Analysis of Urban Storm Drainage*. Hydraulics Research, Wallingford.

Institution of Civil Engineers (1996). *Floods and Reservoir Safety*. Thomas Telford, London.

Linsley, R. K., Kohler, M. A. and Paulhus, J. L. H. (1988). *Hydrology for Engineers*. McGraw Hill, London.

Moore, R. J. (1985). The Probability-distributed Principle and Runoff Production at Point and Basin Scales. *Hydrological Science Journal*, **30**, 263–297.

Nash, J. E. (1959). Systematic Determination of Unit Hydrograph Parameters. *Journal Geophysical Research*, **64**, 111–120.

Nelder, J. A. and Mead, R. A. (1965). A Simplex Method for Function Minimization. *Computer Journal*, **7**, 308–313.

NERC (1975). *Flood Studies Report*. Natural Environment Research Council, London.

Saghafian, B. (1998). A Revised Distributed Time Area Technique for rainfall–runoff Routing. *Proceedings of the BHS International Conference on Hydrology in a Changing Environment*. Wiley, Chichester.

Smith, G. D. (1978). *Numerical Solution of Partial Differential Equations: Finite Difference Methods*. Clarendon Press, Oxford.

WaPUG (1998). *Code of Practice for the Hydraulic Modelling of Sewer Systems*. Wastewater Planning Users' Group, London.

Wedgwood, O. (1995). rainfall–runoff models — over-complex and out of date? *Proceedings of the BHS Fifth National Hydrology Symposium*, Edinburgh.

WRc (2001). *Sewers for Scotland*. WRc, Marlow.

Chapter 8

The analysis and prediction of flows

This chapter is concerned with the estimation of flows mainly based on the analysis of past flow records. The estimation of design or future flows is needed for a variety of purposes, including flood protection and water abstraction. The design of flood protection measures obviously requires the estimation of extreme maximum flows for various return periods: in particular, the design of dam spillways involves the estimation of a 'probable maximum flood'. Likewise, water abstraction from rivers requires data on minimum flows. However, the design and operation of hydro-electric plants are less concerned with extremely high or low flows, but depend more on data over a range of moderate flows.

8.1 Flood prediction

8.1.1 Maximum observed floods in various countries

Although the length of most flow records is too short to extrapolate directly to the return periods of interest to the flood engineer, some information can be obtained from the aggregation of the maximum observed flood peaks from different countries. In a recent project, some 95 countries provided flood-peak data from about 1400 gauging stations. These were compared using an index calculated as:

$$k = 10\left\{1 - \left[\frac{(\log Q - 6)}{(\log A - 8)}\right]\right\} \qquad (8.1)$$

This parameter therefore represents the severity of the flood (Q) in relation to the catchment area (A). It was found (Rodier and Roche, 1984; Herschy, 2001) that for catchment areas greater than 100 km^2, the value of k for the highest 41 rivers exceeded 5·75 (although none of these were from European stations). It was also found that there was generally a logarithmic relationship between Q and A of the form:

$$Q = 500A^{0.43}$$

(8.2)

8.1.2   General principles of flood prediction

The prediction of extreme events from an observed flow record obviously requires a record long enough to include fairly rare events. At the same time, the time interval of the series needs to be fine enough to avoid excessive smoothing of peak flows. The length of a flow record may be enhanced by generating flows from a long rainfall record using a rainfall–runoff model or by combining records from several gauging stations. Since only high flows are of interest in flood estimation, a lot of unnecessary calculation may be needed if the whole record is used, which may require some compromise on the time-step or the complexity of the model. A common approach to estimating extreme flows is to abstract single values, such as annual maxima, for each year and to produce a frequency distribution for these values, although this means that much potentially valuable data are not used. An alternative approach for flood estimation is to select specific storm events, but it is important to select appropriate initial conditions, as this is a major determinant of the effect of such an event. It also may not be clear whether the severity of the flood is related to the peak flow or the duration of the storm.

Flood prediction usually involves the construction of a *flood-frequency curve* showing the relationship between maximum flows and frequency of occurrence. The analysis is normally based on the annual maxima, that is, the maximum of the series of mean daily flows within each year. To avoid complications with multiple peaks around the end of the year, a hydrological year (starting on 1 October) is normally used, rather than a calendar year. The use of an arbitrary period of a year for selecting maxima can present difficulties where storm peaks are poorly defined or where two severe storms occur in one year. A more representative picture of extreme flows is provided by a *peaks over threshold* (POT) series, which includes all the peaks greater than a certain threshold value. The threshold is normally chosen such that, on average, about three or four peaks occur in each year, although some years may contain more than this, while other years may contain no peaks at all.

The frequency of a flood event can be measured either in terms of a return period or an exceedance probability. In the case of an annual maxima time-series, the return period of a given flood peak is the average interval between the years containing a peak greater than or equal to the specified value of peak flow. However, this is not a true return period because of the distortion caused by the use of the year as a unit of time. As noted above, the POT series is a more representative indicator of extreme flows and the return period calculated from a POT series as the average interval between exceedances of the specified peak is closer to the true value.

The exceedance probability is simply the reciprocal of the return period. Thus a 1-in-20 year flood has a probability of being exceeded in any one year of 1/20. The non-exceedance probability is the cumulative probability distribution ($F(Q)$) and is equal to (1 − exceedance probability). Thus:

$$T = \frac{1}{\{1 - F(Q)\}} \qquad (8.3)$$

8.1.3 Estimation of flood return period by plotting

A flood-frequency curve can be constructed by plotting annual maxima data. The data are first ranked in order of magnitude, the largest event having a rank of 1. Each value is then assigned a plotting position reflecting its probability of occurrence, based on its rank (m) and the number of data points (n). The most common formula for the plotting position is Weibull's:

$$p = \frac{m}{n+1} \qquad (8.4)$$

where p is the probability of exceedence. The corresponding return period (the reciprocal of the probability) is:

$$T = \frac{n+1}{m} \qquad (8.5)$$

Plotting the peak flow against the exceedance probability or return period gives the flood-frequency curve. In most cases, the flood return period of interest is greater than the length of the available record and, therefore, the flood–frequency relationship needs to be extrapolated in some way. This leads to inaccuracies because of the non-linear relationship, and any extension to the curve may be highly dependent on one or two extreme floods. The normal approach is to fit one of a number of theoretical distributions to the data, with the assumption that the full range of annual maximum flows in the period of interest follow that distribution.

These distributions are defined by parameters reflecting the mean (or median) and standard deviation or variance of the data and, in some cases, by a third skewness parameter. Examples of appropriate distributions are given in Chapter 3. The normal approach to fitting a distribution curve is to match the parameters or moments of the distribution to those of the sample data. One distribution, which is commonly used for extreme values, is the *Gumbel* distribution, where the annual maximum flow with a return period *T* is given by:

$$Q_T = Q_m - 0{\cdot}45\sigma + 0{\cdot}78\sigma y \qquad (8.6)$$

where:

$$y = -\ln\left[-\ln\left(1 - \frac{1}{T}\right)\right]$$

and Q_m and σ are the mean and standard deviation of the time-series, respectively (Example 8.1).

If the horizontal (return period) scale of the plotting paper is adjusted so that the divisions are proportional to *y* in Equation (8.6), then the Gumbel distribution will plot as a straight line and extrapolation becomes easier. Nevertheless, it is not safe to extend the flood–frequency relationship beyond about twice the length of the observed record.

The main problem associated with fitting distributions to observed data is the assumption of stationarity, i.e. the assumption that the distribution of a time-series remains constant over a significant period of years and that it can be estimated from a relatively short period of data. Special consideration is required where it is suspected that there have been significant trends due, for example, to changes in land use or in climate. Such trends can be identified and removed using the techniques described in Chapter 3.

Example 8.1

The annual maximum flows (m³/s) for a sequence of six years are 6·21, 6·30, 5·07, 6·55, 5·86, 5·33. Plot a flood–frequency curve and estimate the 10-year return period flow using the Gumbel Distribution.

Year	Annual maxima	Ranked series	Rank	Plotting position	Return period
1	6·21	6·55	1	0·14	7·00
2	6·30	6·30	2	0·29	3·50
3	5·07	6·21	3	0·43	2·33
4	6·55	5·86	4	0·57	1·75
5	5·86	5·33	5	0·71	1·40
6	5·33	5·07	6	0·86	1·17

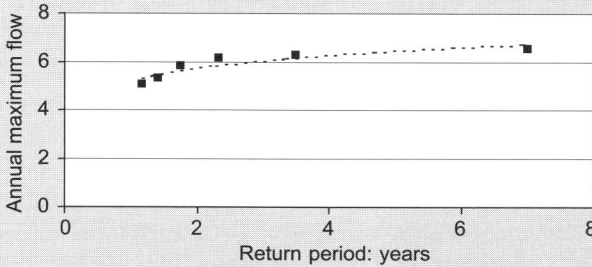

Mean flow $Q_m = 5 \cdot 89$ m³/s

Standard deviation $\sigma = 0 \cdot 53$ m³/s

$$Q_{10} = Q_m - 0 \cdot 45\sigma + 0 \cdot 78\sigma y$$

$$y = -\ln\left[-\ln\left(1 - \frac{1}{T}\right)\right] = -\ln\left[-\ln(0 \cdot 9)\right] = 2 \cdot 25$$

$$Q_{10} = 5 \cdot 89 - 0 \cdot 45 \times 0 \cdot 53 + 0 \cdot 78 \times 0 \cdot 53 \times 2 \cdot 25 = 6 \cdot 58 \text{ m}^3/\text{s}$$

8.1.4 **Flood Estimation Handbook** *method*

The flood–frequency curve used in the *Flood Estimation Handbook* (Robson and Reed, 1999) is based on a non-dimensional *flood growth curve*, which is scaled by an *index flood*. The index flood is the property of a specific site and represents a typical magnitude of a flood at that site. The growth curve is likely to be similar for many catchments with similar hydrological characteristics, and the growth curve for a given catchment can therefore be estimated by combining growth curves from a pool of similar catchments, not necessarily within the same geographic area.

Estimation of the index flood

The *Flood Estimation Handbook* uses the *median annual maximum flood* (*QMED*) as the index flood. *QMED* is defined as the middle-ranking value in a sample of ranked annual maximum flows. Where there is an even number of values, the mean of the two middle values is taken. Since, by definition, half of the dataset exceeds the median value, the probability of exceedance of the median flood is a half and, therefore, the corresponding return period is two years. As before, the annual maxima should be estimated over a hydrological year (starting in October), rather than the calendar year, to avoid a problem of multiple peaks in December and January. Where there are several tied values leading to steps or 'granularity' in the flood–frequency curve, the data may need to be re-abstracted. The advantage of the median value over the mean value is that it is less affected by exceptionally large floods.

Where the available record is less than 13 years, it is recommended that *QMED* is estimated using peaks over a threshold (POT) (Section 8.1.2) rather than annual maxima. *QMED* can be estimated from the weighted combination of two values (Q_i and Q_{i+1}) from the POT series depending on the length of the record, i.e.:

$$QMED = wQ_i + (1-w)Q_{i+1} \qquad (8.7)$$

where w and i are given in Table 8.1 for different record lengths. An example of the calculations is given in Example 8.2.

Where the record length is two years or less, it is unlikely that a reliable estimate of *QMED* can be obtained from the data unless the record can be extended, for example by regression with another nearby (*donor*) catchment. This would require an overlap of at least 12 months between the two records. Examples of appropriate regression models include:

Table 8.1   Coefficients for estimation of QMED from a POT series (reproduced with the permission of the Centre for Ecology and Hydrology, Wallingford)

Record length: years	*i*th postion	Weight, w
1	1	0·602
2	2	0·895
3	2	0·100
4	3	0·298
5	4	0·509
6	5	0·725
7	6	0·945
8	6	0·147
9	7	0·349
10	8	0·557
11	9	0·769
12	10	0·983
13	10	0·185

$$Q_s = a + bQ_d \tag{8.8}$$

or:

$$\ln Q_s = c + d \ln Q_d \tag{8.9}$$

where Q_s and Q_d are the flows at the subject and donor sites, respectively. If the regression explains at least 90% of the variance of the flood peaks at the subject site, it can be considered sufficiently reliable and the estimate of *QMED* can be made based on a POT series. In practice, the values of Q_i and Q_{i+1} can be identified from the donor record and converted to the subject site using the regression relationship.

Example 8.2

A series of peaks over threshold (POT) values (m³/s) were extracted from a time-series of flow as below:

139·3, 160·8, 261·6, 124·4, 162·7, 120·0, 155·8, 141·6, 202·9, 133·4

Estimate the median flow.

The data are ranked in descending order:

Rank	POT series
1	261·6
2	202·9
3	162·7
4	160·8
5	155·8
6	141·6
7	139·3
8	133·4
9	124·4
10	120·0

From Table 8.1 for a record length of 10, $i = 8$ and $w = 0.557$:

$$QMED = wQ_i + (1 - w)Q_{i+1}$$

$$= 0.557Q_8 + 0.443Q_9$$

$$= 0.557 \times 133.4 + 0.443 \times 124.4$$

$$= 129.4 \text{ m}^3/\text{s}$$

If there is no suitable flow record available, or where a quick preliminary assessment is required, *QMED* can be estimated from regression equations with certain catchment descriptors. It should be recognised that estimates using such equations will be much less reliable than estimates based on flow records, and this procedure should not be used for major schemes or where there is a possible threat to life. The regression equation proposed by the *Flood Estimation Handbook* for *QMED* is:

$$QMED = 1.172 AREA^{AE} \left(\frac{SAAR}{1000} \right)^{1.560} FARL^{2.642} \left(\frac{SPRHOST}{100} \right)^{1.211} 0.0198^{RESHOST}$$

$$(8.10)$$

AE and *RESHOST* are coefficients based on the catchment area and the soil type respectively, and are given by:

$$AE = 1 - 0.015 \ln \left(\frac{AREA}{0.5} \right)$$

$$RESHOST = BFIHOST + 1.30 \left(\frac{SPRHOST}{100} \right) - 0.987 \qquad (8.11)$$

AREA is the catchment drainage area in km^2, and *SPRHOST* and *BFIHOST* are estimates of standard percentage runoff and baseflow index respectively, using the HOST soil classification (see Chapter 6). *FARL* is an index of flood attenuation due to reservoirs and *SAAR* is the Standard Average Annual Rainfall, which can be obtained from a published map.

The above equations apply only to rural catchments: where the catchment is significantly urbanised (i.e. where the parameter *URBEXT* (see Section 5.7.4) exceeds 0.025), the above estimate of *QMED* should be multiplied by a factor *UAF* given by:

$$UAF = PRUAF(1 + URBEXT)^{0.83} \qquad (8.12)$$

where:

$$PRUAF = 1 + 0.615 \times URBEXT \left(\frac{70}{SPRHOST} - 1 \right)$$

As noted in earlier chapters, the *Flood Estimation Handbook* calculates these parameters using digital data on a 1 km grid, although the user should check that the estimates of *AREA*, *URBEXT* and *FARL* are reasonable and up to date.

Because of the relative unreliability of estimates of *QMED* derived using catchment characteristics, it is recommended that they should be adjusted by reference to a suitable nearby site (*donor site*), where *QMED* can be estimated from a reasonably long record of flow. The ideal donor site would be situated just upstream or downstream of the subject site with a similar size, land use and topography. If possible, neither the subject nor donor catchments should have any major reservoirs that could cause significant attenuation or any significant urban areas.

The adjustment procedure can be summarised by the following equation:

$$QMED_{s,adj} = QMED_{s,cds} \left(\frac{QMED_{g,obs}}{QMED_{g,cds}} \right) \qquad (8.13)$$

The subscripts s and g refer to the *subject* site and *gauged* (donor) site and cds and obs refer to estimates obtained using catchment descriptors and observed data, respectively.

If more than one suitable donor catchment is available, a weighted average of the estimates from each donor site can be used. If no suitable donor site can be found, one or more *analogue* sites may be used instead. An analogue site is a gauged site which is not in the same geographic area as the subject site but is hydrologically quite similar, specifically in terms of the five parameters used to estimate *QMED*. In view of the greater uncertainty regarding the use of analogue sites, it is recommended that several such sites are used.

Estimation of the growth curve and flood–frequency curve

The growth curve represents the frequencies or return periods of given floods in relation to the index flood, determined as described in the previous section. By normalising the floods by the index flood, the growth curves for different-sized catchments tend towards a single curve, provided the catchments are hydrologically similar. As has been mentioned, the estimation of long period floods is more reliable if a *pooled growth curve* is used. A pooled growth curve is determined using the curves from a number of selected catchments, chosen for their similar hydrological characteristics rather than their geographical location. Greatest weight is applied to those stations which are most similar to the subject site, or which have the longest record. Pooling data from several stations effectively increases the length of record available, although the value of the data is obviously not as great as data from the subject site. The size of the *pooling group* is normally selected to give a total record length of at least five times the return period being considered. Only sites that are essentially rural and have records with at least eight annual maxima are normally included in the pooling group.

The initial selection of the pool of sites is made on the basis of three key catchment features: size, wetness and soil properties. The parameters used are *AREA*, *SAAR* (Section 4.6.7) and *BFIHOST* (Section 5.7.1). The *Flood Estimation Handbook* software ranks the selected sites in terms of their closeness to the subject site in *size–wetness–soils space*. The closeness is measured by the difference between the values of *AREA*, *SAAR* and *BFIHOST* for the subject site and a donor site, i.e.:

$$d = \sqrt{\frac{1}{2}\left(\frac{\ln A_s - \ln A_d}{\sigma \ln A}\right)^2 + \left(\frac{\ln S_s - \ln S_d}{\sigma \ln S}\right)^2 + \left(\frac{B_s - B_d}{\sigma b}\right)^2}$$

$$= \sqrt{\frac{1}{2}\left(\frac{\ln A_s - \ln A_d}{1\cdot 34}\right)^2 + \left(\frac{\ln S_s - \ln S_d}{0\cdot 38}\right)^2 + \left(\frac{B_s - B_d}{0\cdot 15}\right)^2} \tag{8.14}$$

The parameters *A*, *S* and *B* represent *AREA*, *SAAR* and *BFIHOST*, respectively, and the suffixes s and d represent the subject and donor sites. The denominators in the equation are the standard deviations of the respective parameters for the 698 sites used in the study. The factor of a half was introduced because it was felt that the *AREA* was exerting too large an influence on the similarity distance.

The sites are then sorted in order of their closeness and a cut-off is made when the aggregate record length is adequate (normally five times the required return period). A *similarity ranking factor* is then calculated as:

$$S_i = 1 - \frac{1}{n_{\text{total}}}\sum_{j-1}^{i-1} n_j$$

$$= S_{i-1} - \frac{n_{i-1}}{n_{\text{total}}} \tag{8.15}$$

where n_{total} is the total number of station years of records in the pooling group and n_i is the record length of the *i*th most similar site. The value of the similarity ranking factor will therefore vary from 1 for the most similar site (which will often be the subject site) to almost zero for the least similar, but will be weighted according to the length of record.

The user has an opportunity to review the selection in the light of further factors, such as location, record length and seasonality. A check should also be made to see if the distribution of annual maxima from an individual station is significantly different from the group average in terms of its mean, variability and skewness.

Having selected the pooling group, the distribution of annual maxima for each site is derived from the records. Fitting a distribution curve to the annual maxima data is carried out by matching the first, second and third moments (mean, variance and skewness) of the data as described above. The *Flood Estimation Handbook* recommends the use of L-moments rather than conventional moments, as they have been found to be more robust where the data are highly skewed. As described in Chapter 3, L-moments are based on linear combinations of the data (hence the name) and correspond to the three moments mentioned above. The first three sample moments are calculated as below:

$$l_1 = \frac{\sum_{j=1}^{n} x_j}{n}$$

$$l_2 = 2b_1 - l_1$$

$$l_3 = 6b_2 - 6b_1 + l_1 \qquad (8.16)$$

where:

$$b_1 = \frac{\sum_{j=2}^{n} \frac{(j-1)}{(n-1)} x_j}{n}$$

$$b_2 = \frac{\sum_{j=3}^{n} \frac{(j-1)(j-2)}{(n-1)(n-2)} x_j}{n}$$

The moments are normalised by reference to the mean or variance, the L-moment ratios being defined as:

$$t_2 = \frac{l_2}{l_1}$$

$$t_3 = \frac{l_3}{l_2} \qquad (8.17)$$

Example 8.3

Using the data from Example 8.1, calculate the first three L-moments and the moment ratios.

The data are first sorted into ascending order:

Rank, j	Ranked data (x_j)	$\dfrac{(j-1)x}{(n-1)}$	$\dfrac{(j-1)(j-2)x}{(n-1)(n-2)}$
1	5·07		
2	5·33	1·066	
3	5·86	2·344	0·586
4	6·21	3·726	1·863
5	6·30	5·040	3·780
6	6·55	6·550	6·550
Σ	30·25	18·726	12·779

$$l_1 = \frac{\sum_{j=1}^{n} x_j}{n} = \frac{30\cdot25}{6} = 5\cdot887$$

where:

$$b_1 = \frac{\sum_{j=2}^{n} \dfrac{(j-1)}{(n-1)} x_j}{n} = \frac{18\cdot726}{6} = 3\cdot121$$

$$b_2 = \frac{\sum_{j=3}^{n} \dfrac{(j-1)(j-2)}{(n-1)(n-2)} x_j}{n} = \frac{12\cdot779}{6} = 2\cdot130$$

$$l_1 = \frac{30 \cdot 25}{6} = 5 \cdot 887 \quad (= l_1)$$

$$l_2 = 2b_1 - l_1 = 2 \times 3 \cdot 121 - 5 \cdot 887 = 0 \cdot 355$$

$$l_3 = 6b_2 - 6b_1 + l_1 = 6 \times 2 \cdot 130 - 6 \times 3 \cdot 121 + 5 \cdot 887 = -0 \cdot 060$$

(As a comparison, the variance of the data is 0·339 and the skewness is −0·489.)

Moment ratios:

$$t_2 = \frac{l_2}{l_1} = \frac{0 \cdot 355}{5 \cdot 887} = 0 \cdot 060$$

$$t_3 = \frac{l_3}{l_2} = \frac{-0 \cdot 060}{0 \cdot 355} = -0 \cdot 169$$

The moment ratios for the pooled group are formed as the weighted average of the ratios for the individual sites in the group, i.e.:

$$t_{2-\text{pooled}} = \frac{\displaystyle\sum_{i=1}^{M} w_i t_2}{\displaystyle\sum_{i=1}^{M} w_i}$$

$$t_{3-\text{pooled}} = \frac{\displaystyle\sum_{i=1}^{M} w_i t_3}{\displaystyle\sum_{i=1}^{M} w_i} \tag{8.18}$$

The weighting factor (w) depends on the record length (n) and the similarity ranking (S) (Equation (8.15)) of the site and is given by:

$$w_i = S_i n_i \qquad (8.19)$$

The distribution recommended by the *Flood Estimation Handbook* for the flood frequency curve is the *Generalised Logistic* distribution, which is described by:

$$x_r = 1 + \frac{\beta}{k}\left\{1 - (T-1)^{-k}\right\} \qquad (8.20)$$

where:

$$k = -t_3$$

$$\beta = \frac{t_2 k \sin \pi k}{k\pi(k + t_2) - t_2 \sin \pi k}$$

and t_2 and t_3 are the second and third L-moment ratios. Fitting this distribution to the pooled moment ratios then yields the pooled growth curve.

Finally, the flood-frequency curve is constructed as the product of the index flood (*QMED*) and the growth curve (x_r), i.e.:

$$Q_r = QMED x_r \qquad (8.21)$$

8.2   Flow estimation for reservoirs

An estimation of extreme flows is required for the design and assessment of dam spillways. The spillway capacity for a dam is based on the *probable maximum flood* (PMF), which can be regarded as the largest flood that might ever occur. It is normally estimated using the unit hydrograph method (described in Chapter 7) and applying a notional *probable maximum precipitation* (PMP), as described in Chapter 4. The statistical methods of flood prediction described in Section 8.1 above are not normally used for estimating the PMF because it is necessary to have a complete hydrograph which is routed through the reservoir. There is also

Example 8.4

A pooling group of suitable sites is selected in order to produce a flood–frequency curve. The individual hydrological characteristics, moment ratios and record lengths are given below (in order of similarity to the subject site). Estimate the pooled moment ratios.

Site	AREA	SAAR	BFIHOST	Variance ratio, t_2	Skewness ratio, t_3	Record length: years
1 (subject)	550	655	0·550	0·156	0·180	15
2	397	604	0·562	0·199	0·175	25
3	355	623	0·570	0·171	0·208	18
4	345	688	0·590	0·112	0·157	38
5	709	710	0·493	0·148	0·235	25
6	443	610	0·611	0·160	0·201	39
7	295	675	0·485	0·144	0·145	17
8	598	690	0·445	0·139	0·110	18

For site 2, the distance d is given by:

$$d = \sqrt{\frac{1}{2}\left(\frac{\ln A_s - \ln A_d}{\sigma \ln A}\right)^2 + \left(\frac{\ln S_s - \ln S_d}{\sigma \ln S}\right)^2 + \left(\frac{B_s - B_d}{\sigma b}\right)^2}$$

$$= \sqrt{\frac{1}{2}\left(\frac{\ln 397 - \ln 550}{1\cdot34}\right)^2 + \left(\frac{\ln 604 - \ln 655}{0\cdot38}\right)^2 + \left(\frac{0\cdot562 - 0\cdot550}{0\cdot15}\right)^2} = 0\cdot134$$

The similarity factor S is given by:

$$S_i = S_{i-1} - \frac{n_{i-1}}{n_{\text{total}}} = 1\cdot0 - \frac{15}{195} = 0\cdot923$$

The weight (w) is given by:

$$w = nS = 25 \times 0.923 = 23.08$$

The contributions of station 2 to the pooled moment ratios are:

$$\frac{w \times t_2}{\sum w} = \frac{23.08 \times 0.199}{111.29} = 0.041$$

and:

$$\frac{w \times t_3}{\sum w} = \frac{23.08 \times 0.175}{111.29} = 0.036$$

The results for the remaining stations are shown in the table below:

Site	Record length	Dist.	Sim'ty	Weight, (w)	Var. ratio, t_2	Skew. ratio, t_3	$\dfrac{w \times t_2}{\sum w}$	$\dfrac{w \times t_3}{\sum w}$
1	15	0.000	1.000	15.00	0.156	0.180	0.021	0.024
2	25	0.134	0.923	23.08	0.199	0.175	0.041	0.036
3	18	0.155	0.795	14.31	0.171	0.208	0.022	0.027
4	38	0.236	0.703	26.70	0.112	0.157	0.027	0.038
5	25	0.319	0.508	19.80	0.148	0.235	0.026	0.042
6	39	0.335	0.308	7.69	0.160	0.201	0.011	0.014
7	17	0.393	0.179	3.05	0.144	0.145	0.004	0.004
8	18	0.617	0.092	1.66	0.139	0.110	0.002	0.002
Total	195			111.29			0.155	0.186

Pooled moment ratios

$t_2 = 0.155$
$t_3 = 0.186$

Example 8.5

Using the pooled moment ratios from Example 8.4, estimate the 25-year return period growth factor and flood. The index flood is 10·5 m³/s.

From generalised logistic distribution

$$x_r = 1 + \frac{\beta}{k}\left\{1 - (T-1)^{-k}\right\}$$

where:

$$k = -t_3$$

$$\beta = \frac{t_2 k \sin \pi k}{k\pi(k+t_2) - t_2 \sin \pi k}$$

Hence:

$$k = -0\cdot186$$

$$\beta = \frac{0\cdot155(-0\cdot186)\sin\pi(-0\cdot186)}{-0\cdot186\pi(-0\cdot186+0\cdot155)-0\cdot155\sin\pi(-0\cdot186)} = 0\cdot154$$

$$x_{25} = 1 + \frac{0\cdot155}{-0\cdot186}\left\{1-(25-1)^{0\cdot186}\right\} = 1\cdot665$$

25-year flood:

$$Q_{25} = 1\cdot665 \times 10\cdot5 = 17\cdot5 \text{ m}^3/\text{s}$$

the difficulty of assigning a return period to the flood event: values of 10^6 or 10^7 years have been suggested in the *Flood Estimation Handbook* depending on the size of the catchment and the amount of snowmelt. In any event, any flood–frequency curve will be highly unreliable because of the very small length of record in relation to the notional return period.

However, a quick method for an initial assessment of the *PMF* was proposed in the *Flood Studies Report* (NERC, 1975) and is also contained in the *Flood Estimation Handbook*. It uses the following regression equation:

$$PMF = 0.629AREA^{0.937} S1085^{0.328} SOIL^{0.471} (1 + URBAN_{FSR})^{2.04} SAAR^{0.319}$$

(8.22)

S1085 is a measure of the channel slope and is calculated as the slope between a point 10% of the channel length and a point 85% of the channel length measured from the outfall. This estimate should not be regarded as an alternative to the unit hydrograph method (for example, it does not allow for reservoir routing) and should only be used for an initial estimate for simple catchments.

8.3   Low flow estimation

The estimation of very low flows is often required where water is to be abstracted directly from a river channel or where a pollutant discharge is proposed. As in the case of flood estimation, if there is a sufficiently long flow record from a suitable gauging station, the best estimate is obtained using the flow records. Where the flow record is limited or nonexistent, other methods may be used.

A low flow event can be described in terms of either:

- a threshold discharge
- an accumulated volume
- a length of time below a certain threshold
- a rate of recession.

A low flow event is not normally an instantaneous minimum annual flow, but is averaged over a period of time, such as a week or a month. A method of low flow estimation used in the United Kingdom (Gustard *et al.*, 1992) uses the following flow parameters:

Mean flow (MF)	Mean of a record of average daily flows.
$Q_{95}(1)$	The flow exceeded by 95% of the record of average daily flows.
MAM(7)	The mean annual seven day minimum flow, i.e. the mean of the lowest annual flows averaged over seven days.

8.3.1 Estimating low flows from flow records

Low flows can be estimated from *flow–duration curves* (FDCs), which show the proportion of time a given flow was equalled or exceeded. Typical examples are given in Figure 8.1(a). They are constructed from a histogram of average daily flows, which is then integrated to give a cumulative distribution. The cumulative probabilities are subtracted from 1 to give the probabilities of exceedence. The horizontal axis of a flow–duration curve is usually scaled in terms of percentage of time or probability, and the vertical axis has units of discharge (e.g. m^3/s). Since the distribution of flows is usually skewed and often follows a log-normal distribution, the flow–duration curve can be linearised if the horizontal axis is plotted using a normal probability scale and the vertical axis is plotted using a logarithmic scale. Where several catchments of different sizes are being compared, it is also more convenient to normalise the flows by dividing by the mean flow or product of area multiplied by the rainfall. In fact, a better estimate of a flow–duration curve can be obtained by pooling the flow–duration curves from a number of similar adjacent catchments to produce a regional flow–duration curve.

Although the flow–duration curve contains no information about the sequencing of flows, it does indicate useful characteristics of the catchment. In particular, it describes the degree of storage within the catchment, which controls the amount of base flow. Catchments with little storage, such as those of the uplands of northern England and Scotland will typically show large extremes of flow and will have steep flow–duration curves. In contrast, many catchments in the south of England on chalk aquifers will have a much smaller variation in flow and much flatter flow–duration curves. Figure 8.1 compares the typical hydrographs and flow–duration curves of the River Test in Hampshire and the Black Cart River in south-west Scotland. The former, being on a chalk catchment, is largely fed by groundwater and shows a relatively small annual variation in flow. A similar effect is evident where there is a high degree of reservoir storage in a catchment. However, the Black Cart is a much more responsive river with greater fluctuations in flow because of its steep, relatively impermeable, catchment.

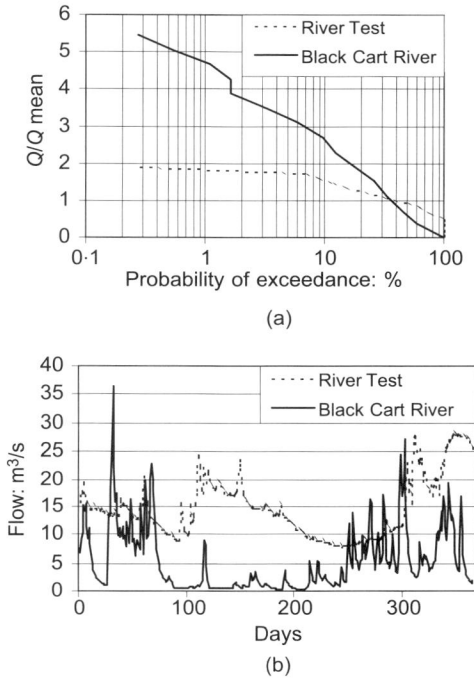

Figure 8.1   Comparison of flow–duration curves and hydrographs of two rivers: (a) flow–duration curves; (b) typical annual hydrographs

8.3.2   Estimating low flows from ungauged catchments

Where there are no flow records, an estimate of the low flow characteristics of a catchment can be obtained using the method proposed by Gustard *et al.* (1992). This involves, firstly, estimating Q_{95} (1) and *MAM*(7), as well as ordinates of the flow–duration curve, as a percentage of the mean flow.

Estimation of mean flow

The mean flow can be estimated either by a water balance approach (Section 5.5.1) or by using regression equations. The basis of the water balance method is that:

$$Mean\ flow = AARD \times Catchment\ area \times C / Record\ length \qquad (8.23)$$

where:

$AARD$ = *Standard Average Annual Rainfall (SAAR) – Losses*

Losses = $r \times PE$

The constant (C) reflects the units used for the area, rainfall depth and record length, and r is an empirical coefficient. For units of km^2, mm and years respectively, the value of C is $3 \cdot 17 \times 10^{-5}$. A regression exercise involving 687 UK catchments (Holmes *et al.*, 2002) provided the following relationship:

$$r = 0 \cdot 00061 SAAR + 0 \cdot 475 \qquad \text{(for } SAAR < 850 \text{ mm)}$$

$$r = 1 \cdot 0 \qquad \text{(for } SAAR => 850 \text{ mm)} \qquad (8.24)$$

$SAAR$ is estimated from a published map and PE is the potential evaporation calculated from the Penman-Monteith or similar equation (see Chapter 5). The value of the constant r increases with water availability. For catchments with average rainfall greater than 850 mm, it is assumed that actual evaporation is equal to potential evaporation, since evaporation is unlikely to be limited by soil moisture deficit.

Alternatively, the mean flow can be estimated from a regression equation based on catchment characteristics:

$$MF = 2 \cdot 70 \times 10^{-7} AREA^{1 \cdot 02} SAAR^{1 \cdot 82} PE^{-0 \cdot 284} \qquad (8.25)$$

In general, the water balance approach is to be preferred as it is more reliable and can incorporate local data.

Estimation of $Q_{95}(1)$ and MAM(7)

The 95 percentile daily flow ($Q_{95}(1)$) and the seven-day minimum flow ($MAM(7)$), are both calculated relative to the mean flow. Where flow records exist, it is recommended that $Q_{95}(1)$ and $MAM(7)$ are estimated from the baseflow index (BFI) (Section 5.7.1), if necessary using a relationship with a nearby station having a longer record. The proposed regression equations for $Q_{95}(1)$ and $MAM(7)$ are:

$$Q_{95}(1) = 44 \cdot 1 BFI^{1 \cdot 43} SAAR^{-0 \cdot 033} AREA^{0 \cdot 0342} \qquad (8.26)$$

and:

$$MAM(7) = 190 \cdot 99 BFI^{1 \cdot 52} SAAR^{-0 \cdot 199} \tag{8.27}$$

Where no flow records are available, it is possible to estimate $Q_{95}(1)$ and $MAM(7)$ using the Hydrology of Soil Types (HOST) classification (Table 5.3). It was found that the 29 HOST soil types (plus urban and lake surfaces) could be grouped into 12 *low-flow groups* of soil type which were significant in the estimation of $Q_{95}(1)$ and $MAM(7)$. The results are summarised in Table 8.2

The distribution of the low-flow HOST groups across the United Kingdom is shown in Figure 8.2 (Gustard *et al.*, 1992), and Figure 8.3 shows the distribution of the $Q_{95}(1)$ ratio (colour versions are available in the original references).

Table 8.2   Estimation of $Q_{95}(1)$ and MAM(7) from soil type (reproduced with the permission of the Centre for Ecology and Hydrology, Wallingford)

Low-flow HOST group	HOST classes	$Q_{95}(1)$: % of MF	MAM(7): % of MF
1	1	40·8	50·8
2	29	31·9	40·3
3	2,4	65·7	71·3
4	3	25	27·5
5	5,12	49	53·4
6	6,7,8,9,10	6·5	1·4
7	13,15,16,17,18,20,21,23	10·7	12·4
8	19,22,24	1·1	0·1
9	14	15·0	14·4
10	11,25,26,27	6·8	5·9
11	97 (urban)	29·4	33·8
12	98 (lake)	65·1	49·6

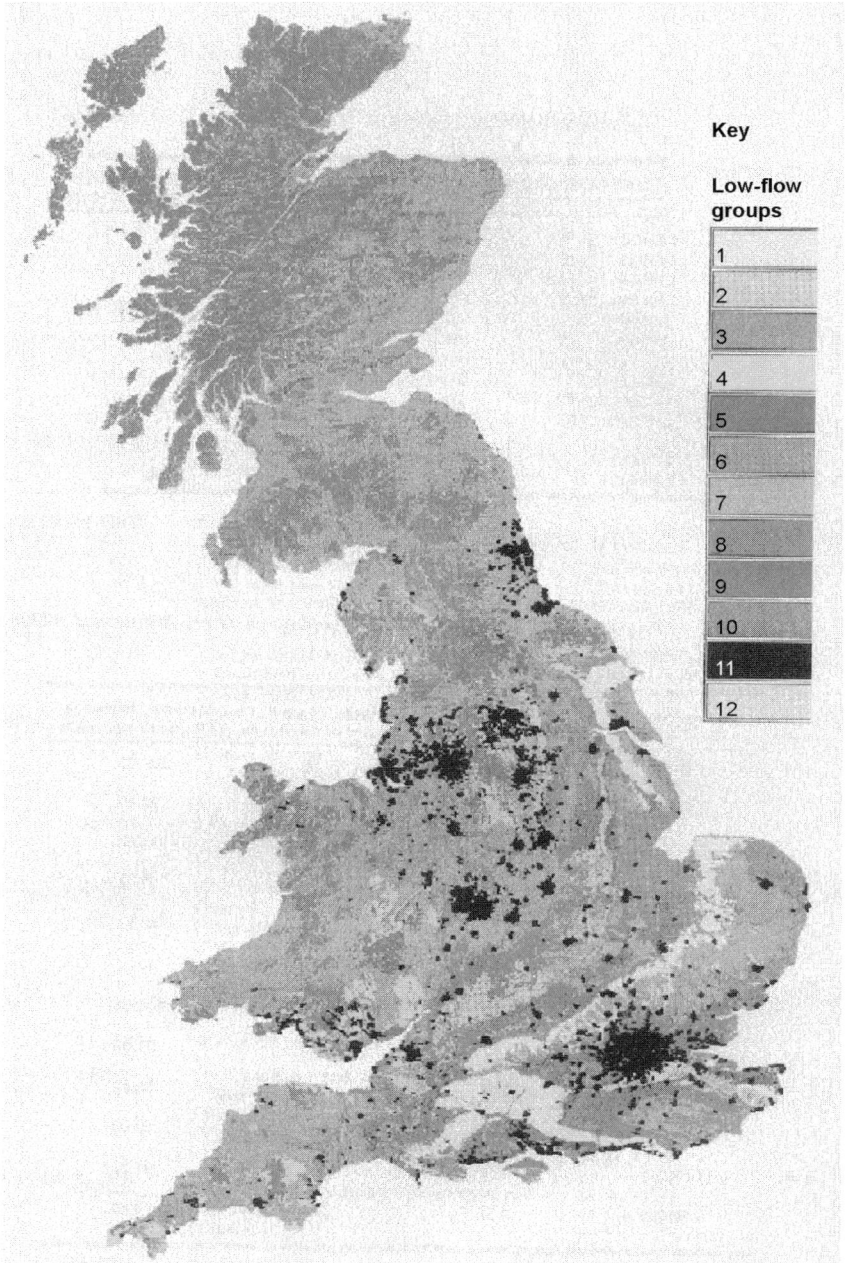

Figure 8.2 Distribution of low-flow HOST soil types (reproduced with the permission of the The Macaulay Land Use Research Institute)

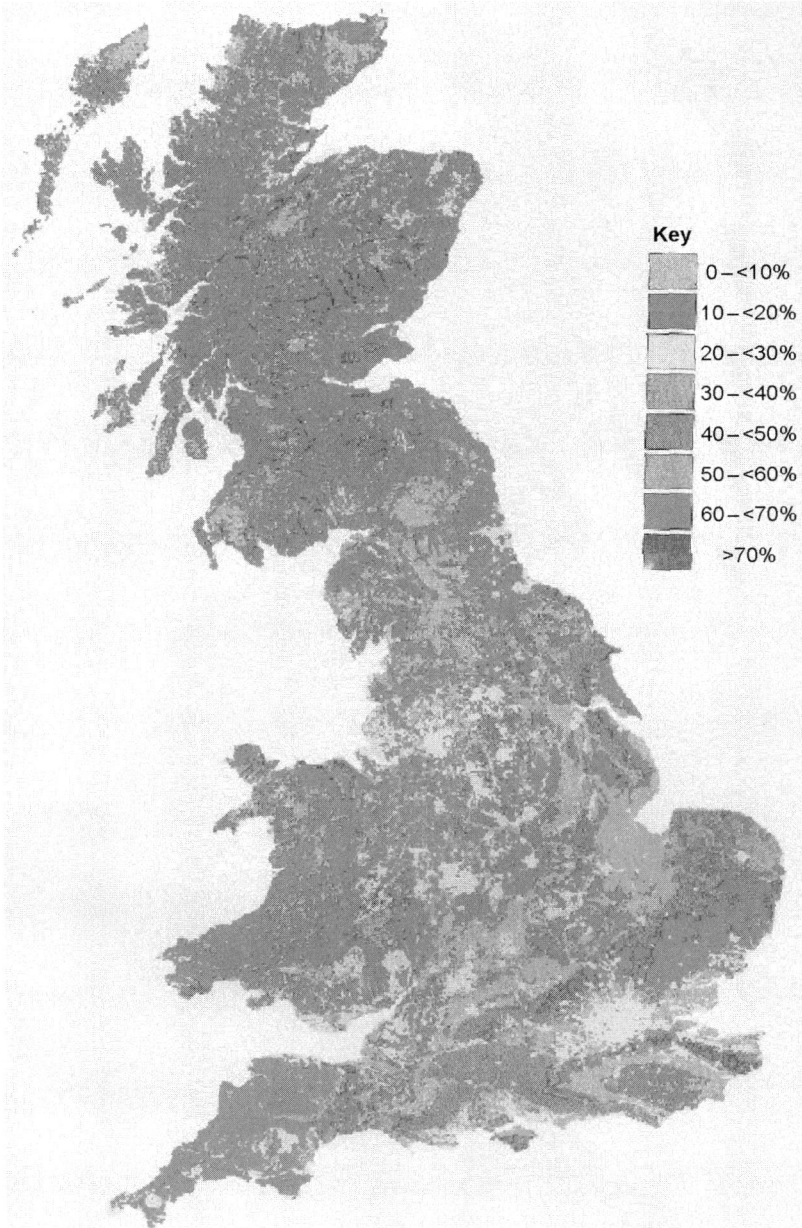

Figure 8.3   Distribution of $Q_{95}(1)$ as a percentage of mean flow (from Low Flow Estimation in the United Kingdom, reproduced with the permission of the The Macaulay Land Use Research Institute)

Estimating ordinates of the flow–duration curve

The actual shape of the flow–duration curves for the 845 UK catchments investigated was found to depend mainly on the value of Q_{95} (1) (Table 8.3). The individual flow–duration curves were pooled into 19 groups according to the value of Q_{95} (1)/ *mean flow*, and a mean curve produced for each group. These are shown in Figure 8.4.

If some flow readings are available for a site, an improved estimate of the flow–duration curve can be obtained by comparison with a suitable nearby gauged catchment with a well-established flow–duration curve. Ideally, the analogue site should have a similar soil type. The percentile of the flow at the gauged station, which occurred at the same time as the flow measurement was carried out, is estimated from the flow–duration curve. This is then plotted against the observed flow and, if sufficient measurements are carried out, a flow–duration curve can be drawn.

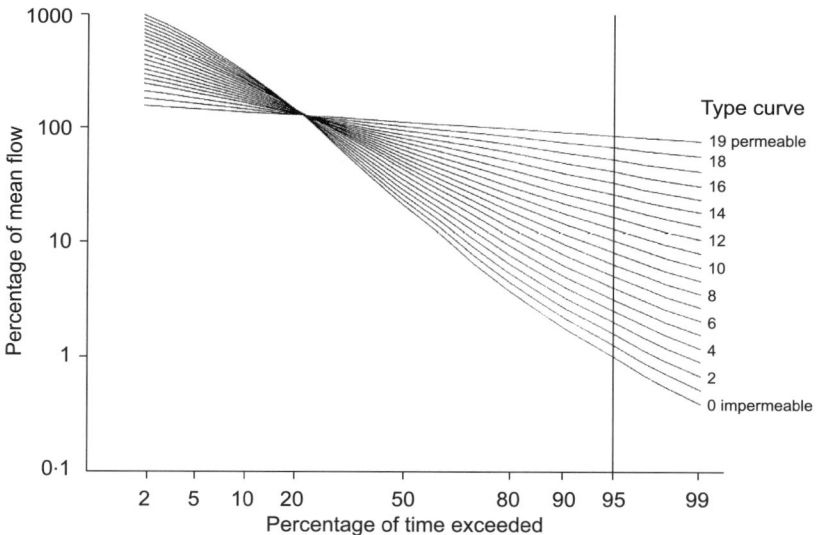

Figure 8.4   Mean flow–duration curves for groups of UK catchments (reproduced with the permission of the Centre for Ecology and Hydrology, Wallingford)

The analysis and prediction of flows

Table 8.3 Flow–duration type curves (percentage of mean flow) (reproduced with the permission of the Centre for Ecology and Hydrology, Wallingford) (continued overleaf)

$Q_{95}(1)$	1·00	1·26	1·58	2·00	2·51	3·16	3·98	5·01	6·30	7·94
Type curve	0	1	2	3	4	5	6	7	8	9
Percentile										
2	975·70	904·17	838·77	776·04	719·91	667·48	618·22	572·53	520·00	472·29
5	577·26	1534·08	511·37	480·48	452·42	425·82	400·44	376·64	350·65	326·46
50	20·49	22·69	25·10	27·86	30·82	34·11	37·81	41·82	45·10	48·64
80	3·70	4·42	5·27	6·33	7·54	9·00	10·77	12·86	15·20	17·98
90	1·73	2·13	2·62	3·25	3·99	4·92	6·07	7·47	9·16	11·22
95	1·00	1·26	1·58	2·00	2·51	3·16	3·98	5·01	6·30	7·94
99	0·38	0·51	0·67	0·88	1·16	1·53	2·02	2·65	3·46	4·52

Table 8.3 *continued*

$Q_{95}(1)$	10·00	12·57	15·83	19·93	25·13	31·64	39·81	50·13	63·12	79·43
Type curve	10	11	12	13	14	15	16	17	18	19
Percentile										
2	428·96	389·60	353·86	321·39	291·65	264·89	240·09	206·89	178·28	153·69
5	303·93	282·96	263·44	245·26	228·19	212·45	197·49	176·99	158·62	142·20
50	52·46	56·57	61·01	65·79	71·00	76·57	82·60	89·91	97·86	106·49
80	21·25	25·13	29·71	35·12	41·58	49·16	58·08	67·82	79·21	92·46
90	13·75	16·86	20·66	25·32	31·09	38·10	46·67	56·95	69·50	84·77
95	10·00	12·57	15·83	19·93	25·13	31·64	39·81	50·13	63·12	79·43
99	5·89	7·69	10·03	13·08	17·11	22·32	29·13	39·00	52·22	69·85

Example 8.6

Estimate the mean flow and $Q_{95}(1)$, *MAM*(7) and flow–duration curve for a catchment with an area of 103 km², a *SAAR* of 1500 mm/year with a potential evaporation of 250 mm/year. The proportion of low flow soil types are given in the table below:

Low-flow host group	Percentage
6	9
7	28
9	34
10	29

From Equation (8.23):

  Mean flow = AARD × *Catchment area* × *C/Record length*

where:

  AARD = Standard Average Annual Rainfall (*SAAR*) – *Losses*
  Losses = r × *PE*

For *SAAR* = 1500 mm/year, *r* = 1

  ∴ *Losses* = 1 × 250 = 250 mm/year
  AARD = 1500 – 250 = 1250 mm/year
  Mean flow = 1250 × 103 × 3·17 × 10^{-5} = 4·08 m³/s

From Table 8.2:

  $Q_{95}(1)$ = 0·09 × 6·5 + 0·28 × 10·7 + 0·34 × 15·0 + 0·29 × 6·8
  = 10·65% (of *mean flow*)
  = 0·1065 × 4·08 = 0·43 m³/s

  MAM(7) = 0·09 × 1·4 + 0·28 × 12·4 + 0·34 × 14·4 + 0·29 × 5·9
  = 10·21% (of *mean flow*)
  = 0·1021 × 4·08 = 0·42 m³/s

For $Q_{95}(1)$ = 10·65%, the type 10 curve from Table 8.3 is appropriate:

Percentile	Percentage of *mean flow*	Flow: m³/s
2	428·96	17·50
5	303·93	12·40
50	52·46	2·14
80	21·25	0·87
90	13·75	0·56
95	10·00	0·41
99	5·89	0·24

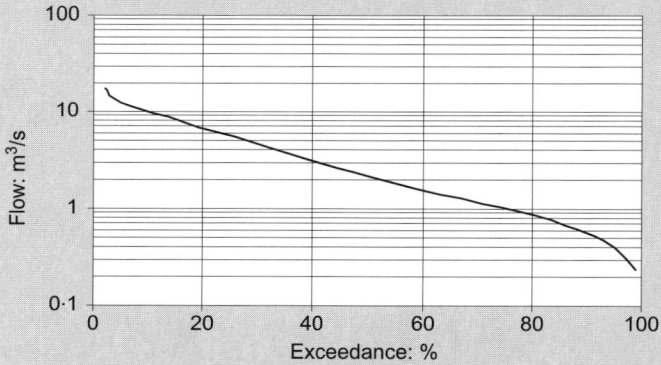

8.3.3   Estimation of low–flow frequency curve

An alternative approach to low-flow estimation using observed data is to construct a flow–frequency curve in a similar way to a flood–frequency curve described in Section 8.1. A low-flow frequency curve indicates the probability of a low flow of a particular duration occurring in any year. For instance, the annual minima of the average daily flow time-series could be ranked from the highest to the lowest, and plotted using a plotting position calculated, for example, from:

$$P_i = \frac{i - 0.44}{N + 0.12} \tag{8.28}$$

The return period (T) is given by:

$$T_i = \frac{1}{1 - P_i} \tag{8.29}$$

It has been found that a *Weibull* distribution given by:

$$F(x) = 1 - \exp\left(\frac{x^\gamma}{\vartheta}\right) \tag{8.30}$$

provides a good fit for annual minima data, where γ and ϑ are parameters which define, respectively, the scale and shape of the distribution. As before, better estimates are obtained if flow–frequency curves are based on pooled data.

The low-flow frequency curve for a seven-day duration ($MAM(7)$) (or other durations) can be estimated from standard curves in a similar manner to the flow–duration curve (Gustard *et al.*, 1992) described in the preceding section.

8.3.4 Estimation of low flow using recession curves

The recession part of a hydrograph is analogous to the flow through an orifice in a tank under a varying head, and tends to follow an exponential relationship of the form (Section 6.3.1):

$$Q(t) = Q_0 e^{-kt} \tag{8.31}$$

where $Q(t)$ is the flow at time t, Q_0 is the initial flow and k is a depletion constant depending on the nature of the catchment. Where extensive dry periods occur, the river flows can be predicted for several weeks or months by extrapolating the recession curve.

From Equation (8.31) it can be seen that the ratio of successive flows at a uniform time-interval is:

$$\frac{Q(t + \Delta t)}{Q(t)} = \frac{Q_0 e^{-kt} e^{-k\Delta t}}{Q_0 e^{-kt}} = e^{-k\Delta t} \tag{8.32}$$

which is a function of the depletion constant and the time interval only. The constant k can therefore be estimated by plotting ratios of successive flows on a log scale.

Curran (1990) showed that the value of $Q_{95}(1)$ is related to the mean flow (MF) by:

$$Q_{95}(1) = A \times MF \times e^{-BkT} \tag{8.33}$$

where A and B are empirical constants, and T is the recession period. The recession period T was estimated as the geometric mean length of a dry spell defined by the accumulated rainfall and evaporation. For 39 gauging stations in Scotland, the regression constants A and B were 0·383 and 1·73 respectively.

In the case of many arid and tropical regions, there is a pronounced dry period lasting for several months, during which a recession curve becomes well established. Mansell and Johnston (1994) showed that the recession constant for small watercourses in Zimbabwe was related to various catchment characteristics and, thus, it was possible to predict the flow during the dry season based on the flow at the end of the wet season.

8.4   Flow estimation for hydropower

In the design of hydropower plants, the significance of very low or very high flows is much less. Hydro-electric turbines are designed to operate at a specific flow rate (for example, 1·3 times the mean flow) and are unable to accept flows much greater than this value. In addition, they do not operate efficiently at very low flows and are usually shut down when the flow is less than a specific minimum value. The output of a hydro-electric turbine depends mainly on the shape of the flow–duration curve between these maximum and minimum flow rates (Department of Energy, 1983). Allowance may also need to be made for any compensation flow, i.e. a minimum flow which is required in the channel at all times to preserve fish and other aquatic life, or for 'freshets', which are discrete volumes of water released at certain times for the same purpose.

Consider a turbine operating in a river with a flow–duration curve as shown in Figure 8.5, from which a compensation flow has been deducted. The design flow rate for the turbine is Q_d and it is assumed that the turbine can operate at flow rates of between, say, $0{\cdot}25Q_d$ and Q_d. The power (kW) generated by a hydraulic turbine is given by:

$$P = \frac{\rho g \eta Q H}{1000} \qquad (8.34)$$

where Q (m^3/s) and H (m) are the flow rate and head respectively, η is the turbine/generator efficiency, ρ is the density of water (= 1000 kg/m^3) and g is the gravitational acceleration (9·81 m/s^2). The energy (kWh) produced by the turbine is the integral of power with time, i.e.:

$$E = g \eta H \int Q dt \qquad (8.35)$$

(assuming the head does not vary).

The integral of flow with time is the area under the flow–duration curve. However, the area is bounded by the upper and lower operating flow rates. Since the plant continues to operate at flows in excess of the upper limit (although the excess flows are not used), this forms a horizontal boundary, while below the lower limit it is assumed that the plant does not produce any output at all and so this forms a vertical boundary (Figure 8.5). The value of annual energy produced is therefore:

$$E(kW) = 3600 \times 24 \times 365 \times g \eta H A \qquad (8.36)$$

where A is the shaded area under the flow–duration curve (assuming that the flow rate is measured in m^3/s).

It can be seen that for a given value of mean flow, the area under the flow–duration curve is maximised when the slope of the curve is a minimum, i.e. when there is a high degree of storage within the

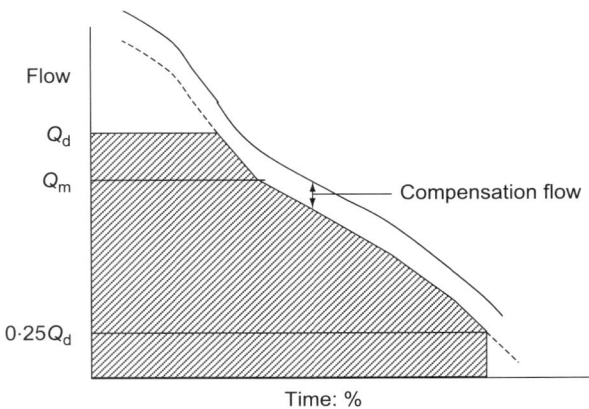

Figure 8.5   Estimation of flow for a hydro-electric turbine

catchment, either within the ground or as a result of artificial reservoirs. The required reservoir storage volume can be estimated by routing typical inflow hydrographs through reservoirs with varying storage and outflow characteristics, and estimating the energy generated using the outflow hydrographs.

Example 8.7

The following average monthly flows (m³/s) were recorded at a gauging station whose catchment area is 13 km². It is proposed to construct a hydro-electric power station at a site further down the river where the catchment area is 19 km². The turbine is to operate between 0·7 m³/s and 2·8 m³/s. Construct the flow–duration curve for the hydro-electric site and estimate the average annual energy production, allowing for 0·6 m³/s compensation flow. The net head at the site is 35 m and the overall efficiency is 70%:

2·163, 1·628, 0·579, 1·280, 0·277, 0·326, 0·304, 1·837, 2·725, 2·196, 2·015, 0·829

The flows for the hydro-electric site are estimated by scaling by the catchment areas (i.e. 19/13):

Gauging data	Flow at site
2·163	3·162
1·628	2·380
0·579	0·846
1·280	1·871
0·277	0·405
0·326	0·476
0·304	0·444
1·837	2·685
2·725	3·982
2·196	3·209
2·015	2·945
0·829	1·211

These flows are then allocated to classes and the cumulative percentage exceeding flows are calculated and the flow–duration curve constructed:

Class upper limit	Frequency	Cumulative frequency	Cumulative percentage
0	—	12	100·0
0·5	3	9	75·0
1	1	8	66·7
1·5	1	7	58·3
2	1	6	50·0
2·5	1	5	41·7
3	2	3	25·0
3·5	2	1	8·3
4	1	0	0·0

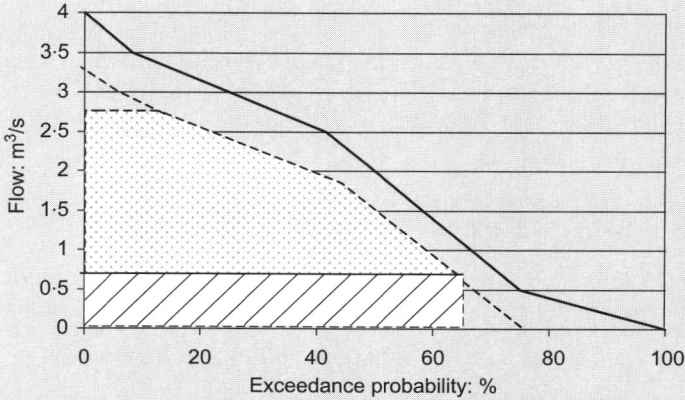

Average flow = shaded area = 0·85 m^3/s

Annual energy = 3600 × 24 × 365 × 9·81 × 0·7 × 35 × 0·85
= 6443 × 10^6 KWh
= 6443 GWh

8.5   Summary

This chapter describes methods for estimating return periods of flood events, either by manual analysis of individual flow records or by the analysis of records from a pool of similar catchments using the *Flood Estimation Handbook* method. It also describes the procedure for estimating various low-flow parameters and the use of the flow–duration curve in estimating hydro-electric power.

8.6   References

Curran, J. C. (1990). Low Flow Estimation Based on River Recession Rate. *Journal Institute of Water and Environmental Management,* **4**.

Department of Energy (1983). *The Development of Small Scale Hydro-Electric Power Plants.* University of Salford.

Gustard, A., Bullock, A. and Dixon, J. M.. (1992). *Low Flow Estimation in the United Kingdom.* Institute of Hydrology, Wallingford.

Herschy, R. (2001). *The world's maximum observed floods.* IAHS.

Holmes, M. G. R., Young, A. R., Gustard, A. and Drew, R. (2002). A New Approach to estimating Mean Flow in the UK. *Hydrology and Earth System Sciences,* **64**, No. 4, 709–720.

Mansell, M. G. and Johnston, J. C. (1994). Recession Characteristics of Small Streams in the Eastern Districts of Zimbabwe. *Proceedings Institution of Civil Engineers (Water and Maritime Division),* **106**, 71–79.

Natural Environment Research Council (1975). *Flood Studies Report.* NERC.

Robson, A. and Reed, D. (1999). *Flood Estimation Handbook Vol. 3 — Statistical Procedures for Flood Estimation.* Institute of Hydrology, Wallingford.

Rodier, J. A. and Roche, N. J. (1984). *World Catalogue of Maximum Observed Floods.* IAHS.

Chapter 9

Hydrological management

This chapter is concerned with the ways in which human intervention has affected the natural flows and hydrological processes. This intervention includes the abstraction and recharge of groundwater resources, the use of reservoirs for storage and flood alleviation, and the provision of drainage systems in urban areas.

9.1 Groundwater management

9.1.1 Pumping from wells

Where water is pumped out of an aquifer, there is a lowering (drawdown) of the water table in the vicinity of the well. The flow rate from a well under equilibrium conditions is related to the drawdown as well as to the permeability of the aquifer (Figure 9.1(a)). Applying Darcy's law (see Chapter 6) for a cylindrical surface at a distance r from a well in a deep semi-unconfined aquifer gives:

$$Q = \frac{2\pi r h K \mathrm{d}h}{\mathrm{d}r} \tag{9.1}$$

Integrating from r_1 to r_2 where the depths of the aquifer are h_1 and h_2 respectively:

$$Q = \frac{\pi K\left(h_1^2 - h_2^2\right)}{\ln\left(\dfrac{r_1}{r_2}\right)} \tag{9.2}$$

However, at the start of pumping, the flow rates will be higher than this because part of the flow will come from the storage in the immediate vicinity of the well. Furthermore, adjustments need to be made if there are boundary effects, such as other wells or streams, which may act as sources.

Where the aquifer is confined by an impermeable layer above, the corresponding equation is:

$$Q = \frac{2\pi K b (s_1 - s_2)}{\ln\left(\dfrac{r_1}{r_2}\right)} \tag{9.3}$$

where b is the thickness of the aquifer, and s_1 and s_2 are the drawdowns (i.e. the depths of the water surface at r_1 and r_2, respectively, below the undisturbed level — Figure 9.1(b)).

9.1.2 Aquifer recharge

It is possible to increase the yield of an aquifer by artificially inducing water to enter it. This can be achieved by, for example, constructing storage ponds or reservoirs on permeable soils, or by storing water and then releasing it into channels which allows percolation to the aquifer, or by reverse pumping in wells. Aquifer storage may be considered as an alternative to conventional reservoirs, with the advantage of reduced evaporation and protection against pollution.

9.1.3 Land drainage

The installation of land drains can have a significant effect on the hydrological characteristics of a catchment. The techniques used in drainage vary and include ploughing into ridges and furrows and 'mole drainage', where a hole is formed by dragging a metal bullet through the soil, as well as the use of porous pipes and open ditches. Such measures are designed to remove water from the upper saturated zone through gravity. Land drainage is sometimes used where there is a high groundwater level in a permeable soil, especially where it is in a basin. Alternatively, it may be used where the saturation is a result of the low permeability of the soil and where lower soil layers may be relatively dry.

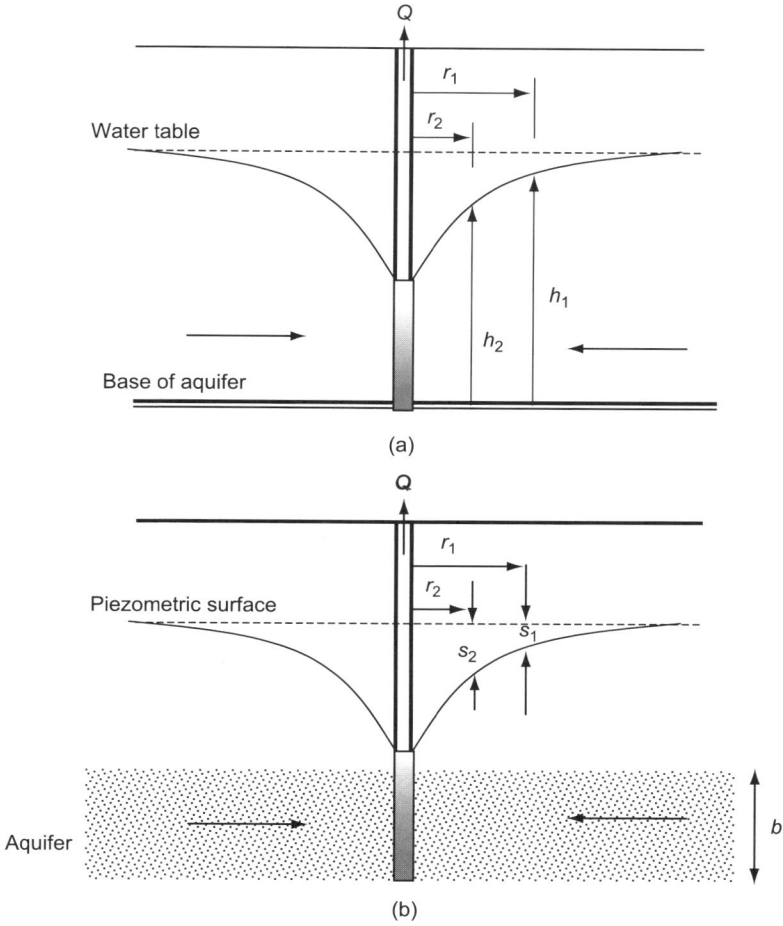

Figure 9.1   Well hydraulics in aquifers: (a) semi-unconfined aquifer; and (b) confined aquifer

As noted in Chapter 1, the effect of land drainage depends on the storm characteristics and antecedent conditions. Where the rainfall is not sufficient to produce saturated conditions, the effect of drainage is to reduce peak flows: for higher rainfall levels it will tend to increase peak flows (Robinson, 1990).

9.2 Water abstraction

9.2.1 Reservoirs

Reservoirs are constructed for a variety of purposes, including flood alleviation, water abstraction, hydro-electric power generation and irrigation. In some cases, a dam will serve more than one purpose, although this will often lead to conflicts or compromises in the design and operation of the dam. However, the basic functions of a reservoir are storage and flow attenuation. The amount of attenuation depends on the capacity of the spillway and the topography of the reservoir basin. The design criteria for a dam may be one of the following:

- the peak outflow for an extreme storm event
- the required storage volume for hydro-electricity generation
- the required draw-off rate or yield.

The peak outflow is determined by routing a given inflow hydrograph through the reservoir as described in Chapter 6. Alternatively, for small detention ponds, the required storage volume (above the outlet level) can be estimated from the area between the inflow and outflow hydrographs. For hydropower reservoirs, the energy generated is estimated from the flow–duration curve based on the outflow hydrograph as described in Chapter 8.

For water-supply reservoirs, the storage volume of a reservoir below the spillway level is important, and this can be calculated from the areas of contours at various levels and a graph of volume versus level can be drawn. The volume below the level of the lowest draw-off is known as the dead storage (Figure 9.2) (in practice there may be draw-offs at different levels). The volume between the lowest draw-off and the spillway level (normal maximum level) is the useful (or live) storage of the reservoir, and the volume above the spillway level is the flood storage. In addition, up to 2% or 3% extra water may be stored in the banks of the reservoir, which may be released as the water level falls, depending on the porosity of the soil and rock.

For storage reservoirs, the important criteria is the maximum safe rate of abstraction or *yield*. This will depend on the useful storage volume in relation to the variation in the inflow. The safe yield can be estimated using the *mass curve* technique. A mass curve is a plot of the cumulative volume of inflow (or outflow) against time, and a typical mass curve for two years of seasonal inflow is shown in Figure 9.3. The steep sections of the curve correspond to high inflows and the flatter sections to low flows.

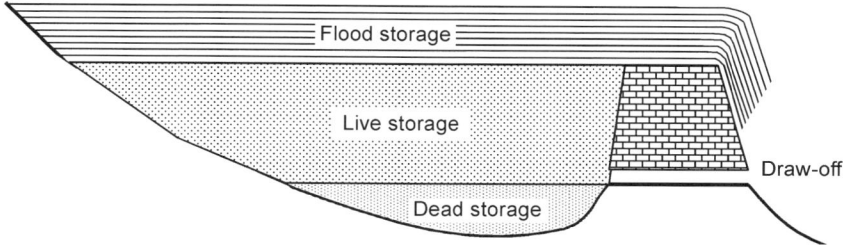

Figure 9.2   Storage zones in a reservoir

The gradient of a straight line joining the ends of the curve represents the average inflow and is the maximum uniform outflow which can be withdrawn over the period of the record. Assume water is abstracted at a uniform rate represented by, say, the slope of the line AB, which is drawn at a tangent to the mass curve. If the reservoir is empty at A, then the vertical distance between the line AB and the inflow mass curve at any time is the volume in storage (*S*) at that time. If a line parallel to AB is drawn at a tangent to the mass curve at C, the vertical distance between the two lines represents the total storage volume (S_{max}) required to collect all the inflow and to avoid any spill (for a yield represented by the gradient of AB) for the first year. It can be seen that the reservoir storage then reduces until point D and then starts to fill again. Therefore, the net storage required for the first year is the vertical distance between the lines passing through C and D. The difference between S_{max} and this volume represents the volume spilled in that year. It can also be seen that, as the gradient of the lines increases (i.e. the yield increases), the vertical distance between the two tangent lines increases and, hence, a greater storage volume is required. Where the combination of gradient (draw-off rate) and storage (vertical distance between the tangents) is such that the tangents do not enclose the inflow mass curve, the reservoir will become empty where the lower line cuts the mass curve.

The mass curve shows that the greater the variability in the inflow, the greater is the required storage. For each subsequent year, a further set of tangents can be drawn and, thus, a storage volume for a given yield can be estimated for each year of the record. Alternatively, for a given storage volume, a maximum yield can be estimated, for any given year, from the gradient of the upper and lower tangents at a vertical distance apart corresponding to the storage volume. Therefore, for a specific reservoir, with a given inflow hydrograph, a series of annual safe yields can be estimated. These can be plotted as a distribution and the reservoir

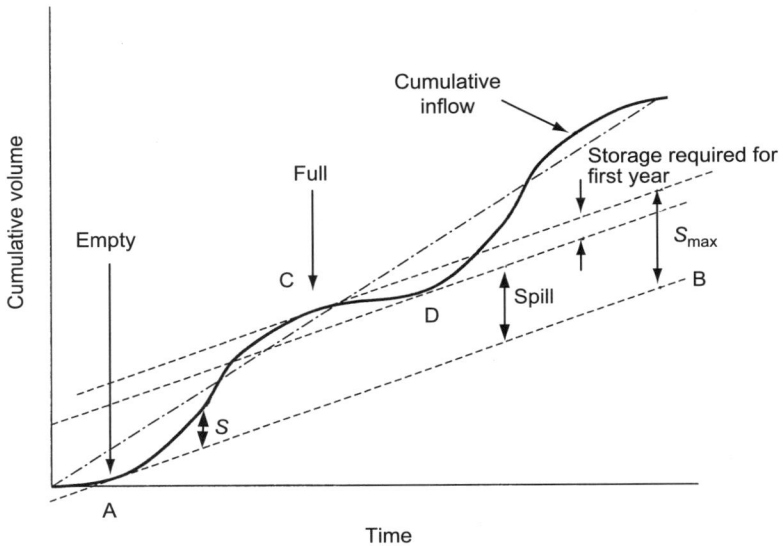

Figure 9.3   Mass curve technique

capacity for a given probability of given safe yield can be estimated. Another approach is to assume a given draw-off for a specified reservoir and to count the number of years when the storage volume was inadequate and, thereby, estimate the probability of failure for a given yield. In either case, it should be recognised that the initial starting condition of the reservoir may have a significant effect on the result, particularly where the flow record is short.

As described in Section 2.6.2, the design and operation of a water supply system is a balance between its storage capacity, its firm yield and the probability of failure. A change in the pattern of inflow, for example, can be met either by increasing the storage, reducing the yield, or by accepting a greater probability of failure to meet that yield.

Fewer large dams are being constructed now than in the past either for hydro-electric power or for water supply. This is partly because of a decreasing number of suitable sites and partly because of the growing public opposition to large dams for the reasons outlined in Chapter 1. In order to overcome this opposition and to establish good practice in the evaluation of such schemes, the World Commission on Dams issued guidelines for good practice (World Commission on Dams, 2000), which are reproduced in Appendix 9.1.

9.2.2   Run of river water abstraction

The impact of widespread, small-scale abstraction of water from rivers is more difficult to assess than that of a large point abstraction, such as a reservoir. There are over 48 000 active abstraction licences in England and Wales (Barker and Kirmond, 1998), and although the local impact of each of the abstractions may be limited, their cumulative impact within a catchment may be significant.

The policy in the United Kingdom is that the minimum sustainable flow in the river and the proportion of the flow above this minimum flow, which is available for abstraction, depend on an environmental weighting for each river, which is based on its physical characteristics, ecology and fisheries. For a single, large abstraction, a fixed proportion of the instantaneous flow (less the minimum flow) may be licensed, but where there are large numbers of relatively small abstractions at fixed rates, a system of *tranches* may be used. A number of threshold flows are determined and a tranche of abstractions may be a proportion of the interval between successive threshold flows. The aim of the abstraction policy is to preserve the variability of the flow regime, as well as to ensure that flows do not fall below a minimum acceptable level, while allowing the abstractor the maximum flexibility to make optimum use of their licences whenever water is available.

9.3   Drainage systems

9.3.1   General theory of pipe flow

Flow through pipes or closed conduits that are flowing full, results from a difference in head that is manifested as a difference in pressure, whereas, in open channels (or where the pipe is not full), the difference in head results from the difference in elevation. Another obvious difference between flow in closed conduits and open channels is that, in the former, the cross-section of flow is defined by the shape of the pipe, whereas in open channels, the cross-section is not uniquely defined even for a given flow rate.

The rate of volume flow in a pipe, usually expressed in m^3/s or l/s, depends on the size of the pipe, the roughness of its internal walls, the viscosity of the water (which is normally taken as about $1 \times 10^{-6}\,Ns/m^2$) and the *hydraulic gradient*. The hydraulic gradient represents the head loss per length of pipe due to the friction between the water and the pipe wall.

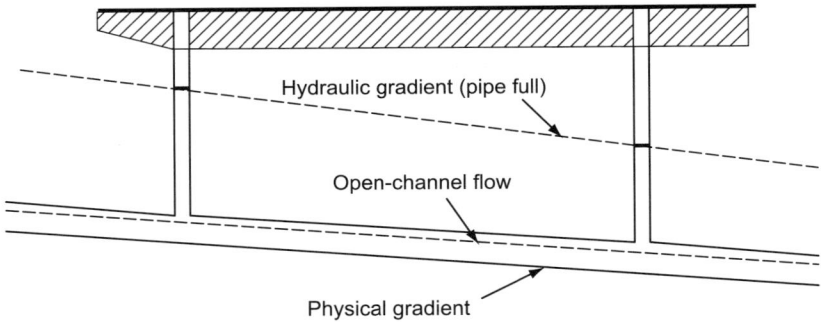

Figure 9.4   Hydraulic and physical gradients in pipe flow

For a pipe flowing partly full, it can be taken as the physical gradient of the pipe, while, for a pipe flowing full, it is approximately the slope of a line joining the free-water surfaces in real or imaginary manholes or stand-pipes (Figure 9.4).

There is no complete theoretical analysis of flow in pipes, especially turbulent flow, which is the type of flow found under normal conditions. Many semi-empirical formulae have been proposed, but in most cases they are only valid for a limited range of flow conditions. The most widely used formula is that developed by Colebrook and White:

$$Q = -2A\sqrt{2gds}\ \log\left(\frac{k}{3\cdot7d} + \frac{2\cdot51\nu}{d\sqrt{2gds}}\right) \tag{9.4}$$

where A is the area of pipe, d is the pipe diameter, s is the hydraulic gradient (h_L/L), k is the pipe roughness, and ν is the viscosity.

In the Colebrook-White formula, the pipe roughness coefficient k corresponds to the mean diameter of sand grains, which, if attached to the pipe wall, would give the same hydraulic roughness. Examples of recommended values for the roughness parameter are given in Table 9.1.

Where there is sediment on the pipe invert, allowance should be made for the roughness of the sediment and a common empirical formula is:

$$k_s = 1\cdot5D_{65} \tag{9.5}$$

where D_{65} is the 65 percentile particle size. The overall roughness can be calculated according to the proportion of the perimeter covered by

Table 9.1   Recommended pipe roughness coefficient (k) (Barr, 1994) — assuming new and clean condition

Pipe material	k value: mm
Copper and plastic	0·003
Steel (uncoated)	0·03
Clayware	0·3
Concrete	0·6

sediment. In old sewers, allowance also needs to be made for bricks and large debris on the pipe invert.

Where pipes are flowing partly full, as in most sewers, they can be considered as open channels and the routing of unsteady flow through such pipes can be carried out in the same manner as described for channels (see Section 6.3.5).

9.3.2   Flow measurement in pipes

Where a pipe is flowing full, the average velocity can be measured using ultrasonic transducers mounted on the outside of the pipe in a similar way to measuring the velocity in an open channel (Figure 9.5). Each transducer acts alternately as a transmitter and receiver, sending and receiving pulses of sound. The sound will naturally travel faster in the downstream direction than in the upstream direction, and the difference in the upstream and downstream transit times can be used to estimate the velocity of the water. Allowance needs to be made for the nature and thickness of the pipe wall in measuring the transit time.

The transducers can either be arranged on the same or opposite sides of the pipe, but, for horizontal pipes, they should both be on the same horizontal plane. If the transducers are mounted in a vertical plane there may be interference to the signal because of sediment at the bottom of the pipe or air bubbles at the top. There should be a length of at least 10 pipe diameters of straight undisturbed flow upstream of the transducers and 5 diameters downstream.

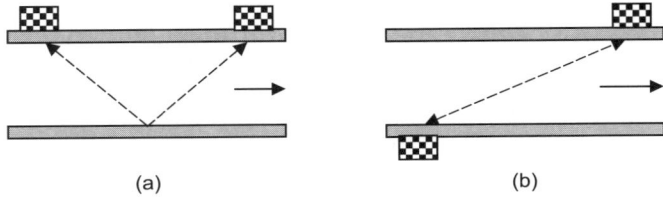

Figure 9.5 Ultrasonic pipeflow measurement: (a)reflection (V) configuration; and (b) through (Z) configuration

Flow in partly full pipes, such as sewers, can be measured using the ultrasonic devices based on the Doppler frequency shift, which are used in open channels (see Section 6.4.2).

9.3.3 Flow regulating devices

Various devices are used to attenuate or modify the flow in pipes, especially in urban drainage systems.

Orifice plates

An orifice plate is a plate with a hole less than the pipe diameter, through which the flow passes. It is usually positioned in a chamber at the entrance to a pipe so that it can be modified if necessary. When the flow exceeds the capacity of the orifice, the chamber acts as a storage reservoir and the water level rises. Using the energy equation, it can be shown that the flow though an orifice can be given by:

$$Q = C_d A_0 \sqrt{(2gh)} \tag{9.6}$$

where C_d is a coefficient of discharge, A_0 is the area of the orifice, and h is the height of the free surface above the orifice.

Where the orifice is not drowned, the value of C_d is between 0·57 and 0·6; where the orifice is drowned the value of C_d can be estimated from:

$$C_d = \frac{1}{1 \cdot 7 - \dfrac{A_0}{A}} \tag{9.7}$$

where A is the area of the pipe.

Throttle pipes

A throttle pipe is a length of pipe with a reduced cross-section, which provides a flow control. Where the pipe is short or has a large diameter or gradient, the pipe acts as an orifice, as described above, and the flow rate is controlled by the nature of the inlet. For longer or smaller pipes, the capacity is limited by the friction loss along the pipe and can be estimated from one of the standard pipe formulae, such as the Colebrook-White Formula (Equation (9.4)). A common use of throttle pipes is in conjunction with overflows (see below), where the reduced capacity of the throttle causes surcharging and spill over the weir. Normally, throttle pipes are not less than 200 mm in diameter.

Penstocks

A penstock is an adjustable gate that causes a reduction in the flow area in the same way as an orifice plate. However, it has the advantage that the opening can be adjusted by raising or lowering the gate, either manually or automatically, possibly as part of a real-time control system. The equation for flow through a penstock is similar to that for an orifice, except that the coefficient of discharge is estimated differently.

Vortex regulators

A vortex regulator constricts flow by creating a vortex with rotating flow and an air core which increases in size as the flow increases. The vortex is created by introducing the flow tangentially into the regulator, and at high flows the air core occupies most of the outlet, acting as a reduced orifice. The shape of the head–discharge curve depends on the geometry of the regulator and the downstream conditions. A typical example is shown in Figure 9.6. The distinctive kink in the curve is due to the formation of a stable vortex and only occurs during increasing flow. The main advantage of a vortex regulator is that smaller units can be used than with orifice plates without the danger of blockage due to debris.

Tanks

Storage tanks are provided to reduce peak-flow and to allow the settlement of sediment. A degree of storage exists in any urban drainage system, by virtue of the volume of the pipes, and this can be exploited by increasing the pipe diameter over a given length. Alternatively, a special tank can be constructed on the line of the sewer. Both of these solutions are termed 'on-line storage', distinguishing them from 'off-line storage',

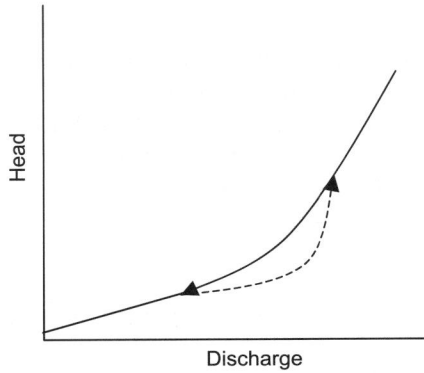

Figure 9.6   Head–discharge relationship for a vortex regulator

where the flow does not normally pass through the storage (Figure 9.7). A recent development in off-line storage is the use of prefabricated plastic cellular units, which are buried underground and provide a measure of flood relief.

Normally, a storage system will include some kind of throttling device and often an overflow device. The characteristics of such a storage system depend on the storage volume and the size of the throttle. Maximum attenuation results from a large storage volume together with a small throttle area.

The estimation of the outflow hydrograph for given storage systems with a given inflow hydrograph can be carried out using the reservoir routing procedure described earlier in Section 6.3.5.

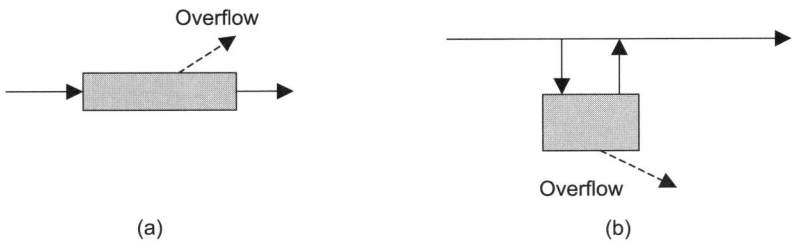

Figure 9.7   Storage tanks: (a) on-line storage; and (b) off-line storage

Stormwater overflows

The function of a stormwater overflow is to divert flows in excess of a certain limit into an appropriate watercourse. This permits the capacity of the sewer and any pumping or treatment facilities downstream of the overflow to be limited. The lower the value of the setting, the greater the saving in capacity, but the greater the frequency of spilling.

In general, an effective overflow structure should satisfy the following criteria:

- it should not come into operation until the design flow setting has been reached
- there should be no increase in flow to treatment while the overflow is operating
- the maximum amount of polluting material (both in sediment and floating debris) should be passed to treatment
- the risk of blocking and silting should be minimised.

There are several different designs available for overflow structures. Some rely on the principle of stilling to allow heavy solids to settle and floating material to rise to the surface, while others use the vortex principle. Figure 9.8 shows an example of a *stilling pond* overflow structure, which consists of a rectangular chamber with a transverse weir at a distance above the outlet. It can be seen that the diameter of the outlet is less than that of the inlet and, therefore, when the flow reaches a certain level, the outlet will be surcharged and the water level in the chamber will begin to rise. When it reaches the level of the weir, spilling will commence and will continue until the flow recedes. The large chamber allows for heavier sediment to settle and to be carried through the continuation pipe,

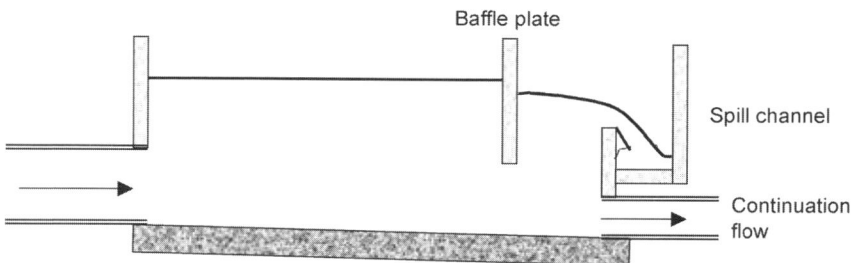

Figure 9.8   Stilling pond overflow

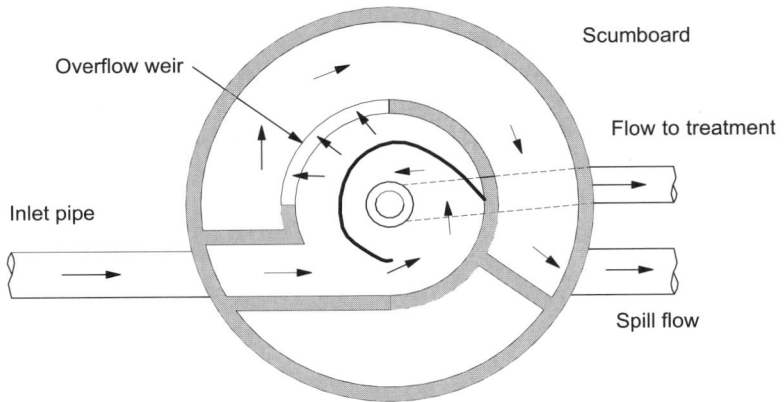

Figure 9.9   Vortex overflow

while the lighter material is trapped behind a baffle and passes through the continuation pipe when the flow recedes. The *side-spilling* overflow operates on a similar principle, except that the flow spills over a side weir.

An alternative design uses a vortex principle, as shown in Figure 9.9. Flow enters a circular chamber and forms a vortex above the central outlet, thus encouraging sediment to pass through to the continuation flow. During a spill event, excess flow passes over a weir into the outer part of the chamber. A scumboard also concentrates the floating debris in the centre. The vortex principle has been developed into several patented devices which use complex spiral flow patterns. The dimensions of the overflow chamber are based on the diameter of the inlet pipe (*D*) which may be determined from:

$$D = K \times Q^{0.4} \tag{9.8}$$

where Q is the design discharge (m³/s) which should be based on a storm with a return period of one year, and K is a constant given in Table 9.2 (Balmforth *et al.*, 1995).

Traditionally, overflows were designed to operate at flows exceeding six times the *Dry Weather Flow*. Further research has led to the development of a more refined design formula (known as *Formula A*):

$$\text{Flow setting} = DWF + 1360 \times Population + 2 \times Industrial\ effluent \quad \text{(l/day)} \tag{9.9}$$

Table 9.2   Flow constant K

	Total efficiency: %				
Flow ratio	20	40	60	80	90
5	1·27	1·47	1·60	1·72	—
10	1·13	1·37	1·52	1·66	1·83
20	0·825	1·19	1·34	1·50	1·65
30	0·815	1·02	1·18	1·33	1·43

where *DWF* is the Dry Weather Flow (usually calculated on the basis of water consumption). The above formula is derived on the basis of a dilution of 8:1. In other words, the mean flow in the receiving watercourse should be at least eight times the *DWF*. In some cases, where there is not adequate dilution, it may be necessary to increase the setting or to provide additional storage following Table 9.3 (Examples 9.1 and 9.2).

Figure 9.10 shows the typical performance of a stormwater overflow during a storm event. The river flow generally rises rapidly but recedes

Table 9.3   Recommended overflow settings and storage

Streamflow/ DWF	Percentage of pollutant to stream	Overflow setting	Tank volume: l/head
8	100	Formula A	—
6	75	Formula A+455P	0 (sewer storage) or 40
4	50	Formula A	40
2	25	Formula A	80
1	12	Formula A	120

Example 9.1

Calculate the setting for an overflow on a combined sewer which serves a town with a population of 15 000. The average Dry Weather Flow is 250 l/day and there is an industrial discharge of 750 m³/day. If the overflow spills into a water course where the mean daily flow is 0.17 m³/s, estimate the volume of storage required.

From Formula A:

Setting = 15 000 × 250 + 15 000 × 1360 + 2 × 750 × 10^3 l/day

= 25 650 × 10^3 l/day

= 297 l/s

$$DWF = \frac{15\ 000 \times 250}{24 \times 3600} = 43 \cdot 4 \text{ l/s}$$

$$\text{Dilution} = \frac{170}{43 \cdot 4} = 3 \cdot 9$$

From Table 9.3 use tank with volume = 40 l/person:

Volume = 0·04 × 15 000 = 600 m³

Example 9.2

The inlet flow hydrograph at the overflow in Example 9.1 was observed as below. Estimate the volume of effluent spilled and the duration of the spill (assuming that the continuation flow does not increase at flows above the overflow setting).

Time: min	Flow: l/s
0	40
10	95
20	210
30	350
40	465
50	325
60	210
70	110

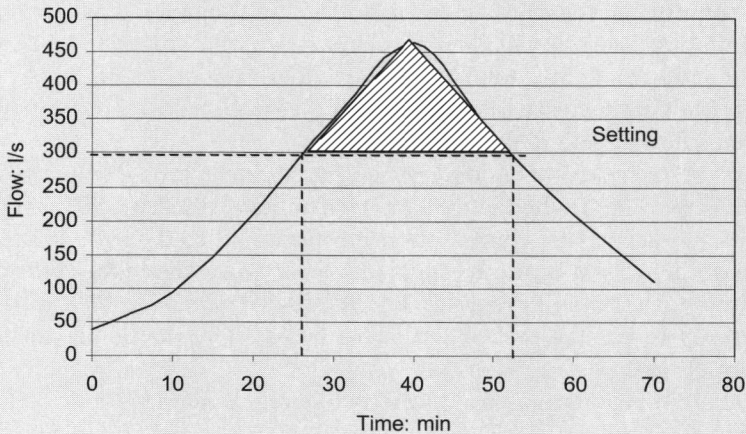

The volume of spill is the area of the hydrograph above the setting flow, which can, in this case, be approximated to a triangle.

Volume = $\frac{1}{2} \times (52 - 27) \times 60 \times (465 - 297) = 126\ 000\ l = 126\ m^3$

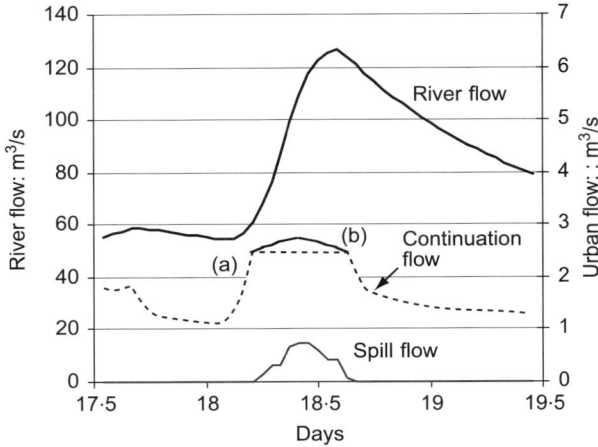

Figure 9.10   Typical storm overflow performance

slowly. The continuation flow (plotted on the right-hand axis) increases to the overflow setting, after which, in theory, it remains constant (curve (a)). However, in practice, the water level in the chamber will continue to rise slightly as the flow increases after spilling and, therefore, the continuation flow curve will also increase slightly (curve (b)). It can be seen that the ratio of the total urban flow (continuation + spill flow) to that of the receiving water can be relatively high at the start of the storm before the river flow has started to increase.

For the purposes of modelling, a simple linear model is often used, which is represented by Figure 9.11 (Hydraulics Research, 1983). At flows below the setting, the continuation flow is equal to the inflow, and the slope of the inflow–outflow line is unity. At flows above the setting, the continuation flow still increases, but at a lower rate given by the slope a, which represents the proportion of the excess flow that is passed to the continuation flow.

The main problems associated with overflows include:

- operation at incorrect flows
- siltation and or blockage
- visual pollution (e.g. toilet paper, etc.).

Unnecessary pollution may be caused if the overflow operates at little more than dry weather flow. Even if the overflow is operating correctly,

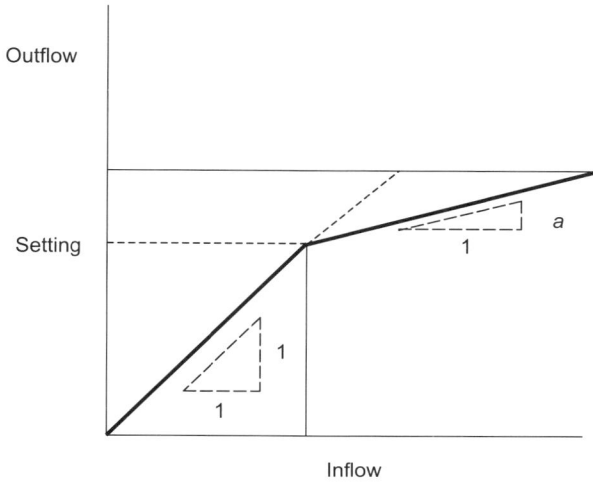

Figure 9.11 Idealising the performance of a stormwater overflow (Hydraulics Research, 1983)

there will often be biological or chemical pollution where the spill occurs before the flow in the receiving water has started to increase. However, if the setting is too high, there may be inadequate relief to the downstream sewerage system. The main public objections associated with overflows are usually the visual pollution of toilet paper and other sanitary material, which can be discharged from an overflow if there is inefficient solids separation. The visual pollution can be mitigated to a certain extent by the use of screens, typically with 6 mm bar spacing or apertures. However, screens can make the situation worse unless they are properly designed and maintained. Special devices are available that use a siphon outlet to periodically reverse the flow across the screen and carry the screened material to the treatment flow.

Inverted siphons

Inverted siphons are U-shaped lengths of pipe, which are sometimes used where a gravity sewer has to cross under a river or a road in a cutting. Unlike the rest of the drainage system, an inverted siphon is full regardless of the flow rate. At low flows, the velocity will be very low, with ideal conditions for settlement of sediment, which is the major problem with inverted siphons. The problem can be mitigated in a number of ways. Many siphons consist of more than one pipe with

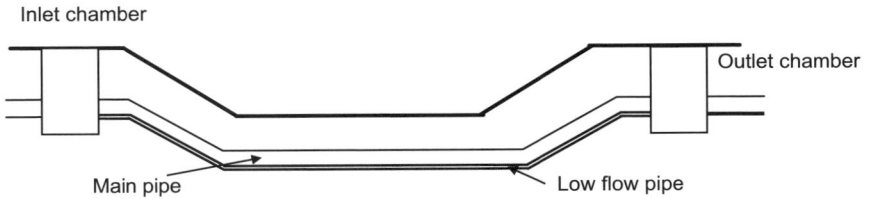

Figure 9.12   Multiple pipe inverted siphon

different diameters (Figure 9.12). The smallest pipe carries the low flows while higher flows are carried by one or more larger pipes and also flush sediment from the smaller pipe.

An alternative approach to reduce the problem of siltation is an air cushion siphon (Figure 9.13). This relies on a pocket of air, which is trapped in the horizontal length of pipe. This reduces the cross-sectional area of the flow and, hence, increases the mean velocity. Even at high flows, the flow does not occupy the whole pipe cross-section.

Other methods that can be used to reduce the problem of siltation include small penstocks upstream to dam the flow, which can then be released to create a flushing wave, or wash out chambers, which can be used to pump out the silt. However, as a rule, inverted siphons should be avoided if possible.

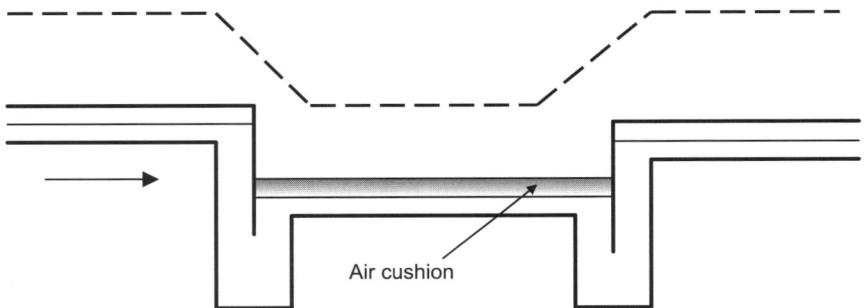

Figure 9.13   Air cushion inverted siphon

9.3.4 Pumped systems

Pumping is sometimes required in urban drainage systems where there is insufficient fall to provide gravity drainage and it may also be used to enhance natural drainage in low-lying areas. A typical pumping system consists of an inlet well at the pumping station followed by a rising main which discharges into a gravity pipe (Figure 9.14).

Hydraulically, the function of a pump is to add energy to a system. The performance of a pumped system depends on the characteristics of the pump and the pumping main. Both characteristics can be expressed in terms of the relationship between head and flow rate (or discharge) (Figure 9.15). The head can be defined as energy per unit weight, but can also be measured in terms of a height of water above a datum. There are two main types of head used in connection with pumping. The *static head* is simply the vertical height through which the water is raised, while the *friction head* is the head lost due to the friction resulting from the flow through the pipe. The friction head is therefore dependent on the velocity (or flow rate), as well as the pipe characteristics, and can be calculated using one of the standard pipe equations, such as the Colebrook-White equation (Equation (9.4)). The pipe characteristic curve (Figure 9.15(a)) therefore combines a constant with a varying function of flow.

The pump characteristic curve (Figure 9.15((b)) shows how the output varies with the head faced by the pump. At one extreme, if the head faced by the pump is very high (i.e. a long, very small pipe or a high static head), the output will be small. At the other extreme, if the pump was discharging straight into the atmosphere, the head would be negligible and the flow would be a maximum. The actual flow could clearly be anywhere between these points, depending on the pipe characteristics.

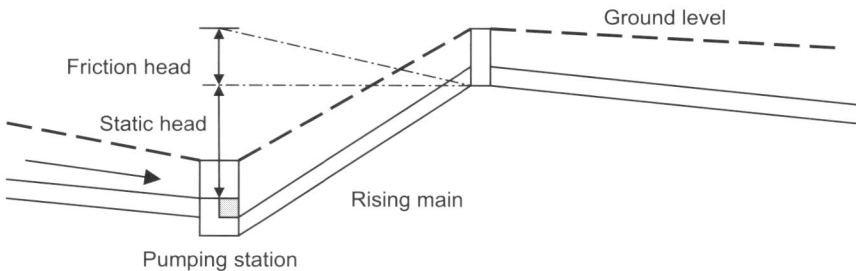

Figure 9.14   General arrangement of a pumping scheme

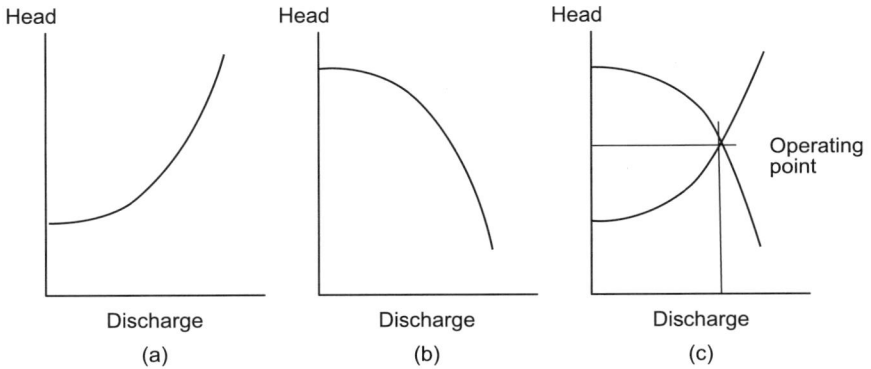

Figure 9.15   Pump system characteristic curves: (a) pipe; (b) pump; and (c) system

Therefore, the actual operating point for a pump will be at the intersection of the pump and pipe characteristic curves (Figure 9.15((c)).

In practice, pumps are not normally run continuously. The inlet well is usually allowed to fill to a fixed level, at which an automatic switch starts the pump, which then runs until the water level in the well falls to another fixed level. Furthermore, two or three pumps of varying sizes are generally used in parallel, so that wide variations in flow can be accepted and to allow for maintenance and repairs. The flow in the rising main is therefore intermittent and is determined by the capacity of the pumps being used.

9.3.5   Culverts

The analysis of culverts carrying natural watercourses under roads and other obstructions, involves a complex interaction of pipe flow and open channel flow, with various flow patterns according to the nature of the channel and the design water levels (Ramsbottam *et al.*, 1997). For any culvert, a number of different types of flow profile can occur depending on the flow rate and the slope and roughness of the channel (Figure 9.16(a)). The key to the type of flow profile is the location of the control, which, as described in Section 6.3.2, is the location with the lowest capacity and where the water level is uniquely defined for a given flow rate. In general, the control is likely to be either at the inlet, which will then function like an orifice, or downstream of the outlet, for example, where uniform flow

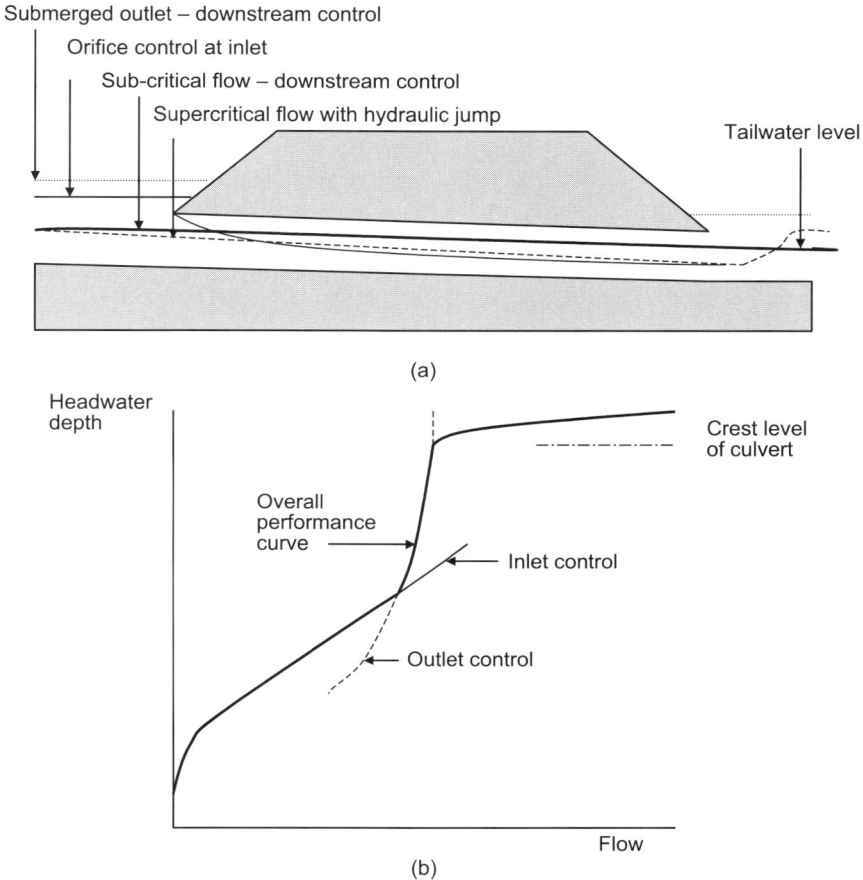

Submerged outlet – downstream control

Orifice control at inlet

Sub-critical flow – downstream control

Supercritical flow with hydraulic jump

Tailwater level

(a)

Headwater depth

Crest level of culvert

Overall performance curve

Inlet control

Outlet control

Flow

(b)

Figure 9.16 (a) Examples of flow through a culvert; and (b) performance curve for a typical culvert (Ramsbottom et al., 1997)

conditions exist or where the water level is controlled by a downstream weir. In fact, control may switch from the inlet to the outlet as the flow changes (Figure 9.16(b)). The most common type of profile is the sub-critical flow with downstream control (Figure 9.16(a)) and, in general, culverts should be designed for this profile. The preliminary design procedure involves assessing the tailwater level for the design flow using the Manning equation (Equation (6.27)) and selecting a culvert size based on the size of the natural channel. The invert level of the culvert should be at or just below the bed level of the watercourse to encourage sediment to cover the bed of the culvert. The upstream water level is then calculated by adding the head loss in the culvert together with entrance and exit head losses to the downstream water level. The size of the culvert is adjusted if this water level is not acceptable in relation to the level of the banks etc. The velocity of flow (V) inside the culvert should be in the range $0.75\,\mathrm{m/s} > V > 2.0\,\mathrm{m/s}$ to avoid both silting and scouring problems. More detailed design may be necessary, using the non-uniform flow equation (Section 6.3.4) with different profiles.

Care should taken in the design of the inlet and outlet structures to provide a smooth transition between the flow regimes inside and outside the pipe. Since the velocities inside the culvert are usually higher than in the natural channel, there is a risk of scouring at the outlet unless adequate protection is provided (Figure 9.17). Many flooding problems have resulted from poorly designed or inadequate culverts and, in particular, from screens which are often used to prevent children or debris from entering the pipes. Screens should only be used where there are large items of debris, where there is a bend or obstruction in the culvert, or where the culvert is long or difficult to clear. If screens are used, they should be sloping and should have good access at night or in poor weather (Figure 9.18). Provision may also need to be made for access for small mammals through the culvert.

9.4   Flood alleviation measures

Flooding in urban areas can be attributed, in general, to two main sources:

- high flows in a river passing through the urban area as a result of excessive rainfall on the upstream catchment
- high runoff due to short intensive rainfall within the urban area.

9.4.1   *Flooding due to high river flows*

Flooding is a natural (and desirable) process in most rivers, as it uses the natural storage of the floodplain to attenuate peak flows as well as replenishing nutrients in the soil. It is therefore neither desirable nor feasible to eliminate the risk of flooding entirely. Flooding from high river flows can be mitigated either by the provision of enhanced storage or by improvements in the capacity or conveyance of the channels (ICE, 2001), or by specific protection measures for properties.

Example 9.3

Estimate the size of a culvert for the channel with a cross section given below. The length of the culvert is 40 m and the invert levels at the inlet and outlet are 35·16 m AD and 35·0 m AD respectively. The design flow is 25·0 m³/s and the Manning friction coefficient can be taken as 0·014 for the culvert and 0·05 for the channel.

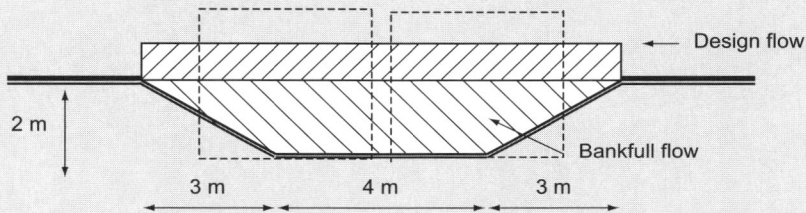

Top width = 10·00 m

Area = $4 \times 2 + 3 \times 2 = 14·00$ m^2

Wetted perimeter = $4 + 2\sqrt{(2^2 + 3^2)} = 11·20$ m

Gradient = $\dfrac{35·16 - 35·0}{40} = 0·00400$

Bankfull flow = $\dfrac{A \times m^{2/3} i^{0·5}}{n} = 14\left(\dfrac{14}{11·2}\right)^{2/3} \dfrac{\sqrt{0·004}}{0·05} = 20·55$ m^3/s

Design flow is greater than bankfull flow, therefore design water level will be, say, x above top of bank. Assuming flow is contained with width of bank, the wetted perimeter is as for bankfull, i.e.:

$$Q = \dfrac{A}{n} \times \dfrac{A^{2/3}}{P^{2/3}} \sqrt{i} = \dfrac{A^{5/3}}{nP^{2/3}} \sqrt{i}$$

$$A^{5/3} = \dfrac{Qnp^{2/3}}{\sqrt{i}} = \dfrac{25 \times 0·05 \times 11·2^{2/3}}{\sqrt{0·004}} = 98·4$$

Flow area A = 15·75 m^2

Height above bank $x = \dfrac{15 \cdot 75 - 14 \cdot 0}{10} = 0 \cdot 17\,\text{m}$

Tailwater depth $d_1 = 2 \cdot 0 + 0 \cdot 17 = 2 \cdot 17$ m

Tailwater level $= 35 \cdot 0 + 2 \cdot 17 = 37 \cdot 17$ m AD

Culvert cross-section

Area $= 15 \cdot 75$ m^2

Height $=$ Tailwater depth $+$ Freeboard $= 2 \cdot 17 + 0 \cdot 5$ (say) $= 2 \cdot 67$ m

Width $= \dfrac{15 \cdot 75}{2 \cdot 17} = 7 \cdot 24$ m

Actual height say 2·70 m

Actual width say 7·20 m (2No. 3·6 m barrels)

Flow area $= 2 \cdot 17 \times 7 \cdot 2 = 15 \cdot 66$ m^2

Wetted perimeter $= 4 \times 2 \cdot 17 + 7 \cdot 2 = 15 \cdot 90$ m

Hydraulic radius $= \dfrac{15 \cdot 66}{15 \cdot 90} = 0 \cdot 98$

Upstream water level

Friction head loss $i = \left(\dfrac{Qn}{Am^{2/3}} \right)^2 = \left(\dfrac{25 \times 0 \cdot 014}{15 \cdot 66 \times 0 \cdot 98^{2/3}} \right)^2 = 5 \cdot 15 \times 10^{-4}$

Velocity $= \dfrac{35}{(2 \cdot 17 \times 7 \cdot 2)} = 1 \cdot 60$ m/s

Entry head loss $= \dfrac{1 \cdot 5 \times 1 \cdot 6^2}{2g} = 0 \cdot 20$ m

Water level difference $= 5 \cdot 15 \times 10^{-4} \times 40 + 0 \cdot 20 = 0 \cdot 22$ m

Upstream water level $= 37 \cdot 17 + 0 \cdot 22 = 37 \cdot 39$ m AD

Figure 9.17   Typical culvert with downstream apron

Figure 9.18   Typical screens for a double culvert

Enhanced storage

Water is stored in all parts of a catchment and river system, and this storage can be enhanced in several ways. Large flood-storage reservoirs are often used with provision for alternative amenity use when not flooded. In some cases, the joint use of water-supply reservoirs for flood alleviation may be feasible, although there are usually operational problems. Catchment storage can also be improved by using small-scale dispersed measures which might involve, for example, forestation or the use of vegetation to reduce the speed of runoff and encourage infiltration or the removal of flood banks to allow farm land to flood. The storage of water within river channels can be encouraged by restoring them to a more natural state, thus slowing down the speed of flow. Where this approach is proposed, it is important that adequate provision is made for upstream storage due to the backing up of the flow. In extreme cases it may be possible to clear a floodplain of development within an urban area. Although this is obviously an expensive option, there may be considerable environmental and amenity benefits.

Improved conveyance

The options for improving the channel conveyance have traditionally meant the construction of flood walls or embankments along the channel to keep the flow within the channel or diversion channels or tunnels in parallel with the channel. The major problem with this approach is that such measures do little to reduce flows and merely pass the problem further downstream. Embankments and walls also attract criticism because of the loss of amenity and visual intrusion. The use of temporary or demountable flood barriers overcomes this problem although these rely on being erected in advance of any flood. Automatic floating flood-wall systems are now being developed to avoid this difficulty.

Protection of properties

Although many flooding events cover extensive areas, often only a small number of properties are affected. It may be possible to protect these properties with small ring-dykes and, in some cases, it may be more economic to consider 'flood-proofing' individual properties. This would involve, for example, raising the floor level or locating all electrics above flood level and providing non-return valves on drains.

Flood management

Flood mitigation can also be achieved by flood management rather than flood control. Flood management includes the restriction of development on floodplains and improved flood-warning systems.

9.4.2   *Flooding from urban runoff*

Urban runoff has traditionally being carried by piped drainage systems. Historically, this developed from the culverting of natural watercourses which were used for foul and surface-water drainage. Other watercourses, which carried only surface water, were often lined with concrete. Since runoff occurs relatively quickly, such drainage systems are typically designed for short-duration storms with return periods of five years or less. Because of the high storage within sewage systems in relation to the volume of flow, such systems are often able to deal with flows with much higher return periods. However, most urban-drainage infrastructure was constructed over 100 years ago and may have reached its effective capacity. There is frequently a need to provide for increased development in towns without having to carry out expensive and disruptive alterations to the existing surface-water drainage systems. In addition, the effects of climate change are likely to increase the pressure on drainage systems (Table 9.4).

In general, the performance of an existing drainage system can be improved through either increased capacity, flow attenuation or flow reduction.

Increased capacity

This is the traditional way of dealing with the problem and may involve either the *replacement* of a length of pipe or its *duplication*. In either case, given that most sewers tend to follow street patterns, it will involve considerable cost and disruption, and, as in the case of channel improvements described above, can merely result in the transfer of the problem further downstream. It may, however, be justified if the pipe is in poor structural condition. Where there are minor defects in a drainage system, its capacity can be significantly improved by *relining*, which usually involves coating the inside of the pipe with a layer of epoxy mortar or similar material. Although this slightly reduces the diameter of the pipe, it reduces the friction coefficient substantially and the procedure has the advantage that it can be carried out with minimum disruption on the surface.

Table 9.4   Likely effects of climate change on urban drainage (Balmforth, 2002)

Likely climate change	Direct effect	Impact on urban drainage systems
More frequent, intense rainfall events, especially in winter	Increased volume and rate of runoff	Reduction in the level of protection against flooding more frequent overland flood flow
Reduced antecedent dry weather periods	Increased surface wetness and soil moisture, increasing volume and rate of runoff	Reduction in level of protection against flooding, more frequent overland flood flow
Prolonged periods of wet weather, especially in winter	Available storage remains full for long periods while inflow exceeds outflow. Soil moisture increases. River levels rise	Surface flooding will occur as drainage facilities are overtopped
Variation in total rainfall between years	Rise in water-table levels during years of exceptionally high rainfall	Increase in infiltration into piped systems, reduction in capacity of infiltration drainage
Prolonged periods of dry weather, especially in summer	Reduced baseflow in rivers. Increased accumulation of pollutants and sediments on surface areas	Increased pollution impact from surface water outfalls and CSO discharges. Increased sedimentation in sewerage systems
Sea level rise	Reduced capacity of outfalls	Backing up of sewer levels in storm events, reducing level of protection against flooding

Increased attenuation

Where it is not practicable to increase the capacity of a system, problems of surcharging can be reduced by providing storage within the drainage system, which attenuates the peak flow. Any drainage system has a certain amount of inbuilt storage, and this can be enhanced by simply increasing the size of the pipes or by providing tanks either on the line of the sewer or off-line, as described in Section 9.3.3.

Flow reduction

An alternative method of dealing with undercapacity is to reduce the peak flow by means of an overflow structure, as described in Section 9.3.3, or by enhancing natural infiltration.

9.4.3   Sustainable urban drainage

It has been recognised for some time that urban drainage systems modify the natural drainage paths to such an extent that they can be a cause of flooding, as well as being very expensive and disruptive to construct. Conventional drainage systems also requires large amounts of water, which may not be available in many developing countries and do not make use of the nutrients within the effluent. Recently, increased attention has been focused on *Sustainable Urban Drainage Systems* (SUDS), which work by replicating and enhancing natural drainage mechanisms (CIRIA, 2000) and provide *inter alia*:

- attenuation of peak flows by providing storage
- flow reduction by infiltration to the groundwater
- some natural biological purification.

The main advantages of SUDS are that they enhance water quality and provide habitats for wildlife in urban areas, as well as encouraging natural groundwater recharge. It is also argued that the capital costs of SUDS are lower than for conventional schemes and the maintenance costs are no higher.

Sustainable drainage measures are becoming more popular and are typically used where a commercial or residential development in an urban area would otherwise increase flows in an existing sewer. The criteria for such a development might be that the attenuation and/or flow reduction should be such as to cause no increase in peak flow after the development is complete. Although the sustainable drainage measures can be used in almost any situation, where the rainfall is high relative to evaporation, or where the soil type is predominately clay, the benefits of such measures may be limited to attenuation and treatment rather than infiltration. There is also the possible risk of contamination of the groundwater.

Sustainable drainage measures can be broadly classified into three main groups, as follows.

Local source control measures

On the principle of dealing with rainwater as close to source as possible, some degree of attenuation can be achieved by providing, for example, *storage butts* connected to down-pipes from buildings. Provision needs to be made for the overflow from these tanks and, in practice, it is difficult to introduce a uniform policy when dealing with a large number of owners and occupiers. A more widely used measure is *permeable surfaces*, which are designed to allow water to pass through to the underlying sub-base material (Figure 9.19). The surface material can be either porous itself or can consist of non-porous slabs, where water can enter through the joints. Water is stored and conveyed through the sub-base or is allowed to infiltrate into the ground. The main design issues in permeable pavements are the structural strength of the pavement, the storage volume and flow capacity of the sub-base, and the provision of either an overflow or surface storage for rainfall in excess of the design storm. In addition to providing attenuation of runoff, some flow is lost through evaporation from the surface and there is often some improvement in the quality of the runoff due to filtration and biological activity.

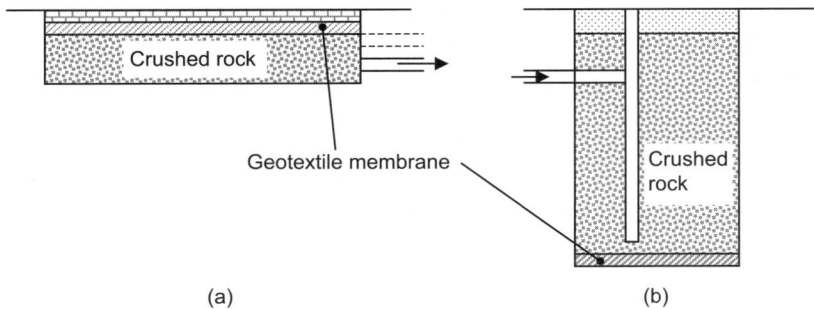

Figure 9.19   Source control measures: (a) permeable surface; and (b) soakaway

Runoff can also be diverted to *soakaways* or *infiltration ditches*, which have a large storage capacity and therefore provide attenuation and gradual infiltration into the surrounding soil. They are either chambers or ditches, which are generally filled with rubble or stones, and are normally designed for an appropriate 10-year return period storm. The main considerations in the design and construction of these structures are:

- the infiltration capacity of the surrounding soil
- possible groundwater pollution
- the presence of groundwater — the bottom of soakaways, etc., should be above groundwater at all times
- the proximity to buildings — they should not be constructed within 5 m of any building.

As with other source control devices, there are water-quality benefits in addition to flow reduction and attenuation, particularly where the flow path length to the water table is maximised.

Conveyance systems

In place of rigid impermeable pipes, sustainable drainage systems tend to use either wide, open ditches (swales) or filter drains. A *filter drain* comprises a perforated or porous pipe surrounded by a suitable filter or granular material. This material may be exposed at the surface or may be capped with turf or topsoil. The pipe is normally only provided over the last few metres before the outlet to maximise the treatment and attenuation benefits of the drains. Filter drains are traditionally used adjacent to roads to intercept surface water from the surrounding ground, and to prevent it from entering the pavement or sub-base and to remove water from the sub-grade. They are also used in a similar way in conjunction with permeable pavements. Sediment traps should be provided where pipes discharge into filter drains. Typically, filter drains are designed to store 10 mm of rainfall from the contributing area and a high-level pipe is provided as an overflow.

Swales are wide shallow ditches normally located adjacent to residential and small commercial developments and paved areas (Figure 9.20). The swale should be wide enough for the depth of flow not to exceed 0·1 m with a velocity less than 0·3 m/s. The longitudinal gradient is therefore usually less than 1 in 50 and the side slope no steeper than 1 in 40 to allow for mechanical grass-cutting. *Filter strips* are often provided in-between hard surfaces and swales to trap some of the sediment and

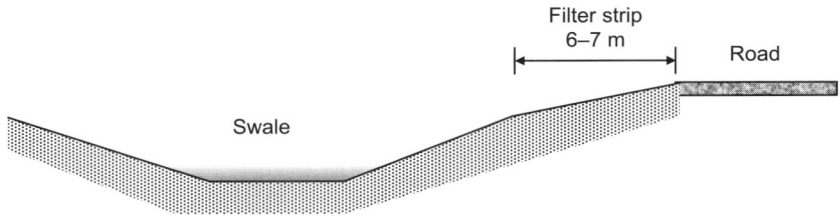

Figure 9.20   Typical SUDS conveyance system

pollutants. Such filter strips are typically 6 to 7 m wide, with a slope of 1 in 20. In common with other sustainable drainage measures, swales can provide attenuation, infiltration and treatment benefits.

Regional storage measures

Storage ponds can be used to provide infiltration, attenuation and treatment, as well as enhancing the amenity and providing wildlife habitats. *Infiltration ponds* have gently sloping sides and a flat grass-covered bottom, although they are not generally linear in plan. Since their primary purpose is infiltration, an essential design consideration is the infiltration capacity of the soil and level of the water table. They are normally designed such that they should be half empty within 24 h of a storm, with a return period of up to two years. An overflow to a suitable ditch or watercourse should be provided, and consideration should be given to the risk of freezing which might prevent infiltration in winter. *Detention ponds* are primarily temporary attenuation structures that are dry, except following storm events. However, the storage does also allow for the settlement of coarse silts and some degree of improvement in water quality. The maximum depth of detention ponds is about 3 m and the outlet is normally designed so that the pond empties in 24 h. A settling basin may need to be provided at the inlet if the inflow has a high-sediment load. Access provision needs to be made for the mechanical removal of the sediment and, because the water level can vary quite rapidly, safety is an important issue. *Retention ponds* are permanently wet ponds, 1 to 2 m deep with wetland vegetation around the edge. As well as providing biological treatment, the vegetation provides a wildlife habitat and reduces the flow velocity, increasing sedimentation. The volume of the pond is chosen to give a retention time of 14 to 21 days, with an average depth of

1 to 2 m. A shallow bench of 25–50% of the surface area of the pond should be provided around the edge for the vegetation. The length-to-width ratio of the pond should be at least 3:1 to minimise the risk of short-circuiting of flows. In some cases, the pond may be designed almost entirely as a wetland, in which case the volume will be much larger and the area of open water much less.

9.5 Catchment and river-basin planning

It is clear that humans have had a major impact on the natural hydrological processes, either as part of their normal activities, e.g. agriculture, as deliberate measures to enhance the supply of water, or to mitigate the effects of excessive flows. With the growth in the world's population, these impacts are reaching a point where there is a real possibility of major political conflict unless the activities are planned or managed in some way. It is logical to use the river basin or catchment as a basic unit of planning, although in many cases this will cross political or administrative boundaries. Nevertheless, it is obvious that what happens in one part of a catchment may well have effects in other parts of the catchment some distance away.

9.5.1 EU Water Framework Directive

The concept of catchment planning has now been formally recognise in Europe with the introduction of the Water Framework Directive (European Community, 2000), which requires European Union countries to introduce integrated, sustainable river-basin management plans. The Directive proposes a system of management of the natural water-environment based around natural river-basin districts, and introduces the concept of 'programmes of measures' which are required to achieve at least 'good' status for most water bodies, including coastal waters and underground water, by specific dates. The steps can be summarised as:

- publish river-basin management plans and establish programmes of measures
- programmes of measures to be operational
- environmental objectives to be met.

The Directive lists the following major components of a river-basin plan:

1. A map of the extent and condition of surface water and groundwater bodies.
2. A summary of the impact of human activity on such water bodies in terms of diffuse and point pollution and abstraction, etc.
3. A map of protected areas.
4. A map of monitoring networks.
5. A list of environmental objectives.
6. An economic analysis of water use.
7. A summary of measures taken for the protection of water resources, control of abstraction, control of point source, and other forms of pollution and other measures taken to meet the environmental objectives stated above.
8. A register of more detailed programmes and management plans for particular sub-basins or water types, etc.
9. A summary of public information and consultation measures taken.

Catchment plans are required by the Directive to be updated periodically, with a summary of changes and an assessment of progress towards the achievement of the environmental objectives. More details of the requirements of the Directive are given in Appendix 9.2.

9.5.2 Catchment planning for flood protection in the UK

It could be argued that the Water Framework Directive is focused more on water quality than other aspects of water resources, such as flooding. The need for catchment planning in relation to flooding has been identified in the UK (ICE, 2001). It is estimated that some 60–80 catchment plans will be required in England and Wales based on catchments of between 1000 and 5000 km^2. A catchment plan would assess general levels of flood risk and would identify long-term policies to manage the risk. Each catchment would be divided into a series of sub-catchments for which strategy plans would be developed to identify appropriate solutions. Following the approval of these plans, specific schemes would be designed and evaluated.

A general approach to catchment planning is suggested in Figure 9.21. At the core of the model are the processes occurring on the catchment, which can be summarised as the attenuation of the rainfall input and the various losses which occur. The physical inputs to the catchment

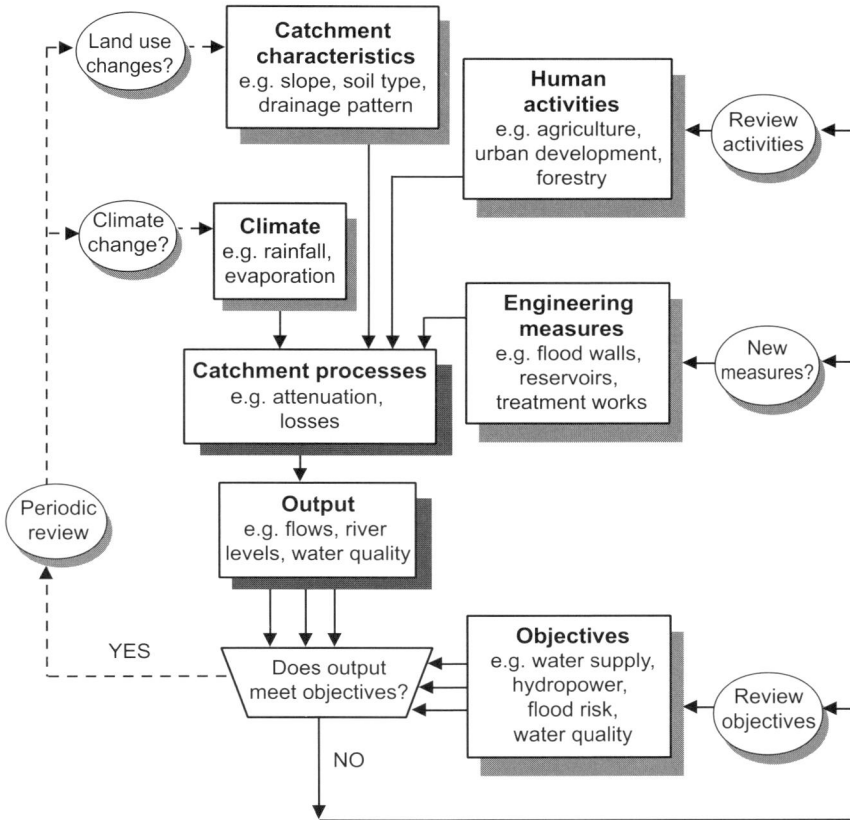

Figure 9.21   The main processes in catchment planning

processes are the climatic variables (mainly rainfall and evapotranspiration), and the physical characteristics of the catchment, such as the size, slope, soil type, drainage pattern, etc. The human inputs to the system are the various activities which impinge on the hydrological processes, such as agriculture and urban land use, together with the specific engineering measures which are put in place to modify the flow of water. The latter could include physical structures, such as reservoirs and flood walls, as well as non-structural procedures, such as planning controls.

The outputs of the process are the river flows, water levels and water quality, etc., which can then each be compared with certain objectives.

The output may be the observed values of these parameters or they may be the result of a mathematical (or physical) model, which enables a range of different input conditions that can be used. Sometimes the impact of the result may depend the nature of the human activity, such as the location of housing in relation to flood levels. The objectives will set certain criteria in terms of the various parameters, such as levels of protection, water-quality standards and minimum flows. In many cases, the standards will involve a probablistic approach with return periods or probabilities assigned to specific values. Where there are deficiencies, a number of possible steps can be taken. Typically, a need may be perceived for new or revised engineering measures to deal with specific problems, such as flooding, but the solution may also involve a review of various activities or it may be necessary to review the objectives themselves.

Where the objectives are met, it will still be necessary to review the plan periodically to allow for changes in land use and possible climate change.

9.5.3   Flood risk and hazard analysis

Flood risk is measured by the probability of a given flood occurring, while the term *hazard* is normally associated with the damage or inconvenience resulting from a given flood. The risk can be expressed either as a return period, which is the average interval between exceedances of a given flow or flood level, or as a probability of occurrence. The return period is the reciprocal of the probability (i.e. a flood with a return period of 100 years has a probability of being equalled or exceeded in any year of 1 in 100 or 0·01), but the term can be misleading in that it might imply that such an event may occur every 100 years. Other things being equal, the probability of a 1 in 100-year flood is the same for any year, even if such an event had occurred in the previous year. However, where there is a trend, such as an increase in river flows, then the probability of a given flow occurring will be greater in the future than it is at present. In the UK, it is recommended that account should be taken of possible increases of up to 20% in design flows for flood defence schemes to reflect the situation by 2050 (Defra, 2001). A 1 in 100-year (1%) event at present, therefore, becomes a 1 in 58-year (1·7%) event in 50 years' time. The recommended standards of protection against river flooding vary from 0·5–2% for intensively developed urban areas to 1–4% for less developed urban areas. Maps indicating areas with a 1% annual risk of flooding have been prepared for England and Wales by the Environment Agency, but they do not take account of climate change or of specific flood defences. In considering land for possible development, planning authorities normally categorise such land according to the potential risk (Table 9.5) and are expected to give priority to development in descending order of the flood zones (Office of the Deputy Prime Minister, 2002).

The hazard associated with a given flood level depends on the level and type of development of the land which is at risk. In the case of agricultural land, the potential hazard of flooding is usually quite low, but flat river-side locations have become natural sites for development and, consequently, the potential damage associated with given flood levels in urban areas has increased dramatically in recent years. Indeed, one logical policy with regard to flood defence might be to reduce the hazard rather than the risk: in other words, to buy and demolish properties that might have an unacceptable risk of flooding rather than invest in flood defences.

It is often difficult for people to accept that, in principle, whatever level of protection is provided, there will always be a finite, albeit very small, risk of flooding. As in many cases, the cost of the flood-mitigation work (which is, in general, relatively easy to estimate) has to be balanced against the

Table 9.5   Planning response to river flood-risk — summarised from Planning and Policy Guidance Note 25 (Office of the Deputy Prime Minister, 2002)

Flood zone	Planning response
Little or no risk ($P < 0.1\%$)	No constraints due to river flooding
Low to medium risk ($0.1 < P < 1\%$)	Suitable for most development. Flood-risk assessment required and some flood-resistant construction and/or warning measures may be required
High risk ($P > 1\%$)	Developed areas may be suitable for development with restrictions as above. Undeveloped areas are generally not suitable for development. Floodplain development should be restricted to recreation or essential transport infrastructure which does not impede water flows or lead to increased flood risk elsewhere

benefit resulting from the work in terms of the reduction in either flood risk or hazard (or both), which are more difficult to estimate. The overall cost of a given flood can be represented by the product or integration of risk and hazard, and the benefit can therefore be regarded as the saving in this cost following some flood-alleviation measure (Pennington-Rowsell and Chatterton, 1977). As discussed in earlier chapters, various techniques can be used to produce a flow–frequency curve indicating the return periods of various flow levels at a certain location (Figure 9.22(a)). Using a hydraulic model of a given drainage system, the flows can be associated with water depths and a depth–frequency curve can be produced (Figure 9.22(b). For any given location, there will also be a relationship between depth (or water level) and damage. At low levels, the hazard might be the relatively minor inconvenience of local flooding around manholes, etc. At higher levels there might be the closure of some roads and flooding of gardens, etc. As the water levels increase the flooding would extend to low-lying properties and, in the extreme, might pose a threat to life. Allowance should also be made for the less-tangible costs associated with flood risk, such as the worry and possible health costs. Clearly, the relationship between flood level and damage will be specific to any given area and a given type of property, but it can generally be represented by a curve, such as Figure 9.22(c).

Starting with a given combination of flow and probability, a value of the damage corresponding to that probability can therefore be estimated to give a point on the curve in Figure 9.22(d). This can be repeated for different probabilities to give a complete curve. If improvements to the drainage system are carried out, the relationship between flow and depth might change to, say, the dotted line in Figure 9.22(b), and the probability hazard curve would then shift to the dotted line in Figure 9.22(d). The area between the two curves in Figure 9.22(d) then represents the reduction in the damage, i.e. the annual benefit, which can be compared with the cost of the scheme.

9.5.4   Sustainability in hydrological management

The concept of sustainability is becoming more important in the appraisal of many forms of investment. The main issues relating to sustainability in terms of water resources can be summarised as (Defra, 2001):

- minimising the social and environmental impact of activities
- ensuring all actions are environmentally neutral or positive, and contribute to biodiversity and other environmental targets

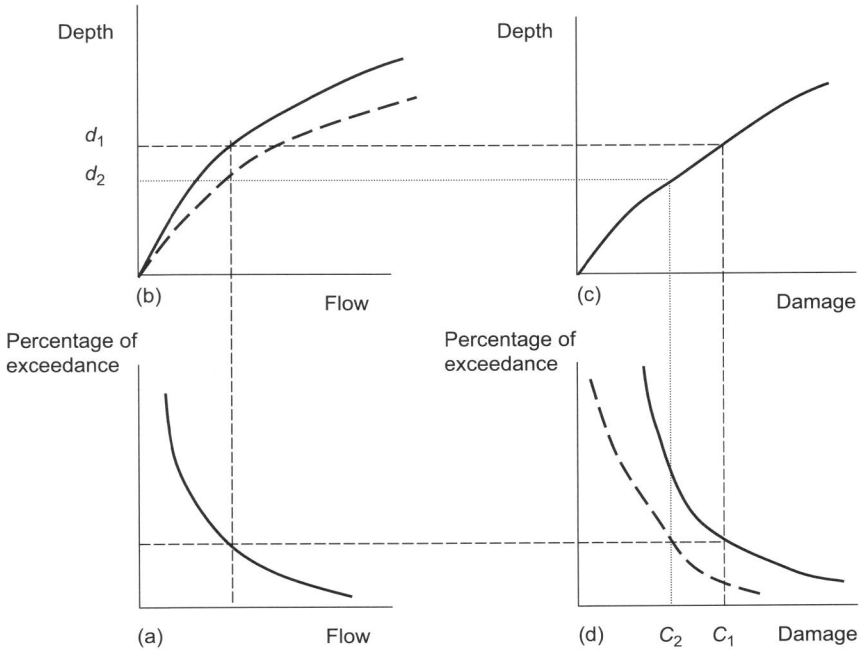

Figure 9.22   Flood risk

- avoiding pollution and the emission of greenhouse gases during construction and throughout the life of the scheme
- using sustainable construction materials
- minimising the use of construction materials and recycling where possible
- being energy efficient in transport and operation activities
- designing for long-term needs
- designing for whole-life approach, addressing the mechanisms of failure and the problems of decommissioning.

9.6   A review of hydrological management objectives

The main objectives of hydrological management are to mitigate the effect of human activities on hydrological processes, and to provide for the sustainable and equitable use of water resources for human needs in terms of water supply, irrigation, hydro-electric power generation and flood mitigation. In Chapter 1, the main human activities which impact on hydrological processes were described as:

- forestation and deforestation
- grazing and agriculture
- urbanisation
- modification of natural channels
- irrigation
- reservoirs
- catchment management
- land drainage.

Many of these practices have the negative effects of reducing the natural attenuation of the Earth's hydrological system and increasing the premature loss of water from the hydrological cycle to the ocean or the atmosphere. The results of this are increasingly being experienced as extreme flood events and water shortages in many parts of the world. These phenomena are now being exacerbated by natural and anthropogenic changes in the world's climate, as well as by the growth in the world's population and the rise in living standards.

Some practices, such as proper catchment management and the judicious construction of reservoirs, preferably on a small scale, can help to mitigate these effects by providing increased attenuation of runoff and reducing losses. Proper catchment management can be applied equally to urban areas as much as to rural areas, using the principles of sustainable urban drainage described above.

It is important to remember that there is still more than enough water in aggregate to supply the whole of the world's population (Table 1.4). The goal of hydrologists must be to reduce the harmful hydrological impact of many of our activities and to overcome the physical, political and economic barriers to the just and equitable use of our water resources.

9.7 References

Balmforth, D. (2002). *Climate Change and SUDS*. Scottish Hydraulics Study Group.

Balmforth, D., Lonsdale, K., Nussey, B. B. and Walsh, M. (1995). A Methodology for Monitoring the Performance of Combined Sewer Overflows. *Journal of CIWEM*, **9**, 510–518.

Barker, I. C. and Kirmond, A. (1998). Managing Surface Water Abstraction. *Proceedings of the BHS International Conference on Hydrology in a Changing Environment*. Wiley.

Barr, D. I. H. (1994). *Tables for the Hydraulic Design of Pipes*. Thomas Telford, London.

British Standards Institution. *BS 3680: Methods of Measurement of Liquid Flow in Open Channels*. BSI, London.

Construction Industry Research and Information Association (2000). *Sustainable Urban Drainage Systems*. CIRIA.

Defra (2001). *Flood and Coastal Defence Project Appraisal Guidance*. MAFF (now Defra).

European Community (2000). *Water Framework Directive 2000/60/EC*.

Hydraulics Research (1983). *Design and Analysis of Urban Storm Drainage*. Hydraulics Research, Wallingford.

ICE (2001). *Learning to Live with Rivers*. Institution of Civil Engineers, London.

Office of the Deputy Prime Minister (2002). *Planning and Policy Guidance Note 25: Development and Flood Risk*. The Stationery Office, London.

Pennington-Rowsell, E. C. and Chatterton, J. B. (1977). T*he Benefits of Flood Alleviation: A Manual of Assessment Techniques*. Saxon House.

Ramsbottam, D., Day, R. *et al.* (1997). *Culvert Design Guide*, CIRIA, London.

Robinson, M. (1990). *Impact of Improved Land Drainage on River Flows*. Institute of Hydrology, Wallingford.

World Commission on Dams (2000). *Dams and development: a new framework for decision-making — the report of the World*. World Commission on Dams. Earthscan, London.

Appendix 9.1
WCD guidelines for good practice

Strategic priority	Guidelines for good practice
1. Gaining public acceptance	1. Stakeholder analysis
	2. Negotiated decision-making processes
	3. Free, prior and informed consent
2. Comprehensive options assessment	4. Strategic impact assessment for environmental, social, health and cultural heritage issues
	5. Project level impact-assessment for environmental, social, health and cultural heritage issues
	6. Multi-critical analysis
	7. Life-cycle assessment
	8. Greenhouse gas emissions
	9. Distributional analysis of projects
	10. Valuation of social and environmental impacts
	11. Improving economic risk-assessment
3. Addressing existing dams	12. Ensuring operating rules reflect social and environmental concerns
	13. Improving reservoir options
4. Sustaining rivers and livelihoods	14. Baseline ecosystem surveys
	15. Environmental flow assessment
	16. Maintaining productive fisheries
5. Recognising entitlements and sharing benefits	17. Baseline social conditions
	18. Impoverishment risk analysis

Strategic priority	Guidelines for good practice
5. *continued*	19. Implementation of the mitigation resettlement and development action plan
	20. Project benefit sharing mechanisms
6. Ensuring compliance	21. Compliance plans
	22. Independent review panels for social and environmental matters
	23. Performance bonds
	24. Trust funds
	25. Integrity pacts

Appendix 9.2
EU Water Framework Directive
Annex VII
River basin management plans

A. River basin management plans shall cover the following elements:
1.   a general description of the characteristics of the river basin district required under Article 5 and Annex II. This shall include:
      1.1 for surface waters:
- mapping of the location and boundaries of water bodies,
- mapping of the ecoregions and surface water body types within the river basin,
- identification of reference conditions for the surface water body types;

      1.2 for groundwaters:
- mapping of the location and boundaries of groundwater bodies;

2.   a summary of significant pressures and impact of human activity on the status of surface water and groundwater, including:
- estimation of point source pollution,
- estimation of diffuse source pollution, including a summary of land use,
- estimation of pressures on the quantitative status of water including abstractions,
- analysis of other impacts of human activity on the status of water;

3.   identification and mapping of protected areas as required by Article 6 and Annex IV:

4.   a map of the monitoring networks established for the purposes of Article 8 and Annex V, and a presentation in map form of the results of the monitoring programmes carried out under those provisions for the status of:
      4.1 surface water (ecological and chemical):
      4.2 groundwater (chemical and quantitative):
      4.3 protected areas:

5.   a list of the environmental objectives established under Article 4 for surface waters, groundwaters and protected areas, including in particular identification of instances where use has been made of Article 4(4), (5), (6) and (7), and the associated information required under that Article:

6.   a summary of the economic analysis of water use as required by Article 5 and Annex III:

7.   a summary of the programme or programmes of measures adopted under Article 11, including the ways in which the objectives established under Article 4 are thereby to be achieved:

7.1   a summary of the measures required to implement Community legislation for the protection of water:

7.2   a report on the practical steps and measures taken to apply the principle of recovery of the costs of water use in accordance with Article 9:

7.3   a summary of the measures taken to meet the requirements of Article 7:

7.4   a summary of the controls on abstraction and impoundment of water, including reference to the registers and identifications of the cases where exemptions have been made under Article 11(3)(c):

7.5   a summary of the controls adopted for point source discharges and other activities with an impact on the status of water in accordance with the provisions of Article 11(3)(g) and 11(3)(i):

7.6   an identification of the cases where direct discharges to groundwater have been authorised in accordance with the provisions of Article 11(3)(j):

7.7   a summary of the measures taken in accordance with Article 16 on priority substances:

7.8   a summary of the measures taken to prevent or reduce the impact of accidental pollution incidents:

7.9   a summary of the measures taken under Article 11(5) for bodies of water which are unlikely to achieve the objectives set out under Article 4:

7.10   details of the supplementary measures identified as necessary in order to meet the environmental objectives established:

7.11   details of the measures taken to avoid increase in pollution of marine waters in accordance with Article 11(6):

8.   a register of any more detailed programmes and management plans for the river basin district dealing with particular sub-basins, sectors, issues or water types, together with a summary of their contents:

9.   a summary of the public information and consultation measures taken, their results and the changes to the plan made as a consequence:

10. a list of competent authorities in accordance with Annex 1.

11. the contact points and procedures for obtaining the background documentation and information referred to in Article 14(1), and in particular details of the control measures adopted in accordance with Article 11(3)(g) and 11(3)(i) and of the actual monitoring data gathered in accordance with Article 8 and Annex V.

B. The first update of the river basin management plan and all subsequent updates shall also include:

1. a summary of any changes or updates since the publication of the previous version of the river basin management plan, including a summary of the reviews to be carried out under Article 4(4), (5), (6) and (7);

2. an assessment of the progress made towards the achievement of the environmental objectives, including presentation of the monitoring results for the period of the previous plan in map form, and an explanation for any environmental objectives which have not been reached;

3. a summary of, and an explanation for, any measures foreseen in the earlier version of the river basin management plan which have not been undertaken;

4. a summary of any additional interim measures adopted under Article 11(5) since the publication of the previous version of the river basin management plan.

Index